Maritime Governance

Michael Roe

Maritime Governance

Speed, Flow, Form, Process

Springer

Michael Roe
Graduate School of Management
University of Plymouth
Plymouth
UK

ISBN 978-3-319-21746-8 ISBN 978-3-319-21747-5 (eBook)
DOI 10.1007/978-3-319-21747-5

Library of Congress Control Number: 2015947406

Springer Cham Heidelberg New York Dordrecht London

Printed on acid-free paper

Springer International Publishing AG Switzerland is part of Springer Science+Business Media (www.springer.com)

Seven social sins: politics without principles; wealth without work; pleasure without conscience; knowledge without character; commerce without morality; science without humanity; and worship without sacrifice.

Mahatma Gandhi

Open to everything happy and sad
Seeing the good when it's all going bad
Seeing the sun when I can't really see
Hoping the sun will at least look at me

Focus on everything better today
All that I needed I never could say
Hold on to people they're slipping away
Hold on to this while it's slipping away

Moby, 2005

Preface and Acknowledgments

> For those with better things to do than immerse themselves in the increasingly recycled waters of Twitter - believed to have passed 700 times through the kidneys of the website – the issue is this: at around midnight on Thursday, (Michael) Fabricant (UK Conservative MP) fired off a tweet in apparent reference to a Channel 4 News debate between the journalists and authors Jasmin Alibhai-Brown and Rod Liddle. He could never appear on a discussion programme with Ms Alibhai-Brown, Fabricant explained (presumably throwing countless telly researchers' plans for Socratic dialogue into disarray). "I would either end up having a brain haemorrhage," he continued, "or be punching her in the throat".
>
> Can I order the brain haemorrhage please? With a side of… but no. No. That was total self-abasement lies. While the knee-jerk response might be to come up with a version of Private Eye's brilliant headline verdict on Rupert Murdoch's diagnosis with prostate cancer some years ago – "Cancer has Murdoch" – the motivation for honking "Brain haemorrhage has a Michael Fabricant" should really have evaporated before you'd worked out where in haemorrhage that eye-catching double sits.
>
> Marina Hyde, *The Guardian*, Saturday 21st June, 2014

This book examines the controversies that surround governance and policy-making in the light of globalisation and with specific reference to the most globalised of all industries—the maritime sector and international shipping in particular. It forms part two of a three-part consideration of the issues that underlie the problems faced by the maritime sector which are manifested in the death, injury, environmental degradation and inefficiency that characterises the industry. In turn, these can be represented as three dimensions.

Dimension 1 is the situation as it exists for maritime governance and policy-making and was considered in detail in the earlier volume—*Maritime Governance and Policy-Making* (Roe 2013). The impact of globalisation upon international shipping was analyzed and the inadequacies of the current hierarchical structure characterised by four features was assessed: the excessive significance still attributed to the nation-state in maritime governance; the domination of anachronistic institutions; the limited range of stakeholders; and the predominant influence of shipowners.

In this volume Dimension 2 focuses upon a fifth characteristic but one which is fundamental to good governance—the need to accommodate dynamic processes and flexibility in governance rather than the domination of stasis and form which is currently the situation. Effective governance does not produce policies for fixed moments in time but allows for the changing industry at which it is directed—and nothing changes quite as much as the heavily globalised maritime sector.

Dimension 3 is something to look forward to in the final of the three volumes and will concentrate upon the need to understand the relationship that exists between policies and their appropriate juxtaposition if they are to maximise effectiveness. Issues such as polycentricism and metagovernance will be considered taking on the argument for dynamic governance made in this volume. But that is for the future.

Traditionally, this is where appreciation for those around me is expressed. In particular, I would like to thank my colleagues at Plymouth University especially in the light of my new existence as semi-retired with the opportunity to focus upon writing and research supervision. Those I am fortunate enough to be supervising whilst writing this book and who have contributed unknowingly to the debate include Xufan Zhang, Xuemuge Wang, Sapna Chacko, Safaa Sindi and Katerina Konsta. In addition, thanks must go to Daria Gritsenko whose contribution has been immense and who introduced me to the delights of Finnish hospitality. Others who have been important include Venus Lun at Hong Kong Polytechnic University without whom nothing would ever have appeared in print and of course to the team at Springer who are a delight to work with. Especially, thanks are also owed to Wanyu Loh in Singapore who has provided unknown (I guess) but extensive support and encouragement over the past few years.

And finally of course, enormous thanks to Liz, Joseph and Siân for making it all worthwhile and possible.

No book of mine could possibly not include a reference to the exploits of Charlton Athletic who have sustained a Championship position and have high hopes for a future in the Premiership sometime in the near future. Meanwhile on a more personal note, thanks to the marvels of modern medicine I am now the proud owner of not only a battery-driven pancreas but also a bluetooth glucose warning system, and two perspex eyes. I have become the personification of reverse logistics and look forward to further plastification in time. Thanks be to God for AAA batteries.

http://en.wikipedia.org/wiki/Implantable_collamer_lens

http://www.cafc.co.uk/

West Hoe, Plymouth
May 2015

Contents

Abbreviations

ACF	Advocacy Coalition Framework
ASEAN	The Association of Southeast Asian Nations
CENSA	Committee of European National Steamship Owners' Association
CMI	Comité International Maritime
EU	European Union
GATT	General Agreement on Tariffs and Trade
GIS	Geographic Information System
ICS	International Chamber of Shipping
IMCO	Inter-Governmental Maritime Consultative Organization
IMF	International Monetary Fund
IMO	International Maritime Organisation
NGO	Non-Governmental Organisation
OECD	Organisation for Economic Cooperation and Development
UK	United Kingdom
UN	United Nations
UNCTAD	United Nations Conference on Trade and Development
USA	United States of America
WTO	World Trade Organisation

List of Figures

Chapter 1
The Beginning

Abstract We left the story of maritime governance acknowledging that much remained to be done, and although many were contributing to resolving global problems and much had been achieved, some fundamental issues still had to be addressed. This book attempts to move the discussion further on and to suggest ways that policy-makers and those responsible for the design of maritime governance can improve upon what we have. We will venture into the dark world of the maritime administrator, shipowner, media company and politician in an attempt to unfathom the inadequacies of maritime governance, digging deep into the philosophical contexts of form, flow, time, speed and process. This chapter proceeds to examine the characteristics and problems that remain with maritime governance, in particular those relating to nation-states, institutions, the narrow definition of stakeholders, shipowner domination and the absence of fluidity in policy-making.

> But in an era of bad faith, the man who does not want to renounce separating true from false is condemned to a certain kind of exile. Albert Camus (1956), quoted in Mooij and De Vos (2003: 30).

> A new scientific truth does not triumph by convincing its opponents and making them see the light, but rather because its opponents eventually die, and a new generation grows up that is familiar with it. Max Planck, *A Scientific Autobiography and Other Papers*, 1949.

> The Palais Stirbey was much older and smaller than the great stucco one with the lions with the blazing eyes... it was built, I should think, early in the nineteenth century, in a charming Regency style: long rooms with ceilings supported by white wooden free-standing pillars. I think with Ionic capitals, and adorned with lustres of many tear-like, glittering drops; and I remember that the parquet floors, during the few moments that these were empty of dancers, had a very slight wave to them, a faint and scarcely discernible warp, like the marquetry of a casket that age has twisted very slightly out of the true. This charm-enhancing blemish, an infinitesimal trace of some long-forgotten earthquake perhaps, gave a wonderful appearance of movement to the interior, something I have hardly ever seen since; a feeling of simultaneous stasis and flux. Patrick Leigh-Fermor (2013: 199).

M. Roe, *Maritime Governance*, DOI 10.1007/978-3-319-21747-5_1

Katie Holmes

Like Katie Holmes in 'Batman Begins' and her troubled relationship with the superhero, we left the story of maritime governance acknowledging that much remained to be done, and although many were contributing to resolving global problems and much had been achieved, some fundamental issues still had to be addressed. This book attempts to move the discussion further on and to suggest ways that policy-makers and those responsible for the design of maritime governance can improve upon what we have, although unlike Katie we cannot necessarily depend upon inestimable talent as well as our good looks and an irresistible taste in silk shirts. Instead, we will again venture into the dark world of the maritime administrator, shipowner, media company and politician in an attempt to unfathom the inadequacies of maritime governance, digging deep into the philosophical contexts of form, flow, time, speed and process. But first, and in the time-honoured way of Danish TV crime dramas, a swift review of what we have already seen.

For those of you with good memories, the advice is to miss the next part and get onto the new plot in Chap. 2; for the rest the story begins here (Fig. 1.1).

Fig. 1.1 Katie Holmes and The Caped Crusader © 2005 Warner Brothers Pictures

What Problems of Maritime Governance?

> The Eastern paradigm looks at ocean wealth as 'value-in-existence', that is the life-giving value of the oceans – and this is something that cannot be mathematically or statistically determined. The national system of accounts… can capture neither the global, planetary dimension of the oceans nor its inestimable value to humanity as the sustainer of life. The Western paradigm aspires to be objective, value-free, based on science, technology and economics; the Eastern paradigm is, in various ways, value-loaded. The Western mind is excessively individualistic; the Eastern world view is holistic, conceiving the individual as an 'illusion', unless integrated into the community in nature, in the universe. The integration of individuality and community has implications for the concept of 'ownership' or 'property'. Thus the Lord Buddha taught, 'it is because people cherish the idea of an ego-personality that they cling to the idea of possession, but since there is no such thing as an *ego* there can be no such things as possessions. When people are able to realize this truth, they will be able to realize the truth of non-duality'. In contrast to the Roman law concept, the Eastern paradigm conceives property as a trust, to be managed responsibly for the good of the community as a whole and with due respect for nature, of which the human community is part. Borgese (1998: 91–92).

To suggest that there is any need to consider changes in maritime governance, there needs to be a case made that something at present could be improved. This is not difficult. The range of failure that maritime policy initiatives continue to display is both substantial and widespread and includes almost all aspects of the industry—all sectors (liner, bulk, ferry); all activities (safety, the environment, security and efficiency); all locations (from the European Union to the USA, and from the Far East and China to the developing countries of Africa); and in particular every part of the jurisdiction and functioning of policy-making and its underlying governance from the international and global down to the local and regional passing on the way through the supranational and national. Perhaps the most indicative and also in some ways the most shocking are the continued problems exhibited by the inadequate functioning of the United Nations International Maritime Organisation (IMO) and its strained relationships with both its supranational (in particular the EU) partners and even with its own national members. This is well documented and covers issues from climate change, environmental policy and safety to issues that stem from the organisational relationship between the IMO and its constituent members (see, e.g. the debate over maritime safety in Tradewinds 2008a, b, c; Lloyd's List 2008, 2009a, b, 2010). In the words of Jordan (2001: 204) in his discussion of the failure of institutions to agree how to approach the problems of governance; 'to all intents and purposes, the dialogue between the two paradigms is essentially one of the deaf'.

There has been considerable commentary on these problems of maritime governance and over many years. See, for example, Sletmo (2001, 2002a, b), Selkou and Roe (2004, 2005), Bloor et al. (2006), Kovats (2006), Roe (2007a, b, c, d, e, 2008a, b, 2009a, b, c, d, 2010a, b, 2013), Roe and Selkou (2006), Van Tatenhove (2008), Sampson and Bloor (2007), De Vivero and Mateos (2010), Van Leeuwen and Van Tatenhove (2010), Baindur and Vegas (2011), Vanelslander (2011), Campanelli (2012) and Wirth (2012: 224, 239); and whilst this does not provide evidence that these governance problems are severe, it is indicative that things are not perhaps straightforward and simple. It is also a trend that can be seen across wider disciplines in their consideration of governance failure. Examples include those analysing the broadest global implications (e.g. Held 1991; Ruggie 1993; Crosby 1996; Stoker 1998; OECD 2000; Jessop 2004; Ramachandran et al. 2009; Borzel and Risse 2010). This in turn raises the issue as to why there has been so little debate about the difficulties of maritime policy-making and the fundamental governance problems that have appeared. If policy-making is problematic, then perhaps something needs to be done (or at least considered). In fact, the structure of maritime governance remains the same as it has been since the 1940s, in turn essentially based upon a framework that was developed from the 1920s and which can be traced back as far as the Treaty of Westphalia signed in 1648. Is it not time that change is considered?

Maritime governance at present has a number of fundamental characteristics that define its operation and structure and which in turn have a major effect upon what can (and cannot) be achieved and by whom. These characteristics can be summarised as follows:

- Nation based,
- Institutionally determined,
- Conservatively defined stakeholders,
- Shipowner dominated and
- A focus on form rather than process.

Each of these issues needs to be addressed if maritime governance is to be appropriate for today's and the future's shipping marketplace. Currently, none are being considered effectively. The nation-state retains its jurisdictional pre-eminence, whilst maritime governance remains essentially institutionally driven with alternative frameworks for policy-making neglected. The role of extended stakeholder involvement is at least understood (see, e.g. recent commentary by the EU on maritime stakeholders). Meanwhile, the ambitions of over-influential shipowners and associated maritime stakeholders is unlikely to change whatever developments in governance occur—these undesirable effects need to be understood and measures taken to produce policies that balance these desires. Major governance revision is not going to remove the significance of shipowners in maritime policy-making, but their ambitions could be accommodated more successfully in policies that address all sides of the environmental, safety, security and efficiency arguments.

At the same time, globalisation centres upon flows—of information, materials, money, etc.—and yet maritime policies are essentially static—designed at one point in time, for a defined situation with an inability to be flexible to accommodate change. Processes—the movement of money, information, materials—dominate the sector and effective governance structures need to accommodate this dynamism, one which takes little account of national borders and acts as the antithesis of the static policies that characterise the maritime sector.

Let us now turn to each of these characteristics in some more detail.

The Characteristics of Maritime Governance

Nation based:

> ...she undertook to devote her untiring active life to getting the Newts accepted as members of the League of Nations. In vain did the statesman explain to the eloquent and energetic lady that Salamanders, having no sovereignty of their own in the world, or their own State territory, could not be members of the League of Nations. Mme Dimimeau began to give currency to the view that the Newts should therefore be granted somewhere their own free territory and their submarine state. This idea, of course was rather unwelcome if not actually opposed; at last, however, a happy solution was arrived at to the effect that the League of Nations should set up a special COMMISSION FOR THE STUDY OF THE NEWT PROBLEM to which two newt delegates would also be invited.
>
> Capek (1936), *War With The Newts*, 235–237.

'The thing exists and no amount of conceptual restructuring can dissolve it'. Nettle's (1968: 559) comment on the nation-state was not entirely popular at the time but may actually have had rather more foresight than envisaged and issues of stateness have remained central to debate ever since (see, e.g. Evans 1997: 62). Despite this, the nation state has been widely seen to be in decline and its political, economic and social influence has lessened over many decades with the rise of globalisation. In governance terms, this has had a marked change on the effectiveness of policy-making and the relationship that exists between increasingly influential global enterprises, the increasingly liberalised consumer and national governments. Many of the maritime policy failures that we see stem from these changes.

Despite this, the nation-state remains central to maritime policy-making forming the most significant jurisdictional element with a key role at the IMO, OECD, UNCTAD, WTO, European Commission, ASEAN and many other policy-making bodies. The inviolability of the state although questioned and threatened remains paramount. The situation is consequently curious. An overtly significant nation-state in terms of maritime policy-making finds itself impotent in terms of maritime governance within an ever-globalised world. This contrasts in particular with a more general political concentration that remains centred upon the nation-state.

The significance of the nation-state in the development of maritime policies has been unrestrained by the spread of globalisation. Shipping is an intensely globalised sector—perhaps more than any other with characteristics of ownership, operation, finance, legality, supply, demand, labour and commodities that can emerge from almost anywhere in the world—and frequently do—as well as changing location with intense and unpredictable speed. The nation-state retains its significant role at the UN (IMO), the EU and of course through the development of domestic shipping policies. This role is as important as it has ever been even though the influence that nation-based decision-making can have over a globalised sector is erratic and minimal. The shipping industry uses this conflict between globalisation and domesticity to its advantage, trading off one jurisdiction against another and involving itself at the different levels as and when it sees fit.

The maritime sector is a classic example of this contradiction but why has the nation-state survived in terms of policy-making? Why does it remain central to governance whilst at the same time largely inadequate in exercising that influence? These questions are fundamental to the nature of the maritime sector and policy-making, policy interpretation and policy implementation—where and how it succeeds and more importantly, where and how it fails.

Although Wright-Mills (1959: 135–136) was an early commentator on the significance of the nation-state, questioning its domination in society and the need for a broader 'sociological imagination' that looks beyond national borders, it was not until the early 1990s that the inadequacies of the nation-state were more widely realised. Walker (1991: 445) emphasises the resilience of the nation-state despite the forces of globalisation. He sees the nation-state as an:

> institution, container of all cultural meaning and site of sovereign jurisdiction over territory, property and abstract space, and consequently over history, possibility and abstract time, that still shapes our capacity to affirm both collective and particular identities. It does so despite all the dislocations, accelerations and contingencies of a world less and less able to recognise itself in the fractured mirror of Cartesian coordinates.

Agnew (1994) suggests that the state is a spatial commodity defined by national boundaries which retains its superiority over other scales (local, regional, global) especially in terms of political sociology, macroeconomics and international relations.

He continues in a later paper to outline the 'Territorial Trap' and analyse the factors that continue to make the state all important in terms of political power, suggesting that the characteristics of bounded territory, the clear decision that remains between domestic and foreign affairs and the widespread view of the nation-state as the geographical container of modern society ensures that the state remains a timeless conception as a 'unique source and arena of political power in the modern world' (Agnew 1999: 503).

Scharpf (1994: 220) considers the role of nation-states in the EU and suggests that the rapid diminution of their powers is unlikely, whilst the EU remains democratically deficient—and little has so far changed. Member states continue to resist erosion of their influence. Meanwhile, Anderson (1996: 133, 135) dismisses ideas that the nation-state is being eroded from below by regionalism and above by globalisation and that it is as a result an anachronism, considering that it lacks plausibility. He suggests that new, postmodern forms of territoriality centring on the nation-state are emerging and that ideas of the death of the state and the emergence of a borderless world are far from the mark (Kaldor 1993; Anderson 1995). States are simply changing their form and function retaining their control over the majority of law and order, education, health, welfare and taxation. They remain the most significant redistributor of resources and wealth and continue to play significant parts in cross-border cooperation (Anderson and O'Dowd 1999: 601).

Brenner (1998: 468) considers that the nation-state—what he defines as a distinctive organisational-territorial locus focussing on capital circulation, class struggle and nationalist/statist ideologies—will always be significant. As such it plays a central role in capitalist territorial organisation accommodating elements of all jurisdictional scales. This focuses upon the mediation of uneven geographical development, itself an essential part of the capitalist model permitting the exploitation of resources and labour. In similar fashion to Harvey's (1981, 1990) vision of the spatial fix and the process of de- and re-territorialisation, the state remains an essential feature. Harvey himself reiterates the continuing significance of the state (Harvey 2001: 29) stressing that far from having its power undermined, it has been in Marx's terms, restructured as 'a committee for managing the common affairs of the whole bourgeoisie' (Marx and Engels 1998: 37). The maritime sector provides inexhaustible examples. Cooper (2000: 23) is firmly convinced:

> The package of national identity, national territory, a national army, a national economy and national democratic institutions has been immensely successful. Economy, law-making and defence may be increasingly embedded in international frameworks, and the borders of territory may be less important but identity and democratic institutions remain primarily national.

Harding (1997: 308) concludes that 'little can happen subnationally without (the national state's) cooperation, acquiescence or benign ignorance' and this could just as easily be applied across other jurisdictions. Le Gales and Harding (1998) are enthusiastic about the state's future, whilst Gordenker and Weiss (1995: 373) stress how government representatives, international officials and academics continue to emphasise the state even in his chosen discipline of transnational cooperation where perhaps a more globalised attitude might have been expected (Waltz 1979; Del Rosso 1995; Morgenthau 2005). Picciotto (1997: 1015) agrees that there are very strong underlying socio-economic forces stimulating globalisation at the expense of the traditional nation-state but even so it is 'misleading to suggest that inexorable tides of global economic flows are eliminating the political structures of nation-states'. Picciotto (1998: 4) continues emphasising that the state remains in Slaughter's (1997: 185) terms the 'primary arena for legitimation and enforcement of societal norms'.

Hirst (1997: 13, 243) is more guarded about the future of the nation-state suggesting that it is losing capacity to deal with international issues such as the environment and economics, but he retains a belief that it remains fundamentally important to broader issues of democratic rights and personal liberty, alongside stabilisation of financial markets, the orchestration of social cohesion and as a guarantor of the rule of law.

Mann (1997: 474) provides two further arguments for why nation-states survive within an increasingly globalised world and which he suggests only 'the most breathless of enthusiasts' could deny:

- State institutions of all types retain causal efficacy because they provide the necessary conditions for social existence, without which society would disintegrate;
- States vary widely in size, characteristics, power, geography, etc. Unless the forces of globalisation can eradicate these differences then they will remain and may grow either sustaining or reformulating the differences that characterise nation-states.

As Opello and Rosow (2004: 2) comment:

> Nation-states, having eclipsed all other types of politico-military rule that have existed on the planet, are and will continue to be for the foreseeable future, the basic building blocks of the global order… The nation-state as a form of politico-military rule has become so ubiquitous that its existence is taken for granted, rarely noticed even by scholars of international relations.

Jessop (2003: 31), along with Zurn (2003: 359), stresses that the nation-state will not just go away even though globalisation has placed enormous pressure upon its legitimacy. Viewing the state as essentially a territorialisation of political power, he considers that this will continue in some form or another and that the state will merely change to accommodate it—spatially, politically, economically, etc. In particular he suggests that the new nation-state has an enhanced role to play. Rather than as an instigator of policy, it has a unique role to act as mediator between the new emerging jurisdictions of power—local, regional, supranational and global—which need some sort of structure to coordinate the diverse governance that they represent. In the maritime sector, examples are beginning to emerge of how member states of the EU can act as mediators through the work of the IMO, the Committee of the Regions and Local Communities within states in the development and implementation of maritime policy. Issues such as territorial integration, social cohesion and social exclusion—for example, in maritime terms coordination across the Mediterranean region, the use of port policies and Motorways of the Sea to encourage cross-national integration and labour policies for seafarers to ensure social inclusion are still ideally handled by (member) nation-states with particularly important roles taken by those richer and more powerful (e.g. in EU maritime terms, Greece, UK, Poland, Cyprus and The Netherlands.).

Jessop (2003: 46) stresses the importance of the nation-state today and in the future:

> While globalization… (has) undermined the effectiveness of the Keynesian national welfare state, a restructured national state remains central to the effective management of the emerging spatio-temporal matrices of capitalism and the emerging forms of post or transnational citizenship. For national states have become even more important arbiters of the movement of state power upward, downward and lateral; they have become even more important meta-governors of the increasingly complex multicentric, multiscalar, multitemporal and multiform world of governance; and they are actively involved in shaping the form of international policy regimes.

To quote Wood (2001: 36):

> However global the economy becomes, it will continue to rely on spatially limited constituent units with a political, and even an economic logic of their own.

Agnew (1994: 56) outlines the state as an example of orthodoxy, the equivalent of a rational individual exercising all the characteristics of free choice. The alternative—anarchy—is simply unimaginable by conservative, organised society in the USA (amongst most other nations) or in the process of undertaking international relations (Ashley 1988; Shimko 1992; Inayatullah and Rupert 1993). The biggest danger of anarchy would come from outside the state borders within which the analyst found themselves. States are 'unitary actors whose nature is determined by their interaction with one another. Each state pursues a calculus of state maximisation relative to the others' thereby best ensuring its survival even at times of intense globalisation (Agnew 1994: 57). The state is so permanent because it existed 'prior to and as a container of society' (Agnew 1994: 59). Its fundamental difference from other societal organisations is that it possesses territory—without this it is lost. We return to the issue of territory later.

Nation-states are widely viewed as rooted in the rise of capitalism and Croucher (2003: 9–10) suggests that the concept of the nation depends upon the structure of the capitalist world economy as without it there can be no freedom of movement of goods or regulation of a market economy. Nation-states provide the opportunity for differences and conformity at the same time and thus form part of the 'ideological superstructure that legitimates and reproduces a particular stage of capitalist development' (Croucher 2003: 9). Consequently although they may change, they will not disappear. Nationalism is a clear manifestation of the significance of the nation-state to capitalism acting as a mechanism of societal elite to maximise the goals of political and economic development. Whilst emphasising the differences between nations and states, a variety of commentators have noted the close relationship that has to exist between capitalism and the nation-state (e.g. Deutsch 1966; Tilly 1975; Giddens 1985; Appadurai 1996).

Lambert (1991: 9) is clear:

> Europe has a lot to answer for. The creation of the nation-state, with its ideology of domination, its centralism, arrogant bureaucracy and latent capacity for repression, must figure high on the list. So must the nurturing and propagation of capitalism, which found in the nation-state an ideal ally, ready to identify a country's fortunes with those of its capitalists.

However, he goes on to note how the nation-state's role in keeping control of capitalism's excesses was dwindling:

> Now capitalism has shifted its ground. Organised worldwide, it escapes those checks and balances built up over the years, in the nation-state framework, by workers' movements and parties of the left. The chances of exerting control at the world level, which would require a political framework and enforceable decisions, are totally remote.

Yeung (1998: 292–293) agrees. He sees the state as performing functions that facilitate capital accumulation and by so-doing legitimise both capitalism and itself. This is what he terms the key argument for the capitalist state (Murray 1971; Jessop 1990). In turn, the expansion of capital beyond its national limits presents a contradiction for the individual state—a global logic to which all states will eventually succumb but which in the intervening period results in a complex system of national and global contradictions and arrangements. Yeung sees the state internationalising itself, losing some traditional capital accumulation functions but gaining others. Consequently, the state will not disappear but reinvent itself as a new beast ready and able to contribute to achieving the capitalist ideal.

Meanwhile, Johnston (1995: 218) reiterates that the main function of the state is to secure conditions of production in both public and private sectors, to regulate patterns of consumption and provide security for the processes of production and exchange. In so-doing, it provides the conditions for making profit and consequently 'ensures the allegiance of the capitalist elite'. The state remains a vital part of the capitalist (shipping) infrastructure without which globalised activities could not take place. In addition, the state remains central to multilateral negotiations (although at least in part because it has generated this framework for them) and in a similar way international legal provisions (Thompson 1999: 149–150).

Duncan and Savage (1989: 181) also consider that nation-states are characterised by inertia and once established through a series of social institutions tend to produce a physical, social, economic and political fixity which persists whatever else happens. This is re-emphasised by Castells (1996) who rejects the end of the nation-state, instead preferring to see it as losing power but not influence.

Globalisation can occur only in conjunction with nation-states (or something that resembles a nation-state) as capital (i.e. shipping) uses national political space in which to generate wealth. Commercial activities still need a national space even in times of increasing globalisation as they have to take place somewhere and that cannot easily be in the few remaining non-places on the globe—international high seas, the Antarctic, in international air-space—and hence, nations are convenient. They are also easily abused, with participants trading off one against another and true allegiance to a national flag against commercially generated loyalty. Thus, policy-making is only effective if it permits capital's (and essentially the shipowner's) globalised excesses to continue.

Walker (1991) was the first to describe the state as 'reified', whereby it is persistently claimed to reflect political reality and its eternal presence is a given and suggesting that it is seen as a series of unalterable units of sovereign space. As a consequence, state formation and disintegration had been dehistoricised and decontextualised. Walker, however, does not see sovereignty and the existence of a nation-state as a 'permanent principle of international order'. The appearance of permanence is just a reflection where convenient, of complex political practices working to sustain continuities and 'shift disruptions and dangers to the margin'. Despite Walker's confidence, we remain fooled by:

> the Euclidean theorems and Cartesian coordinates that have allowed us to situate and naturalise a comfortable home for power and authority. Walker (1991: 459).

As such the nation-state has become 'iconised as the pre-eminent expression of political forms of territorial organisation' (Swyngedouw 2000: 68).

Nation-states have always needed to retain a close link with territory (Johnston 1995: 219). However, the proliferation of globalisation has made the retention of these links more difficult. Shipping is a good example of how states fail to constrain the growth of mobility of people, things, capital and information whilst at the same time attract mobile property to their territory in competition with others (Agnew 1994: 58–60, 1999: 513). In this process, there is a need for the creation and maintenance of state territory, and as such a nation-state to formalise it. In turn, this demands the definition of clear state boundaries which act as the 'geographical container of society' (Agnew 1999: 503).

The issue of boundaries is taken up by Anderson and O'Dowd (1999: 594–595) who emphasise the supreme importance of state borders in the jurisdiction of governance, acting as the framework for all other jurisdictional boundaries and also providing the structure for a range of societal definitions. Controlling state borders in particular provides the key to power in the global system, and the manifestation of the state as a territory within borders is the source of much of the power it retains even within a globalised world. At the same time, it is 'arbitrarily divisive and disruptive of social processes' and 'oversimplifies and hence distorts social realities' (Anderson and O'Dowd 1999: 598).

Nation-states remain vital players in the jurisdictional game of governance albeit with a marked modification of their role and characterised by contradiction and conflict. Maritime transport has felt the impact of these changes and remains heavily entwined with the nation-state as a result. Schrier et al. (1984: 87) actually suggest that the trend is towards global regimes such as the IMO, increasingly dominated by national governments, although the role that those governments play in governance continues to be weakened by extended globalisation.

Institutionally determined:

> Institutionalisation is both a process and a property variable. It is the process by which individual actors transmit what is socially defined as real and at the same time, at any point in the process the meaning of an act can be defined as more or less taken for granted part of this social reality. Institutionalized acts then, must be perceived as both *objective* and *exterior*. Zucker (1977: 728).

> Institutionalization involves the processes by which social processes, obligations, or actualities come to take on a rule like status in social thought and action. Meyer and Rowan (1977: 341).

> (Institutionalisation is) a process in which fluid behaviour gradually solidifies into structures, which subsequently structure the behaviour of actors. Arts and Leroy (2003: 31).

Maritime governance and the development and implementation of maritime policies are essentially institutionally based—and derived from institutions that reflect the industry and its policy needs in the early twentieth century when globalisation was active at a less intense level. This link between institutions and governance is well documented (see, e.g. Oberschall and Leifer 1986: 237; Weingast 1995: 2). Emphasis was placed upon the relationship between those who make up the transactors in government and the institutions that underlie them, an essential part of the rise of institutional economics. Close relationships were also apparent with the state as an institution, something developed further by Clemens and Cook (1999: 442):

> For many political scientists and sociologists, the massively reinforced and embedded array of the state exemplifies the concept of institution.

These institutions—for example, the IMO, UNCTAD, World Trade Organisation (WTO), the EU Commission and ASEAN—are now no longer fit for the task as they reflect a national domination of jurisdictional integrity that in turn is inappropriate for twenty-first-century global shipping (Keohane 2002). However despite this, there has been little debate about how these institutions might adapt or even be replaced. As artificial constructs of a world seeking good governance they are now anachronistic and many of the identifiable maritime policy failures can be traced back to a combination of outdated institutions dominated by their nation-state members. This is combined with a tendency to isomorphic institutionalism, defined by Kostova and Roth (2002: 15) as institutions 'sharing the same environment (employing) similar practices' and thus liable to toxic conformity.

The variety of institutions that can exist is outlined by Mukand and Rodrik (2005: 376), something further stressed by Evans (2001) and Rodrik (2000) as well as Besley (2001). This focus on institutions has been encouraged through the growth of interest in historical institutionalism and its emphasis of the relationship between institutions and policy-making characterised by the work of Evans et al. (1985), March and Olsen (1989), Shepsie (1989), Pierson (1993: 596) and Campbell (1998: 378–379). Meanwhile, institutions have been widely defined (Koelble 1995). Ostrom (1980: 310) suggests that like organisations, they are:

> works of art in which human beings function both as their designers and creators, and as their principal ingredient.

Rather more intriguingly he sees them again like organisations, as:

> Faustian bargains where instruments of evil are used to do no good. Those who have legitimate access to use such instruments of evil have unique opportunities to exploit others and dominate the allocation of values in a society. It is entirely problematic when the use of an instrument of evil may come to dominate social relationships so that rules become oppressive rather than liberating.

Scott (1987: 494) considers them as self-centred:

> technical instruments, designed as means to definite goals. They are judged on engineering premises; they are expendable. Institutions, whether conceived as groups or practices, may be partly engineered, but they also have a 'natural' dimension. They are products of interaction and adaptation; they become the receptacles of group idealism; they are less readily expendable. Selznick (1957: 21–22).

Zucker (1983: 5) sees them rooted in conformity:

> not conformity engendered by sanctions (whether positive or negative), nor conformity resulting from a 'black-box' internalisation process, but conformity rooted in the taken-for-granted aspects of everyday life... institutionalization operates to produce common understandings about what is appropriate and fundamentally, meaningful behaviour.

White (1992: 116) sees them as:

> forced up from counteractions among efforts at control, ... robust articulations of network populations, articulations which draw primarily on structural equivalence. Institutions invoke story-sets across disparate discipline species.

Denzau and North (1994: 4) take a catholic approach suggesting that they are 'the rules of the game of a society and consist of formal and informal constraints constructed to order interpersonal relationships'. O'Riordan et al. (1998: 346) suggest that an institution is:

> an idea that can be clarified only through regular argument, that is through discourse. The notion of institution applies to both organizations with leaders, memberships, clients, resources, and knowledge, and also to socialized ways of looking at the world as shaped by communications, information transfer, and patterns of status and association.

Jones et al. (2003: 153–154) are rather more vocal:

> An institution may be defined as a set of individuals acting according to common rules resulting in collective outcomes. Institutional rules are not neutral, in the sense that different rules often lead to different outcomes (Jackson 1990: 2). These aggregations of individuals interacting according to rules react to information from the environment and come to a collective response.

Ng and Pallis (2010: 2150) define institutions in a maritime context although their comments are equally applicable across other disciplines. They take Hall's (1986: 19) definition:

> the formal rules, compliance procedures, and standard operating practices that structure the relationships between actors in various units of the polity and economy.

which they suggest 'promote efficiency amongst transacting partners, minimise distributional conflicts, and monitor compliance within social spheres'.

Examples of the many other definitions of institutions can be found in Hughes (1936: 180) who was one of the earliest commentators to note the wide variety of interpretations that was possible, Zucker (1977), Ostrom (1980: 310), Bush (1987: 1076), Scott (1987: 494 and 495), North (1993), Koeble (1995: 233–234, 236) and Crawford and Ostrom (1995: 582) who point out a range of other interpretations including those of the *institution as equilibrium* (Von Hayek 1945, 1967; Menger 1963; Riker 1980; Schotter 1981; Calvert 1995), *institution as norm* (Lewis 1969; Ullmann-Margalit 1977; Coleman 1987) and *institution as rule* (Hohfeld 1913; Commons 1968; Shepsie 1975, 1979, 1989; Shepsie and Weingast 1984, 1987; Oakerson and Parks 1988; North 1986, 1990; Ostrom 1986, 1990; Williamson 1985; Knight 1992).

Many of these definitions hint at a much grander role than those adopted for the maritime sector in the rather formalised structures of the UN, European Union, OECD and so on. Informality is an inherent part of institutionalism accommodating the socialised relationships between individuals as a central and vital part of an institution and something notably missing in the institutional framework that dominates the maritime sector.

The importance of institutions within governance is not in doubt. Riker (1980: 432) emphasises the 'force of institutions' seeing them as essentially rules about behaviour derived from language and values:

> Even the priestess in her frenzy probably behaves according to rules and, for certain, her interpreter is constrained by specific conventions. So interpersonal rules, that is, institutions, must affect social outcomes just as much as personal values.

March and Olsen (1984) note how the interest in institutions has increased since 1970 with evidence of considerable research in legislature (Shepsie and Weingast 1983), budgets (Padgett 1981), policy-making (Ashford 1977; Scharpf 1977), local government (Kjellberg 1975), political elites (Robins 1976), the state (Wright 1977), national administration (Skowronek 1982), democracy (Potter 1979), corporatism (Berger 1981; Olsen 1981; and Schmitter and Lehmbruch 1979) amongst many others. Whilst they go on to suggest that institutions had become less important since 1950, Colomar (1995: 74) continues to stress that 'institutions matter'.

Allegret and Dulbecco (2002: 174) summarise the role of institutions, why they are needed and the problems they can generate if their design and operation are inadequate. They describe them as 'behavioural regularities associated with a set of rules, norms and routines'. Schotter (1981: 11) concurs considering institutions as a regularity in social behaviour that is agreed by all members of society that specifies behaviour in specific recurrent situations and is either self-policed or policed by some external authority'. Institutions are thus seen to form an essential part of the market to create harmonious social, political and economic functioning.

Borghese (1998: 132) suggests that institutions play a vital part in addressing global problems but that inappropriate institutions would do nothing but create an institutional gap to which the normal response is violence. She goes on to suggest that the twentieth century has revealed a series of institutional gaps and that institutions 'both national and international, have remained basically static and unchanged'. We return to the static inadequacies of maritime institutions later.

Borghese continues suggesting that there are four principles for the design of institutions that need to be used as a guide:

- Comprehensive—any institution must be effective across all jurisdictions—from local through regional, national and supranational to global. This should reflect the transparency of jurisdictional boundaries which are even more apparent in the maritime sector. Current institutional design reflects a jurisdictional arrangement with clear and strict boundaries between levels generating many of the inadequacies of policy that we have seen.
- Consistent—this demands that regulation and decision-making processes at all jurisdictional levels must be compatible. Innumerable examples of the failure of maritime regulation across jurisdiction (double-hulled tankers; environmental controls; flag-hopping, Port State Control inadequacies, etc.) provide evidence that the current institutional structure does not work.
- Transsectoral—all problems within the maritime sector must be seen as interconnected and holistic and the institutions designed to address them must be the same. Whilst the EU has acknowledged the need for movements in this direction in recent years, there remains little concrete evidence that the wider maritime problems and solutions are considered together. This must include not only consideration of policy problems but also the stakeholders who have an interest in them.
- Participational—refers directly to the involvement of all stakeholders in generating maritime policies and directing maritime governance. We return to this later.

The fact that institutions have a history of failure and inadequacy in all sectors is widely reported. Frankel (1955: 296) notes the institutional rigidity that exists quoting as far back as Veblen (1915: 127) who referred to institutions as 'installations', having been:

> placed and constructed to meet the exigencies of what is now in a degree an obsolete state of the industrial arts (and having changed little since their origin) are all and several, irrelevant, incompetent and impertinent.

Dopfer (1991: 545) is scathing quoting Bush (1987): 'institutionalised behaviors have… a tendency to degenerate and to become encapsulated by dysfunctional ceremony'. Koeble (1995: 232) stresses institutional embeddedness and the consequential stasis noting the inertia that they commonly display (1995: 235). Meanwhile, Slaughter (1997: 183) suggests that the 'new world' promised by the former US President George Bush following the demise of communism in Eastern Europe was a chimera with the UN unable to function independently of the major powers that make it up.

Wuisan et al. (2012: 165) emphasise the importance of institutions to maritime governance and how the IMO has ostensibly failed exemplified by its inability to move quickly to resolve or ameliorate global issues. Shinohara (2005) continues much in the same vein also noting the prominent role of institutionalism in the maritime sector. Meanwhile, Ng and Pallis (2010: 2150–2151) emphasise the function of ports as institutions citing Airriess (2001), Hall (2003), Jacobs (2007), and Jacobs and Hall (2007).

The substantial and delicate relationship between the IMO (at the time of writing IMCO) and the nation-state was made clear by Silverstein (1978: 158):

> It is, of course, a platitude that world order can only be achieved by the sacrifice of a greater or lesser degree of national sovereignty. Perhaps in an ideal world IMCO would have a fleet of ships enforcing some of the Conventions I have mentioned! But that day is not with us and it is the Sovereign States who accept the international agreements who are responsible for their enforcement – not by any means let it be said, a simple task. Colin Goad, IMCO Secretary General (Stockholm, June 1972).

Williams (1987: 2) is positive about the role of the IMO but places the responsibility for inadequacies in maritime policy-making firmly in the court of the nation-state members rather than the institution itself, emphasising the need for them to spend more time dealing with technical rather than political issues. However, Silverstein (1976: 375) attributes the failures of the IMO almost entirely to the creation of an organisation whose member states were 'hyperdependent upon scientific and technological expertise' and which fails to address economic or political issues adequately. The significance of any member state is almost entirely related to its ability to contribute to the technical debate, and although economic issues were specifically written into Article 1 of its original convention (e.g. to remove discriminatory restrictions and unfair shipping practices), they have never been invoked (Silverstein 1976: 373–374). The only overt political discussions have been exemplified by the Mainland China–Taiwan, Israel–Suez Canal and Cuban missile crises and a variety of other seemingly minor seating disagreements. Meanwhile, covert political agendas have always been serious (and harmful). Keohane (2002: 34) stresses the over-dependency of the IMO (amongst other global institutions) upon an elite of technocrats and high government officials with a minimum of democratic control. This is made worse by the inadequate representation of flags of convenience over the years which, whilst abhorrent, are more representative of the maritime sector than many of the established flags that continue to wield power.

Lee et al. (1997: 345–350) provide a detailed discussion of the IMO as a global institution that continues to dominate maritime governance albeit inadequately. Whereas any success of the IMO has been widely believed to be a result of its technical competence exercised through committees and with an aversion to politics, it has been dominated by core private/public shipping interests 'intent upon expanding and protecting global trade and industry' (Lee et al. 1997: 346). In fact, there is serious resistance to any global shipping institution from the industry itself which sees it as a threat to profit. Private shipping and its commercial interests remain effective lobbyists of governmental delegations and also possess some considerable representation themselves through obtaining consultative status at the IMO.

In addition, the tendency for delegates to fail to represent domestic priorities at the IMO contributes to the dislocation between domestic and global maritime policies that has been noted. The conventional state-centric jurisdictional paradigm depends upon two unjustifiable assumptions about global institutions. Those states should be the only significant actors in world politics and that they are unified actors (Keohane and Nye 1972; Silverstein 1976: 377).

Keohane (2002: 36) notes that transnational institutions similar to the IMO could 'invigorate transnational society in the form of networks amongst individuals and non-governmental organisations'. The problems that besiege global institutions might then be addressed and as a consequence the dominance that institutions of this type exhibit might be more justifiable. The current situation is one of institutional centricity combined with structural inadequacy and organisational ineptitude creating a maritime policy-making disaster. Keohane (2002: 245) goes on to explain the difficulties in creating effective global institutions—having to meet high standards of accountability as well as trying to rely upon persuasion rather than coercion and interest-based bargaining. The conflict between nation-state self-interest and global altruism based upon universal values and beliefs makes maritime policy-making difficult. The result will be a need for more, not less, global institutions and ones also redesigned to provide effective maritime governance.

Williams (1987: 3) relates some of the inadequacies of the IMO to the fact that its structure dates from institutions designed before 1914 (Silverstein (1976: 368) suggests 1897 or even 1873!) and consequently bound to be inadequate for a substantially more globalised society. Along with others (e.g. Hughes 1936: 182) who noted the increasing complexity of institutions, and White (1992: 116), he is also conscious of the number and diverse nature and operation of global organisations created each with varying and also commonly overlapping interests in the maritime field—for example the IMO, ILO, GATT (subsequently WTO), International Confederation of Free Trade Unions, World Confederation of Labour, UNCTAD, OECD and so on (Williams 1987: 202). He sees the UN as an organisational system in crisis lacking coordination and authority, with declining standards of management, and decision-making systems unrelated to national strength or financial contributions.

Meanwhile, Silverstein (1978: 160–161) emphasises the issue of slowness of ratification that has always followed the IMO around as a major criticism of its activities. This particularly stems from the failure of national governments to ratify negotiated conventions and amendments with a 'lag time of from five to seven years between passage and final ratification... not uncommon' (Silverstein 1978: 160). The IMO has no power to force member states to comply with decisions, even those to which they have agreed. It has no independent research capacity and relies upon information from member states, private companies and interest groups. Even its financial structure is questionable as fees are related to gross registered tonnage and as a result those who have to contribute the most have the largest fleets, but are not necessarily those with the most significant influence in the industry or at the IMO from a political, economic or social viewpoint. To quote Hobsbawm (1998: 3):

> the *world* does not exist as a political unit at all. Only the so-called *nation-states* exist, although from time to time some of them are powerful enough to have effective global policies or to set-up global institutions for certain special purposes. The United Nations (typically so named) illustrates this problem. It has no power of its own apart from what is made available by its members, and no single policy that cannot be sabotaged by one or more of its members.

Meanwhile, Silverstein (1976: 371) suggests that the average 11-year delay at the IMO between agreement and ratification was largely the effect of disagreement between shipping stakeholders rather than disagreement between nation-states. A rather more cynical view might be that the delay is due to tacit agreement between shipping interests determined to dilute the impact of any maritime regulation.

> Discussions with Norwegian, American, British and other shipping men (*sic*) indicate that the chief support of IMCO as an adjunct of the UN comes from national bureaus, such as our State Department and Maritime Commissions, and their equivalents in other nations, rather than from industry levels. George Horne, Times (London), 18th January, 1949.

Despite these clear inadequacies, there remains an air of permanence about maritime institutions once formed whether global, supranational or national. O'Riordan et al. (1998: 361) quote Keohane and Nye (1972: 55). International institutions... will provide a network of interactions which:

> once established, will be difficult either to eradicate or drastically to rearrange.

As a result, these institutions tend to outlive the decline of the countries which stimulated their creation. Sheldon (1980: 62) suggests that they are in an 'ultrastable 'state, and because change of any sort is threatening, none occurs. Acceptance of change would suggest failure and 'destroy their stability'.

> In this state, keeping constant who they are and what they do is more important than any consideration of the value of this activity in and to the outside world.

Clemens and Cook (1999: 441) are equally as convinced that 'institutions endure', with a 'relative permanence of a distinctly social sort' (Hughes 1936: 180–181; Zucker 1988: 25).

And so to reform? Clearly an institutional problem exists and possibly it lies within the existing framework—or maybe even with the need for new institutions—or perhaps with no institutions at all? There has been much discussion including Angelides and Caiden (1994: 227) along with Peters (2002: 11–12) with an emphasis of needing to incorporate greater 'fluidity', Clemens and Cook (1999: 448–450) who look at the factors that can affect the ability to change, Kovats (2006: 78) who suggests the need for a new global forum, Johnson et al. (2000), Buitelaar et al. (2007: 891) who emphasise the need to 'break through (existing) institutional pathways' something already recognised by Healey (1998) and Healey et al. (2002), and Lazarus (2009: 1158). This focuses of the need for reform of the UN in particular and can be applied in our case more specifically to the IMO. Whilst this is not the only maritime policy-making institution with global impact, it remains the most significant because of its jurisdictional position and consequential influence upon other jurisdictions (and also them upon it). Matheson (2001) provides a full analysis, and although referring specifically to the Security Council, he indicates along with others (e.g. Luck 2005; Krasno 2004; Reisman 1993) that there is considerable support for something at least to be considered. Meanwhile, little is done and the institutional deficit remains—much to the detriment of maritime governance.

Knight (1971: 384) introduces the idea that the influence of authority (commonly in the form of institutions) can be traced in the geographical landscape, and using a similar model the relationship between institutions and policy frameworks can be analysed. He cites Whittesley (1935: 85) who observed that 'deep and widely ramified impress upon the landscape is stamped by the functions of effective central authority'. One might add by ineffective authority as well.

This notion is reinforced by Schwind (1970: 103), and Knight (1971: 384) who notes: 'it is only when ideological considerations and the nature of political control are considered that we can understand many landscape developments'. He quotes examples from Eastern Europe, China, Israel and South Africa.

Three components make up this model of authority and how it might relate to effective impact, and although examples are taken from landscape studies, the principles hold just as true for the relationship that exists between institutional authority, policy implementation and development and could equally be applied to the maritime sector.

- Political goals of all stakeholders need to be clearly defined without which the implications of decisions by authorities may well be misunderstood, deliberately misread or ignored (Hartshorne 1950; Douglas 1968: 16; Kasperson and Minghi 1969: 429). Schat (1969: 258) suggests that we should not concern ourselves so much with 'problems that are tackled by authorities, but (with) the explanation of the way of approach and the contents of the policy of authorities'.
- Agents then need to be identified including an understanding of the distribution of political power, the legislative structure and the political partitioning that exists. These need to all be considered from a mobile rather than static point of view and an understanding reached of how they work both in competition and cooperation.

- Processes need to be understood including the distribution and redistribution of resources, spatial competition and conflict between political units, and the whole impact of symbolisation. The latter refers to the perception of authority and institutional impact by stakeholders.

Keohane (2002: 31) disagrees with the suggestion that states retain the majority of power through the role they play at international institutions and are considered by some 'institutionalists' to wield the only real power in world decision-making. In fact, the decisions by individual states, although highly significant in world terms, are almost always affected by international institutions and their policies, and this is the case with the maritime sector as well. However, the clearly awkward relationship between jurisdictions and the entrenched nature of their structure and operation in a globalised world suggest that the institutional paradox that exists needs to be addressed.

Keohane (2002: 202) along with Borghese (1998: 140–141) goes on to stress that the state retains substantial power within both the national and international/global governance frameworks. The state remains the most important power in maritime governance, but its influence is more nuanced and made more complex by the emergence of new institutional stakeholders including NGOs, TNCs, interest groups, individuals and a range of bodies from seemingly non-maritime areas:

> It is state structures, and the loyalty of people to particular states that enable states to create connections among themselves, handle issues of interdependence, and resist amalgamation, even if it might seem justified on purely functional grounds. (Keohane 2002: 203).

Current maritime institutional structures do nothing to address this curious jurisdictional relationship; an historical accident reflecting priorities from earlier times and now increasingly irrelevant. Parker (2000: 1292) helps to sum it all up:

> The prince of Darkness no longer appears as a personage… but distinguishes himself willingly, even preferably, under the appearance of corporate personalities or institutions. Davidson (1971: xiv).

And

> In Thomas More's *Utopia,* there is a type of person who rather than live in wretched poverty at home, volunteers for slavery in Utopia (More 1988: 102). That is what organization means to them. A steady job, shops with food in them, and a police force that enforces the law: this has its attractions, and anyone who studies organization will understand the importance of certain sorts of predictability. Lucifer would rather 'reign in Hell than serve in Heaven' (More 1988: 263). That is what angelic organization means to him. It means preferring disobedience to the boredom of condescending angels, to the inevitability of hierarchy, the asymmetry of power, and the machine that endlessly manufactures false gods. As Kurt Vonnegut beautifully observes, there is no reason why good cannot triumph as often as evil, and the triumph of anything is a matter of organization. So organization is not one thing, whatever 'the most Holy Hierotheus' might claim. Parker (2000: 1297).

Conservatively defined stakeholders: whilst institutions such as the EU have in recent years recognised the need to expand stakeholder definition in policy-making if maritime governance is to be more effective, the consequences have been very limited in scope. Much is made of the processes of public consultation and involvement in policy development (e.g. through social networking and email), but in practice it is the same interested parties, for a long time part of the policy-making process, that have been involved. The need to widen and deepen stakeholder involvement in maritime governance means taking on-board the opinions and significance of the media, politicians, interest groups of all types and many more if policy-making is to be effective.

The issue of stakeholders is undoubtedly a big one and much has been written—and in some sectors progress made. Collier and Esteban (1999: 176) and Martin (2001: 191) note the growth in recognition of the number, significance and diversity of stakeholders in policy-making, something reaffirmed some years later by Fritz (2010: 2, 4) in considering policy-making in the European Union. The Commission of the European Communities (2001: 14, 2008: 6, 2009: 3, 6) has shown tenacity in attempts at least to consider the stakeholder issue in governance including the maritime sector but elsewhere there is little evidence that the shipping industry has really taken on-board the need to widen and deepen actor involvement. An example of clear evidence of a wider appreciation of the extent of stakeholders comes from Brunn (1998: 121) who some years ago was emphasising the significance of the media as a stakeholder in both state and non-state activities of all sorts. However, an appreciation of the impact upon news and social media and of them upon the shipping industry has so far been minimal and yet from a stakeholder perspective, the impact of Facebook, Twitter and YouTube, the almost ubiquitous use of tablets, smart phones and camera phones is almost inestimable. The maritime sector, along with others, has shown little enthusiasm to expand its stakeholder directory beyond traditional, maritime interests although examples of a more liberal attitude do exist—see for example the work of Braithwaite and Drahos (2000: 476–479) and their understanding of the significance of NGOs, interest groups, the general public and business organisations in addition to the state and more conventional players in the maritime policy-making process. In addition, Furger (1997: 446) suggests that there is a large number of policy-makers that remain unrecognised and which 'cannot be equated to regulators or regulatees' but which nevertheless have a significant impact upon the maritime environment. These 'intermediaries' represent a system of predominantly private and occasionally public self-governance, largely stemming from the maritime sector itself but also including other less obvious players.

Aligica (2006: 79) suggests that any change initiative or project needs to understand the inventory of institutions which have an involvement and in the process to identify the key players. At a more sophisticated level, these players need to be assessed in terms of their contribution, support or opposition and their role within the sector. The maritime industry is no exception. Borzel (2007: 5) identifies a wide range of actors that have been described as stakeholders by a number of authors. These include public actors alone; public and private interests (Kooiman 1993; Mayntz 1993; Scharpf 1993); those in network form (Rhodes 1997; Eising and Kohler-Koch 1999); and those involving private sector interests alone (Streeck and Schmitter 1985). Brunn (1998: 106, 116–117) emphasises the growth in power and range of stakeholders in governance in the discussion of territoriality and statehood. Meanwhile, Freeman (1984: 25) defines stakeholders as 'any group or individual who can affect or is affected by the achievement of the firm's objectives'—which in shipping's terms probably includes almost everyone. Other definitions include Rhenman (1973)—'individuals or groups who depend on the company for the realisation of their personal goals and on whom the company is dependent. In that sense employees, owners, customers, suppliers, creditors, as well as many other groups can all be regarded as stakeholders in the company'; and Thompson (1967)—'those organisations in the environment which make a difference to the organisation in question'.

The importance attached to stakeholders has been reflected in the growth in number of publications in recent years. These include Charan and Freeman (1979), Savage et al. (1991), Hill and Jones (1992), Donaldson and Preston (1995), Rowley (1997), Donaldson (1999), Gioia (1999), Harrison and Freeman (1999), Hemmati (2002), Friedman and Miles (2002), Fletcher et al. (2003) and Friedman and Miles (2006), whilst Sutton (1999: 25) discusses a range of categories of stakeholder (she terms them interest groups) and their relationship to policy-making.

Stakeholders have a close relationship to governance and policy-making and this is as true for the maritime sector as any. Stubbs (2005: 67) is explicit in his consideration of the relationship between multilevel governance (MLG) and stakeholders. MLG is a way of conducting governance that permits an understanding of the complexity at and between jurisdictional levels including incorporating the contributions of institutions both above and below the nation-state as well as the state itself along with all forms of public and private actors and across all types of policy formulator and contributor. It thus avoids the 'very narrow, linear debates' that can characterise governance and policy-making.

Sutherland and Nichols (2006: 6) are even stronger in their conviction of the relationship between stakeholders and governance and even place this relationship within the framework of marine space. They see as fundamental to good governance the recognition and incorporation of all stakeholders and the allocation of priorities between them. They identify a variety of issues that need to be addressed and three that dominate—identification, engagement and managing input. Only then can effective governance have a chance of being implemented. The former was traditionally characterised by a narrow definition, engagement by telling stakeholders what will be done rather than asking them what to do; and the latter by a failure to identify priorities within the maritime community.

Freeman (1984: 26), supported by Altman and Petkus (1994: 39), considers that any form of policy management needs a structured approach for dealing with multiple stakeholders who were involved in multiple issues. The relationship with each stakeholder group would need to be managed including formulation, implementation and monitoring of strategies to address their concerns and interests. Issues of direction/mission, policy choice, resource allocation and the system used to adopt policies for each stakeholder objective need to be included if stakeholders are to form the serious part of governance and policy-making that they should.

Hosseus and Pal (1997: 404) discuss the choice of stakeholders in maritime policy-making, whilst Mason and Gray (1999), Notteboom and Winkelmans (2002), Wang et al. (2004), Pallis (2005–2006), Pallis and Tsiotis (2006, 2008) and Brooks and Pallis (2008) cover related issues in port, land and air transport. Dicken et al. (2001: 91) stress the need to include all stakeholders in policy-making including both agents (e.g. states, labour organizations, regulatory bodies) and what they termed 'non-human intermediaries' (facilities, telecommunications, infrastructure, policy documents, manuals, etc.). Phillips and Orsini (2002) provide a detailed discussion of the relationship between citizen involvement and the policy-making process, focussing in particular on the need for greater stakeholder engagement with governance and policy-makers. They spend a considerable time analysing why this should be the case and the techniques for doing so effectively. Pomeroy and Douvere (2008: 616) in their consideration of marine spatial planning and in particular the marine ecosystem outline why it is important to involve stakeholders in the policy-making process:

- Better understanding of the system under examination;
- Better understanding of the individual relationship with specific issues;
- Understanding the compatibility and conflicts of multiple use objectives;
- Identifying and resolving real and potential conflicts; and
- Discovering existing patterns of interaction.

They emphasise how stakeholder involvement in policy-making means much more than just collecting comments on a completed plan or policy. Stakeholders need to be involved from very early on in policy planning and then their interest and involvement needs to be maintained throughout the process. This would encourage ownership with all the benefits that can bring. In turn, this suggests four phases of policy-making where stakeholders would be important:

- Planning including setting out objectives, priorities and ultimate purposes. These need to be derived in discussion with stakeholders and ultimately confirmed with them.
- Evaluation of plans and policies. Choice of policies should incorporate stakeholders' direct views about choices to be made.
- Implementation. Comanaged application of policies makes their application much more effective and objectives much more likely to be achieved.
- Post-implementation. After policy implementation, to review the whole process and to provide advice for future policies and stakeholder roles and involvement.

Bennett (2000: 876) outlines how currently policy-making and the design of regulations is commonly too simplistic as it excludes innumerable third-party private and public actors who are not the ultimate target of the regulations or policies but have power to influence and be influenced by them. Policy success is dependent upon a whole range of actors whose presence within the policy-making process needs to be enhanced. Governance as a whole needs to incorporate all manner of actors as even if the correct rules, regulations and policies are adopted, the intentions behind them will not otherwise be translated into reality (Vogler 1995: 154).

Bennett (2000: 863) continues emphasising that inputs to maritime policy-making cannot rest only with the IMO and nation-states but must also include shipowners, cargo owners, insurers, classification societies, banks and many others if it is to be relevant and effective.

The importance of the nation-state to the effectiveness of maritime governance is apparent, and Sutton (1999: 26) outlines how state-centred models of stakeholder involvement in policy-making can be a useful way ahead as the state is always a 'powerful actor in its own right'. This is supported by Stalder (2006: 124) who comments on the role of the nation-state as a stakeholder operating at the supranational level:

> we now find all kinds of state institutions, not just national governments, immersed in complex games of alliances, trying to use their position within the network as a whole to advance their own specific goals.

Picciotto (1998: 3) sees the state fragmenting into a range of stakeholders each performing specialised tasks—central banks, competition authorities, utility regulators and agencies supervising health, social services, education, policing, prisons, etc. Each has its own agenda. Each has a role to play in policy-making and not necessarily only in their own specific sector. Macleod and Goodwin (1999: 506) agree identifying:

> a relative decline in the state's direct management and sponsorship of social and economic projects and an analogous engagement of quasi- and non-state actors in a range of public-private partnerships and networks.

One major relationship between the state and stakeholders in governance was:

> not the formal machinery of government, but rather the informal partnership between City Hall and the downtown business elite. This informal partnership and the way it operates constitute the city's regime: it is the means through which major policy decisions are made. (Stone 1989: 3).

MacLeod and Goodwin (1999: 513) go on to outline the concept of 'institutional thickness' as an approach to governance which requires a plethora of actors providing a collective representation (and which reflects a policy of inclusiveness).

To be involved effectively in governance requires a mechanism for stakeholder identification that meets the objectives of the policy-making process. Hosseus and Pal (1997) provide a full discussion on the selection of relevant topics for policy analysis in the shipping sector. Over 140 are listed and the authors claim (rather frighteningly) that this is a substantially restricted version of the full list. These in turn can be interpreted as guidelines for those with an interest in the sector—and thus potential stakeholders.

The choice of stakeholders needs to be comprehensive, something emphasised by Dicken et al. (2001), and Sabatier (1998: 99) suggests that this should always involve all domains and policy subsystems with actors from 'all levels of government within a country and increasingly from international organizations and other countries'. His discussion of advocacy coalition frameworks (ACF) as applied to European policy-making suggests that they should include legislators, administrative agency officials, interest group leaders, researchers and journalists. Examples can be found in Derthick and Quirk (1985), Robyn (1987), Brown and Stewart (1993), Jenkins-Smith and Sabatier (1994), and Zafonte and Sabatier (1997).

Finally, and with particular reference to the discussion on process and maritime governance that is to follow, Sabatier (1998: 102) emphasises the 'hundreds of actors from dozens of organizations seeking to influence the overall policy process over periods of a decade or more', most of which remain excluded from the maritime governance and policy-making framework.

The significance of shipowners: clearly shipowners are very important stakeholders in maritime policy-making and must form a central plank of maritime governance. However at present, their role is so significant as to overwhelm much else. The explosion of tonnage taxation regimes is one example of this whereby relatively profitable shipping companies can gain subsidies through preferential taxation awarded by nation-states desperate to retain a pseudo-domestic shipping industry (Gekara 2010). In order to remain competitive, one country after another has introduced a concessionary scheme of this sort, the like of which is unavailable for any other specific industrial sector (e.g. in other globalised sectors such as the airline industry, satellite TV, international trucking, mobile telephone communications, space exploration). Globalisation has enabled shipowners to trade off national jurisdictions against each other to obtain dubiously justifiable concessions. The role that national flag registers play in the shipping industry is similar in providing opportunities for shipowners to bargain their fleet registration between countries to obtain concessions in taxation, regulation and the like.

This territorial hypocrisy is a characteristic of maritime governance almost always beneficial to the shipowner, manifesting itself in the way that the shipping industry attempts (and largely succeeds) to take advantage of both the national and global framework within which it works. This is the case even when considering wholly domestic shipping activities—and in some ways this is even better evidence of the importance of the hypocrisy that characterises wholly national-based shipping activities which take place within a globalised atmosphere where crewing and flag choices, options for finance, bunkering and insurance and many other decisions are taken in the context of global rather than national standards. In turn, this raises the opportunity of territorial promiscuity that the global shipowner has not been slow to take. Evidence comes from the widespread abuse of maritime policies through flags of convenience, the application of national tonnage tax regimes and the inadequacies of Port State Control which are just some of the ways that the industry trades off policies at national and global level to achieve the best of both worlds (and the worst for the environment, safety, security and competition).

The result is characterised by territorial porosity whereby the impact of national borders can be imposed at will (and taken away) by the shipping sector to maximise profit—either actually or at least by threat. National territory no longer has the meaning it once had and globalisation has created a nightmare for policy-makers condemned to working within a nationally defined framework.

This is not to say that the shipping industry is ambivalent to failures which occur in maritime policy. It is quick to suggest the social advantages that come from clean, secure and safe seas through representatives at the EU, the IMO, through national Ministries and the multitude of ship-owning associations, through a variety of interest and lobbying groups and even through professional associations. However at the same time, it continues to manipulate the globalised maritime governance framework for its own commercial convenience with little true regard for the wider social issues that characterise it. Unlike many other globalised industries, which are positioned in such a way that they can see both short- and long-term benefits in accommodating corporate social responsibility (see, e.g. Bhattacharya and Sen 2004; Kotler and Lee 2005; Quartey and Puplampu 2012), the shipping industry sees (or at least appears to see) almost none.

The factors that lie behind the cause of this state of affairs have been discussed widely by Roe (2007a, b, c, d, e, 2008a, b, 2009a, b, c, d, 2010a, b, 2013), Roe and Selkou (2006), and Selkou and Roe (2004, 2005). In particular, they stem from the wholly inappropriate jurisdictional framework that is central to current maritime governance. This framework emerged over the past 100 years or more and is deeply rooted in the inviolability of the nation-state and the significance that the nation retains regardless of jurisdiction—global, supranational, national, regional, local. The newly globalised world, reflecting continued shifts away from a national focus and towards each of the other jurisdictions, often characterised by linkages that miss out the heavily structured requirement of hierarchical sequencing that remains a feature of maritime governance, has exposed the nation-state and its fragile new existence. Curiously at the same time, the nation-state remains central to the new, globalised maritime governance in that it is the main representative at all global and supranational policy-making institutions. This contradiction between a governance framework designed for nation-state pre-eminence but operating within a highly globalised environment, where nation states have lost their true relevance, provides the opportunity for shipowners to exercise their preference for territorial hypocrisy. Maritime policy derives from this nation-state-dominated framework and consequently clashes with the globalised imperatives of the maritime sector. This generates a series of policy failures which provides the opportunity for the shipping industry to take commercial advantage of structural inadequacies.

The significant role that the nation-state has retained within maritime governance is surprising given the continued acceleration in the importance of globalisation and especially for an industry where the impact of time-space compression has been obvious (Harvey 1981, 1990, 2001). Shipping has played a substantive role in reducing the friction of distance and increasing place utility for all manner of commodities and products, so much so that to suggest that the nation-state has any function in maritime policy-making may sound ludicrous. Where an industry like shipping is characterised by the ability to largely ignore national definitions and boundaries at will whether it be for financial, legal, administrative, operational, social, environmental or any other issues, then it would appear to be nonsensical for the nation-state to claim any realistic influence. However, the structure of maritime governance assumes just that, with nation-states central to policy-making yet peripheral to policy implementation.

Mangat (2001: 9) suggests that governments have been stripped of any sovereignty they once had and now lack not only the ability but also the desire to develop and enforce policy in an area such as maritime. In the terms of Mittelman (1999), states acting as a facilitator of this process actually encourage further globalisation and their own ultimate demise as effective maritime policy-makers handing over responsibility for the market to the shipowner. This post-Westphalian world is one characterised by companies that fly the flag of their 'home' country and for a variety of political, financial, marketing and to a certain extent legal reasons retain a distinct national identity. Shipping companies are prime examples. Simultaneously, they are independent of any specific state, and their domicile is one they have chosen rather than been born to. This in turn generates footloose capital and a market for global activity again epitomised by the markets for seafarers, ship registration, ship taxation regimes and the like. One consequence is the inadequacy of maritime governance, reflected in maritime policy-making failure and in the problems faced by the maritime environment, maritime security, maritime safety and maritime efficiency—and ultimately the pollution, injury and death that results.

Evidence for the influence that nations still seemingly hold in maritime policy comes from Alderton and Winchester (2002: 36) who suggest that a ship's nation state still has exclusive dominion over that ship. However, they also emphasise that the choice of which state has this dominion is now down to the shipowner and as such a competitive market for ship registration has emerged with clear regression to the lowest standard. Globalisation in maritime governance has enabled the shipowner to dictate policy standards to his or her own advantage. Open registries have the possibility to provide vessel registration with almost no conditions because sufficient shipowners welcome the financial opportunities this affords.

Alderton and Winchester (2002: 39) go on to discuss flags of convenience and the failure of national or global policy-makers to provide an adequate response to their deficiencies. The path between the 'flag state and the ship owner is at best, obscure and minimal' and the role of the nation-state in encouraging such registries is in fact a deliberate attempt to minimise the influence of the nation-state in the governance of shipping. Shipowners have welcomed the divorce of national (or in fact any) governance and flag. International attempts at raising standards of registries have in many cases just provided more opportunities for shipowners to flout rules by encouraging disreputable flags to emerge to fill a growing demand for lower standards for shipowners who cannot or choose not to afford the costs of higher standards. Globalised maritime governance has no framework to deal with this because all authority ultimately rests with the nation-state which in turn is regulator of each open registry. Hence, the policeman is also the criminal.

To quote Alderton and Winchester (2002: 43):

> In the context where international regulation is enacted upon a nation by nation basis then it is no wonder that this situation occurs. Where legislation still relies on a state as the analytical model, yet the context itself is irredeemably global, there is always a remainder, a remainder that, due to its sovereign privilege may create an unregulated environment where capital is free to act as it pleases.

The shipowner finds him/herself protected in two ways:

- The whole reason for the existence of open registries is that in terms of regulation they are liberal; their success depends on the maintenance of this. The associated nation-state has no incentive to regulate as to do so would destroy the cash-cow from which they benefit.
- As an intrinsic part of this, the nation-state neither attempts to encroach on the autonomy of the shipowner nor introduces or encourages structures and frameworks that might restrict the power of the flag over the state or shipowner.

The insignificance of the nation-state is recognised by Bauman (2000: 192) citing Hobsbawm (1998: 4–5):

> What we have today is in effect a dual system, the official one of the 'national economy' of states, and the real but largely unofficial one of transnational units and institutions… [Un]like the state with its territory and power, other elements of the 'nation' can be and easily are overridden by the globalization of the economy. Ethnicity and language are the two obvious ones. Take away state power and coercive force, and their relative insignificance is clear.

Bauman goes on to suggest that multinationals (and shipping companies, however, small have many of the same characteristics) have a penchant for small states. Small states are generally weak, and as a result it takes less money to buy them and their favours. Small of course does not mean necessarily spatially small, but it does mean politically and financially small. But with globalisation, the influence that small states can have through providing opportunities for manipulating maritime governance has declined to the point where their purchase is hardly necessary. As national governments increasingly cannot balance books with their own resources, they are forced into global collaboration, of which the maritime sector plays a disproportionate part, something Anthony Giddens calls a juggernaut controlling modernisation. Most governments now happily:

> vie with each other to implore, cajole or seduce the global juggernaut to change track and roll first to the lands (*or ship registry*) they administer. The few among them who are too slow, dim-witted, myopic, or just vainglorious to join in the competition will either find themselves in dire trouble having nothing to boast about when it comes to wooing their 'voting with wallets' electors or be promptly condemned and ostracized by the compliant chorus of world 'opinion' and then showered with bombs or with threats of showering with bombs in order to restore their good sense and prompt them to join or rejoin the ranks. Bauman (2000: 192–193) (*italics added*).

Walton and McKersie's (1965) 'behavioural theory' of social negotiations is cited by Putnam (1988: 433–434) as central to much state-centric policy-making and a considerable amount of international policy-making, and shipping is one commercial activity that is central to this and operates at what they termed a 'two-level game'. Domestic interests (and this includes shipowners) do what they can to pressurise their national governments to adopt policies that are favourable to them. Meanwhile, domestic politicians seek to increase their influence by 'constructing coalitions amongst those groups'. Internationally, these same national politicians aim to 'maximise their own ability to satisfy domestic pressures, whilst minimising the adverse consequences of foreign developments'. Every nation-state appears at both game boards.

The complexities that follow are substantial but regularly form part of shipowners' activities in maritime governance and policy-making. Decisions at one table may conflict with decisions taken by the same interests at another or even the same table but on another occasion. This game played by the shipowning community (amongst others) is commonly seen at the IMO, at the various institutions of the European Union, in discussion with national ministries and so on—with the ultimate consequences reflected in the inconsistency and inadequacy of tonnage tax regimes, delayed or even failed environmental international legislation and the practice of flag-hopping. On some occasions, a clever player will be able to align his or her ambitions at all tables to achieve the ultimate ambition of both national and global maritime policy which whilst seemingly provides benefits on both fronts.

Druckman (1978), Axelrod (1987) and Snyder and Diesing (1977) amongst many others recognise that this two (or even more)-level game has been played across the national/global divide so that at times even the most conflicting policies can be agreed by all depending upon circumstances and the characteristics of those present. The shipping sector finds itself fortunate in that its intensely global nature permits trade-offs to be undertaken much more easily than sectors where national factors dominate. The close links between globalisation, shipowner interests and the decline of the influence of the nation-state are clear.

The somewhat difficult relationship between state and shipowner has been recognised for some time although this does not seem to have made it any easier for national governments to do anything about it—even if they wish to. Strange (1976: 358) emphasised in the 1970s that the 'authority of states over the operators and the market is generally rather weak', something she continued to emphasise in later years (Strange 1996). She contrasts this to other markets and in particular the air transport industry which initially might seem to be comparable with commercial shipping. However, shipping exhibits considerably more flexibility in the opportunities it has to enter and leave ports and in the difficulties of preventing shipping movements. Shipping is characterised by two concepts almost unheard of in air transport—the freedom of the seas and the Master's sole authority over a vessel. Both reflect the independence of shipping, the globalised nature of the business, and the inherent power of the shipowner over much of what the Master decides to do.

Strange continues to emphasise the ambivalent attitude of national governments to the shipping industry and to shipowners in particular. This has stretched across full-scale nationalisation (or at least full-scale national control) of shipping interests at times of war; partial but substantial control of imperial interest at times of colonial expansion; protectionism to support commercial activities through subsidy, trade direction, guarantees for labour markets typified by the US Jones Act (1920); and over-liberalisation of markets where shipping interests emphasise their need for support and can threaten domestic emigration without concessions (e.g. tonnage tax, flag favouritism or even direct financial support). Strange's comment that:

> Considering the multitude of international problems which the transnational operations of ship-operators create, the impact of international law and organization on shipping is still relatively weak. (Strange 1976: 361).

...is as valid today as when written. The weakness of international maritime law is a reflection of the ability of shipping interests to influence its design and application, and with globalisation, this ability rather than decline has intensified. The IMO remains a classic example of this situation with laudable aims continuously thwarted or severely delayed to meet the increasingly significant demands of an industry as globalisation itself becomes more and more significant.

Strange continues with some severe criticism of a number of other international organisations with shipping interests relevant at the time (1976)—UNCTAD, CENSA (Committee of European National Steamship Owners' Association), ICS (International Chamber of Shipping), CMI (Comité International Maritime) and the OECD. These she describes as 'transnational pressure groups or... collective bargaining organisations more than sources of regulatory authority'—something that still rings true today. Many of these global groups are effectively representational bodies for the shipping community (and shipowners in particular) rather than organisations with the prime aim of making the industry cleaner, safer and more efficient.

The distinction between lobbying and regulation is blurred and much more so than in many other sectors, even those with clear global characteristics—air transport, satellite television, telecommunications, space exploration, etc. The result has been almost universally more success in lobbying by international shipping as commercial interests dominate. She sums up:

> ...the world shipping business seems to be heading for decreased efficiency, and for increased inequity and continued instability. It is in a condition of relative anarchy dangerous to the environment and to human life, and potentially very disruptive both to the rest of the world economy and to the political relations between governments – and even perhaps to politics within states. Strange (1976: 364).

Whilst her comments are from 40 years ago, their pertinence remains clear and her ability to see the impact of globalisation upon maritime governance and policy-making is remarkable. This vacuum is a convenience for the international shipping industry and has stemmed from a nation-state focussed governance framework that has become anachronistic in the light of change exemplified by globalisation. Shipowners delight in the increasingly globalised world whilst taking advantage of the lax governance that it permits. In the meantime policy-making frameworks, processes and institutions take their time to catch up. Strange finishes with what might be seen as prophetic comments—that as shipping continues to expand it needs to be regulated more rather than less, but despite this with every technical change, the level of political authority has declined (Strange 1976: 364).

Cerny (1995: 618) provides support in commenting upon the complex relationship between globalisation, the state and the democratic and social aspirations of people. The processes of globalisation have detrimental effects upon the relationship between major commercial interests (e.g. shipping) and society.

> Thus economic globalization contributes not to the suppression of the state by a homogenous world order as such but to the differentiation of the existing national and international political orders, as well. Indeed globalisation leads to a growing disjuncture between the democratic, constitutional and social aspirations of people - which continue to be shaped by and understood through the framework of the territorial state – and the increasingly problematic potential for collective action through state political processes.

Cerny (1995: 618) suggests that collective action through multilateral regimes (e.g. the IMO) might well increase but will operate at least one remove away from democratic accountability. New 'nodes of private and quasi-public economic power' were seen to be crystallising which would ultimately end up as more powerful than the state. This sounds very familiar in the early decades of the twenty-first century and in the context of the global shipping industry where the state has become less effective, the private sector in the form of shipping interests evermore powerful and the whole charade dressed up in the form of quasi-public representation. The state today:

> …is a potentially unstable mix of civil association and enterprise association – of constitutional state, pressure group, and firm – with state actors, no longer so autonomous, feeling their way uneasily in an unfamiliar world. Cerny (1995: 619).

Cerny goes on to cite Andrews (1994: 201) who sees the state as an agent for the 'commodification of the collective, situated in a wider, market-dominated playing field' (Cerny 1995: 620). The nation-state then can be seen increasingly to be free riding on 'opportunities created by the autonomous transnational market structure'.

Cerny emphasises further the governance problems faced by a globalised world where the nation-state retains its policy-making significance at the same time as losing its policy implementing influence. A globalised world retains much of the anarchy that has always characterised governance, but its structural nature will change. The new void in governance that globalisation has generated and of which shipping takes advantage has yet to be filled in any effective way. This cannot be done using the hierarchical structures that characterise shipping governance at the moment as they are ineffective and largely irrelevant. New structures are needed which interact across current hierarchical jurisdictions and which feature complex relationships between actors within which the nation state may have no role to play—'a complex, world-wide evolutionary process of institutional selection' (Cerny 1995).

> Globalization does not mean that the international system is any less structurally anarchic; it merely changes the structural composition of that anarchy from one made up of relations between sovereign states to one made up of relations between functionally differentiated spheres of economic activity, on the one hand, and the institutional structures proliferating in an ad hoc fashion to fill the power void, on the other. Different economic activities – differentiated by their comparative goods/assets structures – increasingly need to be regulated through distinct sets of institutions at different levels organized at different

> optimal scales Such institutions of course, overlap and interact in complex ways, but they no longer sufficiently coincide on a single optimal scale in such a way that they could be efficiently integrated into a multitask hierarchy like the nation-state. Some are essentially private market structures and regimes, some are still public intergovernmental structures, and some are mixed public-private. Cerny (1995: 620–621).

Central to these changes are the private-sector shipowners who as Forsyth (1993: 209) suggests have found themselves conveniently placed within international law that recognises the flag of a vessel as its nationality, not necessarily related to the nationality of the owner, crew or officers. The confusion so caused is one prime example of how the shipowner holds all the cards in globalised governance and acting effectively as transnational corporations engages widely in socially injurious actions (Forsyth 1993: 208; Michalowski and Kramer 1987: 39–45).

Taking the neo-classical argument of Rugman (1982) and the Marxist approaches of Brett (1985), and Jenkins (1988), Picciotto (1991: 46) suggests that the internationalised ownership of capital, which is exemplified by the shipping sector, developed through the creation of corporations which were facilitated by the proliferation of nation-state-backed protectionist regulations. These included tariffs, national procurement policies and national financial protection measures. In particular, foreign-owned capital is commonly considered in a different way to national capital (e.g. tonnage taxation regimes which avoid domestic taxation rules applied to other sectors). However, having been offered and accepted special conditions because of their international characteristics, these same, global investors then become the staunchest of all nation-state defenders. Shipping obtains this significant competitive advantage by exploiting national differences across all jurisdictional levels both politically and economically. Effective maritime policy-making is almost impossible.

Chowdhury (2006: 141) suggests a close and inevitable relationship between capital (in this case shipping and shipowners), globalisation and the nation-state. Each relies upon the other, but at the same time, some elements (notably the shipping community) can take advantage of the situation in which they find themselves and this in turn does nothing to encourage effective governance.

Form rather than process: maritime policies have traditionally focussed on the static rather than the dynamic; on form rather than process. The shipping industry is far from static and the problems which policy-makers are addressing need dynamic approaches if they are to be effective. Curiously, policies are traditionally 'snapshot', relevant to one moment in time and increasingly anachronistic even before they are implemented. Whilst the idea of developing dynamic policies rather than static ones, focussing on processes that are going on rather than their form, on developing a dynamic governance, is a difficult one; it is one that has attracted much attention in many fields already—it is time that the maritime sector entered into this debate.

The issue of dynamism is fundamental to the problems facing maritime governance and the need to overcome the dead hand of stasis is paramount. Process, change, dynamism or whatever else it might be called is consequently the focus of the remaining chapters and will be considered in full in the following pages.

Conclusions

Each of these issues needs to be addressed if maritime governance is to be appropriate for today's and the future's shipping marketplace. Currently, none are considered effectively. The role of extended stakeholder involvement is at least understood [see, e.g. recent commentary by the EU on maritime stakeholders (Commission of the European Communities 2008)]. Meanwhile, the ambitions of over-influential shipowners and associated maritime stakeholders are unlikely to change whatever governance changes are made—these undesirable effects need to be understood and measures taken to produce policies that balance these desires need to be generated. Major governance revision is not going to remove the significance of shipowners in maritime policy-making, but these ambitions could be accommodated more successfully in policies that address all sides of the environmental, safety, security and efficiency arguments.

Meanwhile, the importance of the nation-based bias within current maritime governance cannot be overestimated and the outdated characteristics of the institutions that dominate policy-making remain fundamental. However, it is none of these highly significant characteristics which will be considered in the pages that follow. Instead, it is the nature of policies and the continued focus on static rather than dynamic measures that will be addressed, upon form rather than process. And upon the need to develop maritime governance and policy-making so that it can accommodate change rather than fixation. Fundamentally, this needs to be addressed if any of the other inadequacies are to be resolved.

The remainder of this book will focus on this debate central to maritime governance—how a dynamic industry that needs policies that can accommodate change can emerge from what is essentially a static governance framework based upon an anachronistic institutional structure dominated by commercial and traditional self-interests rather than those of society.

References

Agnew, J. (1994). The territorial trap: The geographical assumptions of international relations theory. *Review of International Political Economy, 1*(1), 53–80.

Agnew, J. (1999). Mapping political power beyond state boundaries: Territory, identity, and movement in world politics. *Millennium: Journal of International Studies, 28*(3), 499–521.

Airriess, C. A. (2001). The regionalization of Hutchison port holdings in mainland China. *Journal of Transport Geography, 9*, 267–278.

Alderton, T., & Winchester, N. (2002). Globalization and de-regulation in the maritime industry. *Marine Policy, 26*(1), 35–43.

Aligica, P. D. (2006). Institutional and stakeholder mapping: Frameworks for policy analysis and institutional change. *Public Organization Review, 6*, 79–90.

Allegret, J.-P., & Dulbecco, P. (2002). Global governance versus domestic governance: What roles for international institutions? *European Journal of Development Research, 14*(2), 173–182.

Altman, J. A., & Petkus, E, Jr. (1994). Towards a stakeholder-based policy process: An application of the social marketing perspective to environmental policy development. *Policy Sciences, 27*, 37–51.

Anderson, J. (1995). The exaggerated death of the nation-state. In J. Anderson, C. Brook, & A. Cochrane (Eds.), *A global world? Reordering political space* (pp. 65–112). Oxford: Oxford University Press.

Anderson, J. (1996). The shifting stage of politics: New medieval and postmodern territorialities? *Environment and Planning D, 14*(2), 133–153.

Anderson, J., & O'Dowd, L. (1999). Borders, border regions and territoriality: Contradictory meanings, changing significance. *Regional Studies, 33*(7), 593–604.

Andrews, D. M. (1994). Mobility and state autonomy: Toward a structural theory of international monetary relations. *International Studies Quarterly, 38*(2), 193–218.

Angelides, C., & Caiden, G. (1994). Adjusting policy thinking to global pragmatics and future problematics. *Public Administration and Development, 14*, 223–239.

Appadurai, A. (1996). *Modernity at large: Cultural dimensions of globalisation*. Minneapolis, MI: University of Minneapolis Press.

Arts, B., & Leroy, P. (2003). *Veranderingen van politiek, vernieuwing van milieubeleid klassieke en postmoderne arrangements*. Nijmegen: Nijmegen University Press. (in Dutch).

Ashford, D. E. (1977). Political science and policy studies: Towards a structural solution. *Policy Studies Journal, 5*, 570–583.

Ashley, R. K. (1988). Untying the sovereign state: A double reading of the anarchy probelmatique. *Millennium, 17*, 227–262.

Axelrod, R. (1987). *The Gamma Paradigm for studying the domestic influence on foreign policy*. Washington, DC: Annual Meeting of the International Studies Association.

Baindur, D., & Vegas, J. (2011). Challenges to implementing motorways of the sea concept—Lessons from the past. *Maritime Policy and Management, 38*(7), 673–690.

Bauman, Z. (2000). *Liquid modernity*. Cambridge: Polity Press.

Bennett, P. (2000). Environmental governance and private actors: Enrolling insurers in international maritime regulation. *Political Geography, 19*, 875–899.

Berger, S. (Ed.). (1981). *Organizing interests in Europe: Pluralism, corporatism and the transformation of politics*. Cambridge: Cambridge University Press.

Besley, T. (2001). Political institutions and policy competition. In G. Kochendorfer-Lucius & B. Pleskovic (Eds.), *Institutional foundations of a market economy* (pp. 102–109). Villa Borsig Workshop Series 2000. Washington, DC: World Bank.

Bhattacharya, C. B., & Sen, S. (2004). Doing better at doing good: When, why and how consumers respond to corporate social initiatives. *California Management Review, 47*(1), 9–24.

Bloor, M., Datta, R., Gilinskiy, Y., & Horlick-Jones, T. (2006). Unicorn among the cedars: On the possibility of effective 'smart regulation' of the globalized shipping industry. *Social and Legal Studies, 15*(4), 534–551.

Borghese, E. M. (1998). *The oceanic circle: Governing the seas as a global resource*. New York: United Nations University Press.

Borzel, T. A. (2007). *European governance—Negotiation and competition in the shadow of hierarchy*. European Union Studies Meeting, Montreal, May 7–10, 2007.

Borzel, T. A., & Risse, T. (2010). Governance without a state. Can it work? *Regulation and Governance, 4*, 113–134.

Braithwaite, J., & Drahos, P. (2000). *Global business regulation*. Cambridge: Cambridge University Press.

Brenner, N. (1998). Between fixity and motion: Accumulation, territorial organization and the historical geography of spatial scales. *Environment and Planning D, 16*(4), 459–481.

Brett, E. A. (1985). *The world economy since the war—The politics of uneven development*. Basingstoke: Macmillan.

Brooks, M. R., & Pallis, A. A. (2008). Assessing port governance models: Prices and performance components. *Maritime Policy and Management, 35*(4), 411–432.

Brown, A. E., & Stewart, J. (1993). Competing advocacy coalitions, policy evolution and airline deregulation. In P. Sabatier & H. Jenkins-Smith (Eds.), *Policy change and learning* (pp. 83–104). Boulder, CO: Westview Press.

Brunn, S. D. (1998). A treaty of silicon for the treaty of westphalia? New territorial dimensions of modern statehood. *Geopolitics, 3*(1), 106–131.

Buitelaar, E., Lagendijk, A., & Jacobs, W. (2007). A theory of institutional change: Illustrated by Dutch city-provinces and Dutch land policy. *Environment and Planning A, 39*, 891–908.

Bush, P. D. (1987). The theory of institutional change. *Journal of Economic Issues, XXI*(3), 1075–1116.

Calvert, R. (1995). Rational actors, equilibrium and social institutions. In J. Knight & I. Sened (Eds.), *Explaining social institutions*. Ann Arbor, MI: University of Michigan Press.

Campanelli, O. (2012). The global governance of maritime piracy. *Journal of Global Policy Governance, 1*, 73–84.

Campbell, J. L. (1998). Institutional analysis and the role of ideas in political economy. *Theory and Society, 27*, 377–409.

Camus, A. (1956) The Fall. London: Vintage.

Capek, K. (1936). *War with the Newts*. London: Geo. Allen and Unwin.

Castells, M. (1996). *The rise of the network society*. Cambridge, MA: Blackwell.

Cerny, P. G. (1995). Globalization and the changing logic of collective action. *International Organization, 49*(4), 595–625.

Charan, R., & Freeman, R. E. (1979). Building bridges with corporate constituents. *Management Review, 68*(11), 8–13.

Chowdhury, K. (2006). Interrogating "newness". Globalization and postcolonial theory in the age of endless war. *Cultural Critique, 62*, 126–161.

Clemens, E. S., & Cook, J. M. (1999). Politics and institutionalism: Explaining durability and change. *Annual Review of Sociology, 25*, 441–466.

Coleman, J. S. (1987). Norms as social capital. In G. Radnitzky & P. Bernholz (Eds.), *Economic imperialism*. New York: Paragon.

Collier, J., & Esteban, R. (1999). Governance in the participative organisation: Freedom, creativity and ethics. *Journal of Business Ethics, 21*(2–3), 173–188.

Colomar, J. M. (1995). Strategies and outcomes in Eastern Europe. *Journal of Democracy, 6*(2), 74–85.

Commission of the European Communities. (2001). *Multi-level governance: Linking and networking the various regional and local levels*. Report of Governance Working Group 4c, Brussels.

Commission of the European Communities. (2008). *Guidelines for an integrated approach to maritime policy: Towards best practice in integrated maritime governance and stakeholder consultation*, COM (2008) 395 final, Brussels.

Commission of the European Communities. (2009). *Towards an integrated maritime policy for better governance in the mediterranean*. Communication from the Commission to the Council and the European Parliament, COM (2009) 466 Final, Brussels.

Commons, J. R. (1968). *Legal foundations of capitalism*. Madison, WI: University of Wisconsin Press.

Cooper, R. (2000). *The post-modern state and the world order*. London: Demos.

Crawford, S. E. S., & Ostrom, E. (1995). A grammar of institutions. *American Political Science Review, 89*(3), 582–600.

Crosby, B. L. (1996). Policy implementation: The organizational challenge. *World Development, 24*(9), 1403–1415.

Croucher, S. L. (2003). Perpetual imagining. Nationhood in a global era. *International Studies Review, 5*, 1–24.

Davidson, G. (1971). *A dictionary of angels, including the fallen angels*. New York: The Free Press.

De Vivero, J. L. S., & Mateos, J. C. R. (2010). Ocean governance in a competitive world. The BRIC countries as emerging maritime powers—Building new geopolitical scenarios. *Marine Policy, 34*, 967–978.

Del Rosso Jr., S. J. (1995). The insecure state (What future for the state?). *Daedalus, 124*, 2.

Denzau, A. T., & North, D. C. (1994). Shared mental models: Ideologies and institutions. *Kyklos, 47*(1), 3–31.

Derthick, M., & Quirk, P. (1985). *The politics of deregulation*. Washington, DC: The Brookings Institution.

Deutsch, K. (1966). *Nationalism and social communication*. New York: MIT Press.

Dicken, P., Kelly, P. F., Olds, K., & Yeung, W. (2001). Chains and networks, territories and scales: Towards a relational framework for analysing the global economy. *Global Networks, 1*(2), 89–112.

Donaldson, T. (1999). Making stakeholder theory whole. *Academy of Management Review, 24*(2), 237–241.

Donaldson, T., & Preston, L. E. (1995). The stakeholder theory of the corporation: Concepts, evidence, and implications. *Academy of Management Review, 20*(1), 65–91.

Dopfer, K. (1991). Toward a theory of economic institutions: Energy and path dependency. *Journal of Economic Issues, XXV*(2), 535–550.

Douglas, J. N. H. (1968). Political geography and administrative areas: A method of assessing the effectiveness of local government areas. In C. A. Fisher (Ed.), *Essays in political geography*. London: Methuen.

Druckman, D. (1978). Boundary role conflict: negotiations as dual responsiveness. In J. W. Zartman (Ed.), *The negotiation process: Theories and applications* (pp. 100–101, 109). Beverly Hills, CA: Sage.

Duncan, S., & Savage, M. (1989). Space, scale and locality. *Antipode, 21*(3), 207–231.

Eising, R., & Kohler-Koch, B. (1999). Introduction. Network Governance in the European Union. In R. Eising & B. Kohler-Koch (Eds.), *Interessenpolitik in Europa*. Baden Baden: Nomos.

Evans, P. (1997). The eclipse of the state? *Reflections on stateness in an era of globalization, World Politics, 50*(1), 62–87.

Evans, P. (2001). *Beyond institutional monocropping: Institutions, capabilities and deliberative development* (Unpublished).

Evans, B., Reuschemeyer, D., & Skopcol, T. (1985). *Bringing the state back in*. Cambridge: Cambridge University Press.

Fletcher, A., Guthrie, J., Steane, P., Roos, G., & Pike, S. (2003). Mapping stakeholder perceptions for a third sector organization. *Journal of Intellectual Capital, 4*(4), 505–527.

Forsyth, C. J. (1993). Transnational corporations: Problems for study in the new international order of maritime shipping. *Maritime Policy and Management, 20*(3), 207–214.

Frankel, M. (1955). Obsolescence and technological change in a maturing economy. *The American Economic Review, 45*(3), 296–319.

Freeman, R. E. (1984). *Strategic management: A stakeholder approach*. London: Pitman.

Friedman, A. L., & Miles, S. (2002). Developing stakeholder theory. *Journal of Management Studies, 39*(1), 2–21.

Friedman, A. L., & Miles, S. (2006). *Stakeholders, theory and practice*. Oxford: Oxford University Press.

Fritz, J. (2010). Towards a 'new form of governance' in science policy relations in the European maritime policy. *Marine Policy, 34*, 1–6.

Furger, F. (1997). Accountability and systems of self-governance: The case of the maritime industry. *Law and Policy, 19*(4), 445–476.

Gekara, V. O. (2010). The stamp of neo-liberalism on the UK tonnage tax and the implications for British seafaring. *Marine Policy, 34*(3), 487–494.

Giddens, A. (1985). *The nation-state and violence*. Cambridge: Polity Press.

Gioia, D. A. (1999). Practicability, paradigms, and problems in stakeholder theorizing. *Academy of Management Review, 24*(2), 228–232.

Gordenker, L., & Weiss, T. G. (1995). Pluralising global governance: Analytical approaches and dimensions. *Third World Quarterly, 16*(3), 357–387.

Hall, P. A. (1986). *Governing the economy: The politics of state intervention in Britain and France*. Cambridge: Polity Press.
Hall, P. V. (2003). Regional institutional convergence? Reflections from Baltimore waterfront. *Economic Geography, 79*, 347–363.
Harding, A. (1997). Urban regimes in a Europe of the cities. *European Urban and Regional Studies, 4*(4), 291–314.
Harrison, J. S., & Freeman, R. E. (1999). Stakeholders, social responsibility, and performance: Empirical evidence and theoretical perspectives. *Academy of Management Journal, 42*(5), 479–485.
Hartshorne, R. (1950). The functional approach in political geography. *Annals of the Association of American Geographers, 40*(2), 95–130.
Harvey, D. (1981). The spatial fix—Hegel, Von Thunen and Marx. *Antipode, 13*(3), 1–12.
Harvey, D. (1990). *The condition of postmodernity*. Cambridge, MA: Blackwell.
Harvey, D. (2001). Globalization and the "spatial fix". *Geographische Review, 2*, 23–30.
Healey, P. (1998). Building institutional capacity through collaborative approaches in urban planning. *Environment and Planning A, 30*, 1531–1546.
Healey, P., Cars, G., Madanipour, A., & de Magalhaes, C. (2002). Transforming governance institutionalist analysis and institutional capacity. In G. Cars, P. Healey, A. Madanipour, & C. de Magalhaes (Eds.), *Transforming governance institutional analysis and institutional capacity* (pp. 6–28). Aldershot: Ashgate.
Held, D. (1991). Democracy, the nation-state and the global system. *Economy and Society, 20*(2), 138–172.
Hemmati, M. (2002). *Multi-stakeholder processes for governance and sustainability: Beyond deadlock and conflict*. London: Earthscan.
Hill, C. W. L., & Jones, T. M. (1992). Stakeholder-agency theory. *Journal of Management Studies, 29*(2), 131–154.
Hirst, P. (1997). *Democracy, civil society and global politics*. London: UCL Press.
Hobsbawm, E. (1998). The nation and globalization. *Constellations, 5*(1), 1–9.
Hohfeld, W. N. (1913). Some fundamental legal concepts as applied in the study of primitive law. *Yale Law Journal, 23*, 16–59.
Hosseus, D., & Pal, L. A. (1997). Anatomy of a policy area: The case of shipping. *Canadian Public Policy, XXIII*(4), 399–415.
Hughes, E. C. (1936). The ecological aspect of institutions. *American Sociological Review, 1*(2), 180–189.
Inayatullah, N., & Rupert, M. E. (1993). Hobbes, Smith and the problem of mixed ontologies in neorealist international political economy. In S. Rosow (Ed.), *The global economy as political space* (pp. 61–85). Boulder, CO: Lynne Rienner.
Jackson, J. E. (Ed.). (1990). *Institutions in American society*. Ann Arbor, MI: University of Michigan Press.
Jacobs, W. (2007). Port competition between Los Angeles and Long Beach: An institutional analysis. *Tijdschrift Voor Economische en Sociale Geografie, 98*, 360–372.
Jacobs, W., & Hall, P. V. (2007). What conditions supply chain strategies in ports? The case of Dubai. *GeoJournal, 68*, 327–342.
Jenkins, R. (1988). *Transnational corporations and uneven development*. London: Routledge.
Jenkins-Smith, H., & Sabatier, P. (1994). Evaluating the advocacy coalition framework. *Journal of Public Policy, 14*, 175–203.
Jessop, B. (1990). *State theory: Putting capitalist states in their place*. Cambridge: Polity Press.
Jessop, B. (2003). The future of the state in an era of globalization. *International Politics and Society, 3*, 30–46.
Jessop, B. (2004). Multilevel governance and multilevel metagovernance. Changes in the EU as integral moments in the transformation and reorientation of contemporary statehood. In I. Bache & M. Flinders (Eds.), *Multi-level governance* (pp. 49–74). Oxford: Oxford University Press.

Johnson, G., Smith, S., & Codling, B. (2000). Microprocesses of institutional change in the context of privatization. *Academy of Management Review, 25*(3), 572–580.

Johnston, R. J. (1995). Territoriality and the state. In G. Benko & U. Strohmayer (Eds.), *Geography, history and social science* (pp. 213–226). Dordrecht: Kluwer.

Jones, B. D., Sulkin, T., & Larsen, H. A. (2003). Policy punctuations in American political institutions. *American Political Science Review, 97*(1), 151–169.

Jordan, A. (2001). The European Union: An evolving system of multi-level governance… or government? *Policy and Politics, 29*(2), 193–208.

Kaldor, M. (1993). Civil society and democratic renewal. In A. Roberts (Ed.), *Power to the People? Economic self-determination and the regions*. London: European Dialogue, Ebert Foundation.

Kasperson, R. E., & Minghi, J. V. (1969). *The structure of political geography*. Chicago, IL: Aldine Publishing Co.

Keohane, R. O. (2002). *Global governance and democratic accountability*, Miliband Lectures, London School of Economics.

Keohane, R. O., & Nye, J. (1972). *Transnational systems and international organizations*. Boston, MA: Harvard University. (Unpublished Seminar Paper, Harvard Center for International Affairs).

Kjellberg, F. (1975). *Political institutionalization*. London: Wiley.

Knight, D. B. (1971). *Impress of authority and ideology on the landscape* (pp. 383–387). LXII: Tijdschrift Voor Economische en Sociale Geografie.

Knight, J. (1992). *Institutions and social conflict*. New York: Cambridge University Press.

Koelble, T. A. (1995). The new institutionalism in political science and sociology. *Comparative Politics, 27*(2), 231–243.

Kooiman, J. (Ed.). (1993). *Modern governance. New government-society interactions*. London: Sage.

Kostova, T., & Roth, K. (2002). Adoption of an organisation practice by subsidiaries of multinational corporations, institutional and relational effects. *Academy of Management Journal, 45*(1), 215–233.

Kotler, P., & Lee, N. (2005). *Corporate social responsibility: Doing the most good for your company and your cause*. Hoboken, NJ: Wiley.

Kovats, L. J. (2006). How flag states lost the plot over shipping's governance. Does a ship need a sovereign? *Maritime Policy and Management, 33*(1), 75–81.

Krasno, J. E. (Ed.). (2004). *United Nations: Confronting the challenges of a global society*. Boulder, CO: Lynne Rienner.

Lambert, J. (1991). Europe: the nation-state dies hard. *Capital and Class, 43*, 9–24. (Spring).

Lazarus, R. J. (2009). Super wicked problems and climate change: Restraining the present to liberate the future. *Cornell Law Review, 94*(5), 1153–1234.

Le Gales, P., & Harding, A. (1998). Cities and states in Europe. *West European Politics, 21*(3), 120–145.

Lee, K., Humphreys, D., & Pugh, M. (1997). 'Privatisation' in the United Nations system: Patterns of influence in three intergovernmental organisations. *Global Society, 11*(3), 339–357.

Leigh-Fermor, P. (2013). *The broken road. From the Iron Gates to Mount Atlas*. London: John Murray.

Lewis, D. K. (1969). *Convention: A philosophical study*. Cambridge, MA: Harvard University Press.

Lloyd's List. (2008). *French owners urge IMO to follow lead on Erika3*, October 13.

Lloyd's List. (2009a). *Is the European Commission really interested in ship safety*, January 14.

Lloyd's List. (2009b). *IMO under fire over fatigue related accidents*, February 19.

Lloyd's List. (2010). *Mordue quits European maritime safety chief role*, January 25.

Luck, E. C. (2005). How not to reform the United Nations. *Global Governance, 11*, 407–414.

MacLeod, G., & Goodwin, M. (1999). Space, scale and state strategy: Rethinking urban and regional governance. *Progress in Human Geography, 23*(4), 503–527.

Mangat, R. (2001). The death of distance? Globalization of international relations. *E-merge, Student Journal of International Affairs*.

Mann, M. (1997). Has globalization ended the rise and rise of the nation-state? *Journal of International Political Economy, 4*(3), 472–496.

March, J. G., & Olsen, J. P. (1984). The new institutionalism: Organizational factors in political life. *The American Political Science Review, 78*(3), 734–749.

March, J. G., & Olsen, J. P. (1989). *Rediscovering institutions: The organizational basis of politics*. New York: Free Press.

Martin, R. (2001). Geography and public policy: The case of the missing agenda. *Progress in Human Geography, 25*(2), 189–210.

Marx, K., & Engels, F. (1998). *The Communist Manifesto*. London: Verso.

Mason, K. J., & Gray, R. (1999). Stakeholders in a hybrid market: The example of air business passenger travel. *European Journal of Marketing, 33*(9/10), 844–858.

Matheson, M. J. (2001). United Nations governance of postconflict societies. *The American Journal of International Law, 95*(1), 76–85.

Mayntz, R. (1993). Policy-Netwerke und die Logik von Verhandlungssystemen. In A. Hertier (Ed.), *Policy analyse. Kritik und Neuorientierung* (Politische Vierteljahresschrift Sonderheft 24, pp. 148–168). Opladen: Westdeutscher Verlag.

Menger, K. (1963). *Problems in economics and sociology*. Urbana, IL: Univeristy of Illinois Press.

Meyer, J. W., & Rowan, B. (1977). Institutional organizations: Formal structure as myth and ceremony. *American Journal of Sociology, 83*, 340–363.

Michalowski, R. J., & Kramer, R. C. (1987). The space between laws: The problem of corporate crime in a transnational context. *Social Problems, 34*(1), 34–53.

Mittelman, J. H. (1999). Rethinking the 'New Regionalism' in the context of globalization. In J. Hettne, A. Inotai., & O. Sunkel (Eds.), *Globalism and the new regionalism* (Vol. 1, pp. 25–53). New Regionalism Series. London: Macmillan.

Mooij, J., & De Vos, V. (2003). *Policy processes: An annotated bibliography on policy processes, with particular emphasis on India*. Working Paper 221, Overseas Development Institute, London.

More, T. (1988). *Utopia*. London: Penguin.

Morgenthau, H. J. (2005). *Politics among nations: The struggle for power and peace*. New York: McGraw Hill.

Mukand, S. W., & Rodrik, D. (2005). In search of the Holy Grail: Policy convergence, experimentation and economic performance. *The American Economic Review, 95*(1), 374–383.

Murray, R. (1971). The internationalization of capital and the nation state. *New Left Review, 67*, 84–108.

Nettl, J. P. (1968). The state as a conceptual variable. *World Politics, 20*(4), 559–592.

Ng, A. K.-Y., & Pallis, A. A. (2010). Port governance reforms in diversified institutional frameworks; generic solutions, implementation asymmetries. *Environment and Planning A, 42*, 2147–2167.

North, D. C. (1986). The new institutional economics. *Journal of Institutional and Theoretical Economics, 142*, 230–237.

North, D. C. (1990). *Institutions: Institutional change and economic performance*. New York: Cambridge University Press.

North, D. C. (1993). *Economic performance through time*. Stockholm, Sweden: Alfred Nobel Memorial Prize Lecture.

Notteboom, T. E., & Winkelmans, W. (2002). *Stakeholder relations management in ports: Dealing with the interplay of forces among stakeholders in a changing competitive environment*. In International Association of Maritime Economists Annual Conference (IAME), Panama.

O'Riordan, T., Cooper, C. L., Jordan, A., Rayner, S., Richards, K. R., Runci, P., & Yoffe, S. (1998). Institutional frameworks for political action. In S. Rayner & E. L. Malone (Eds.), *Human choice and climate change, the societal framework* (Vol. 1, pp. 345–440). Columbus, OH: Battelle Press.

Oakerson, R. J., & Parks, R. B. (1988). Citizen voice and public entrepreneurship: The organizational dynamic of a complex metropolitan county. *Publius, 18*, 91–112.

Oberschall, A., & Leifer, E. M. (1986). Efficiency and social institutions: Economic reasoning in sociology. *Annual Review of Sociology, 12*, 233–253.

OECD. (2000). *Reducing the risk of policy failure. Challenges for regulatory compliance*, Organisation for Economic Cooperation and Development, Paris.

Olsen, J. P. (1981). Organizational integration in government. In P. Nystrom & W. Starbuck (Eds.), *Handbook of organizational design* (Vol. 2). New York: Oxford University Press.

Opello, W. C, Jr., & Rosow, S. J. (2004). *The nation-state and global order: A historical introduction to contemporary politics*. Boulder, CO: Lynne Reinner.

Ostrom, V. (1980). Artisanship and artifact. *Public Administration Review, 40*(4), 309–317.

Ostrom, E. (1986). An agenda for the study of institutions. *Public Choice, 48*, 3–25.

Ostrom, E. (1990). *Governing the commons: The evolution of institutions for collective action*. New York: Cambridge University Press.

Padgett, J. F. (1981). Hierarchy and ecological control in federal budgetary decision-making. *American Journal of Sociology, 87*, 75–129.

Pallis, A. A. (2005–2006). Maritime interest representation in the EU. *European Political Economy Review, 3*(2), 6–28.

Pallis, A. A., & Tsiotis, S. G. P. (2006). *Inside EU-level maritime interest groups: Structures, lobbying practices and governance*. Report of the Jean Monnet Centre in European Port Policy, University of the Aegean, Chios, Greece.

Pallis, A. A., & Tsiotis, S. G. P. (2008). Maritime interests and the EU port services directive. *Trasporti Europei, 38*, 17–31.

Parker, M. (2000). Angelic organization: Hierarchy and the tyranny of Heaven. *Organization Studies, 30*(11), 1281–1299.

Peters, B. G. (2002). *Governance. A garbage can perspective*. Political Science Series 84, Institute for Advanced Studies, Vienna.

Phillips, S. D., & Orsini, M. (2002). *Mapping the links; citizen involvement in policy processes*. CPRN Discussion Paper F 21, Canadian Policy Research Networks, Ottawa.

Picciotto, S. (1991). The internationalization of the state. *Capital and Class, Spring, 43*, 43–63.

Picciotto, S. (1997) Networks in international economic integration, *Northwestern Journal of International Law and Business, 17*(2/3), 1014–1056.

Picciotto, S. (1998). Globalization, liberalization, regulation. In *Conference on Globalization, The Nation-State and Violence, University of Sussex*.

Pierson, P. (1993). When effect becomes change. *World Politics, 45*, 595–628.

Pomeroy, R., & Douvere, F. (2008). The engagement of stakeholders in the marine spatial planning process. *Marine Policy, 32*, 816–822.

Potter, A. L. (1979). *Political institutions, political decay and the Argentine crises of 1930*, Unpublished Doctoral Dissertation, Stanford University, Stanford, CA.

Putnam, R. D. (1988). Diplomacy and domestic politics: The logic of two-level games. *International Organization, 42*(3), 427–460.

Quartey, S. H., & Puplampu, B. B. (2012). Employee health and safety practices: An exploratory and comparative study of the shipping and manufacturing industries in Ghana. *International Journal of Business and Management, 7*(23), 81–95.

Ramachandran, V., Rueda-Sabater, E., & Kraft, R. (2009). Rethinking fundamental principles of global governance: How to represent states and populations in multilateral institutions. *Governance, 22*(3), 341–351.

Reisman, W. M. (1993). The constitutional crisis in the United Nations. *The American Journal of International Law, 87*(1), 83–100.

Rhenman, E. (1973). *Organization theory for long range planning*. New York: Wiley.
Rhodes, R. A. W. (1997). *Understanding governance. Policy networks, governance, reflexivity and accountability*. Buckingham: Open University Press.
Riker, W. (1980). Implications from the disequilibrium of majority rule for the study of institutions. *American Political Science Review, 74*, 432–446.
Robins, R. S. (1976). *Political institutionalization and the integration of elites*. Beverly Hills, CA: Sage.
Robyn, D. (1987). *Braking the special interests: Trucking deregulation and the politics of policy reform*. Chicago, IL: University of Chicago Press.
Rodrik, D. (2000). Institutions for high quality growth: What they are and how to acquire them. *Studies in Comparative Development, 35*(3), 1–33.
Roe, M. S. (2007a). *European Union Maritime Policy. The role of new member states and the need for a review of governance*, International Shipping Management Forum 07, May 10–11, Athens, Greece.
Roe, M. S. (2007b). Shipping, policy and multi-level governance. *Maritime Economics and Logistics, 9*, 84–103.
Roe, M. S. (2007c). *Jurisdiction, shipping policy and the "new" states of the European Union*. International Association of Maritime Economists Annual Conference (IAME), Athens, July 4–6.
Roe, M. S. (2007d). Shipping policy governance: A post-fordist interpretation. In I. Visvikis (Ed.), *Trends and developments in shipping management* (pp. 131–142). Athens: T and T Publishing.
Roe, M. S. (2007e). Shipping policy, governance and jurisdictional friction. *Economic Outlook*, 104, 8–15 January.
Roe, M. (2008a). Safety, security, the environment and shipping—The problem of making effective policies. *WMU Journal of Maritime Affairs, 7*(1), 263–279.
Roe, M. S. (2008b). *State versus the stakeholder: The governance of the maritime sector.* International Association of Maritime Economists Annual Conference (IAME), Dalian, April 2–4.
Roe, M. S. (2009a). *Maritime capitalist liposuction—A postmodern interpretation of maritime governance failure*. International Association of Maritime Economists Annual Conference (IAME), Copenhagen, June 24–26.
Roe, M. S. (2009b). *Policy failure in shipping: A global disaster of epic dimensions*. 2nd International Symposium on Maritime Safety, Security and Environmental Protection, National Technical University of Athens, Athens, Greece, September 17–18.
Roe, M. S. (2009c). Multi-level and polycentric governance: Effective policymaking for shipping. *Maritime Policy and Management, 36*(1), 39–56.
Roe, M. S. (2009d). Maritime governance and policy-making failure in the European Union. *International Journal of Shipping, Transport and Logistics, 1*(1), 1–19.
Roe, M. S. (2010a). *Globalization and maritime policy-making—The case for governance rehabilitation*. International Association of Maritime Economists Annual Conference (IAME), Lisbon, July 7–9.
Roe, M. S. (2010b). Shipping policy and globalization: Jurisdictions, governance and failure. In C. Grammenos (Ed.), *The handbook of maritime economics and business* (pp. 539–556, 2nd ed.). London: Lloyd's List.
Roe, M. S. (2013). *Maritime governance and policy-making*. London: Springer.
Roe, M. S., & Selkou, E. (2006). Multi-level governance, shipping policy and social responsibility. In *International Conference 'Shipping in the Era of Social Responsibility' in honour of the late Professor Basil Metaxas (1925–1996), Argostoli, Kefalonia, Greece*.
Rowley, T. J. (1997). Moving beyond dyadic ties: A network theory of stakeholder influences. *Academy of Management Review, 22*(4), 887–910.
Ruggie, J. G. (1993). Territorality and beyond: Problematizng modernity in international relations. *International Organization, 47*(1), 139–174.

Rugman, A. (Ed.). (1982). *New theories of the multinational enterprize*. London: Croom Helm.
Sabatier, P. A. (1998). The advocacy coalition framework: Revisions and relevance for Europe. *Journal of European Public Policy, 5*(1), 98–130.
Sampson, H., & Bloor, M. (2007). When Jack gets out of the box: The problems of regulating a global industry. *Sociology, 41*(3), 551–569.
Savage, G. T., Nix, T. W., Whitehead, C. J., & Blair, J. D. (1991). Strategies for assessing and managing organizational stakeholders. *Academy of Management Executive, 5*(2), 61–75.
Scharpf, F. W. (1977). Does organization matter? Task structure and interaction in the ministerial bureaucracy. In E. H. Burack & A. R. Negandhi (Eds.), *Organizational design: Theoretical perspectives and empirical findings*. Kent, OH: Kent State University Press.
Scharpf, F. W. (1993). Positiv und negativ Koordination in Verhandlungssystemen. In A. Hertier (Ed.), *Policy-analyse. Kritik und Neuorientierung* (pp. 57–83). PVS Sonderheft 24, Opladen: Westdeutscher Verlag.
Scharpf, F. W. (1994). Community and autonomy: Multi-level policy-making in the European Union. *Journal of European Public Policy, 1*(2), 219–242.
Schat, P. (1969). Political geography: A review. *Tijdschrift Voor Economische en Sociale Geografie, 60*(4), 255–260.
Schmitter, P., & Lehmbruch, G. (1979). *Trends towards corporatist intermediation*. Beverly Hills, CA: Sage.
Schotter, A. (1981). *The economic theory of social institutions*. Cambridge: Cambridge University Press.
Schrier, E., Nadel, E., & Rifas, B. E. (1984). Forces shaping international maritime transport. *World Economy, 7*, 87–102.
Schwind, M. (1970). Die Aufgaben einer Politischen Geografie in neuer Sicht. *Geographische Rundschau, XXII*, 3.
Scott, W. R. (1987). The adolescence of institutional theory. *Administrative Science Quarterly, 32*, 493–511.
Selkou, E., & Roe, M. S. (2004). *Globalization, policy and shipping*. Cheltenham: Edward Elgar.
Selkou, E., & Roe, M. S. (2005). Container shipping policy and models of multi-level governance. *Annals of Maritime Studies, 43*, 45–62.
Selznick, P. (1957). *Leadership in administration*. New York: Harper and Row.
Sheldon, A. (1980). Organizational paradigms: A theory of organizational change. *Organizational Dynamics, 8*(3), 61–80.
Shepsie, K. A. (1975). Congressional committee assignments: An optimization model with institutional constraints. *Public Choice, 22*, 55–78.
Shepsie, K. A. (1979). Institutional arrangements and equilibrium in multidimensional voting models. *American Journal of Polictical Science, 23*, 27–59.
Shepsie, K. A. (1989). Studying institutions some lessons from the rational choice approach. *Journal of Theoretical Politics, 1*, 131–147.
Shepsie, K. A., & Weingast, R. (1983). Institutionalizing majority rule: A social choice theory with policy implications. *American Economic Review, 73*, 357–372.
Shepsie, K. A., & Weingast, R. (1984). When do rules and procedure matter? *Journal of Politics, 46*, 206–221.
Shepsie, K. A., & Weingast, R. (1987). The institutional foundations of committee power. *American Political Science Review, 81*, 85–104.
Shimko, K. (1992). Realism, neo-realism and American liberalism. *Review of Politics, 54*, 281–302.
Shinohara, M. (2005). Quality shipping and incentive schemes: From the perspective of the institutional economics. *Maritime Economics and Logistics, 7*, 281–295.
Silverstein, H. B. (1976). Technological politics and maritime affairs—Comparative participation in the intergovernmental maritime consultative organization. *Journal of Maritime Law and Commerce, 7*, 367–407.
Silverstein, H. B. (1978). *Superships and nation-states*. Boulder, CO: Westview Press.

Skowronek, S. (1982). *Building a New American State*. Cambridge: Cambridge University Press.

Slaughter, A.-M. (1997). The real new world order. *Foreign Affairs, 76*(5), 183–198.

Sletmo, G. K. (2001). The end of national shipping policy? A historical perspective on shipping policy in a global economy. *International Journal of Maritime Economics, 3*, 333–350.

Sletmo, G. K. (2002a). The rise and fall of national shipping policies. In C. Grammenos (Ed.), *The handbook of maritime economics and business* (pp. 471–494). London: Lloyd's of London Press.

Sletmo, G. K. (2002b). *National shipping policy and global markets: A retrospective for the future.* International Association of Maritime Economists Annual Conference (IAME), Panama City, November 13–15.

Snyder, G. H., & Diesing, P. (1977). *Conflict among nations: Bargaining, decision-making and systems structure in international crises*. Princeton, NJ: Princeton University Press.

Stalder, F. (2006). *Manuel Castells*. Cambridge: Polity Press.

Stoker, G. (1998). Governance as theory: Five propositions. *International Social Science Journal, 50*(155), 17–28.

Stone, C. (1989). *Regime politics: Governing Atlanta 1946-88*. Lawrence, KS: University Press of Kansas.

Strange, S. (1976). Who runs world shipping? *International Affairs, 52*(3), 346–367.

Strange, S. (1996). *The retreat of the state. The diffusion of power in the world economy.* Cambridge: Cambridge University Press.

Streeck, W., & Schmitter, P. C. (Eds.). (1985). *Private interest government beyond market and state*. London: Sage.

Stubbs, P. (2005). Stretching concepts too far? *Multi-level governance, policy transfer and the politics of scale in South-East Europe, Southeast European Politics, vi, 2*, 66–87.

Sutherland, M., & Nichols, S. (2006). Issues in the governance of marine spaces. In M. Sutherland (Ed.), *Administering marine spaces: International issues* (FIG Publications No. 36, pp. 6–20). International Federation of Surveyors, Copenhagen, Denmark.

Sutton, R. (1999). *The policy process: An overview*. London: Overseas Development Institute.

Swyngedouw, E. (2000). Authoritarian governance, power and the politics of rescaling. *Environment and Planning D, 18*, 63–76.

Thompson, J. (1967). *Organizations in action*. New York: McGraw-Hill.

Thompson, G. (1999). Introduction: Situating globalization. *International Social Sciences Journal, 160*, 139–152.

Tilly, C. (1975). *The formation of National States in Western Europe*. Princeton, NJ: Princeton University Press.

Tradewinds. (2008a). *Accident process to cast wider net*, February 29.

Tradewinds. (2008b). *A good plan that could go wrong*, February 29.

Tradewinds. (2008c). *Time to reassess the safety regime*, March 14.

Ullmann-Margalit, E. (1977). *The emergence of norms*. Oxford: Clarendon.

Van Leeuwen, J., & Van Tatenhove, J. (2010). The triangle of marine governance in the environmental governance of Dutch offshore platforms. *Marine Policy, 34*, 590–597.

Van Tatenhove, J. (2008). *Innovative forms of marine governance: A reflection*. Paper presented at Dies NatalisWageningen University, Wageningen, Netherlands.

Vanelslander, T. (2011). Port and maritime governance and regulation: Emerging issues. *Maritime Policy and Management, 38*(3), 215–217.

Veblen, T. (1915). *Imperial Germany and the industrial revolution*. London: Macmillan.

Vogler, J. (1995). *The global commons: A regime analysis*. London: Wiley.

Von Hayek, F. (1945). The use of knowledge in society. *American Economic Review, 35*, 319–330.

Von Hayek, F. (1967). Notes on the evolution of rules of conduct. In F. A. Von Hayek (Ed.), *Studies in philosophy, politics and economics*. Chicago, IL: Chicago University Press.

Walker, R. B. J. (1991). State sovereignty and the articulation of political space/time. *Millennium, 20*, 445–461.

Walton, R. E., & McKersie, R. B. (1965). *A behavioral theory of labor negotiations: An analysis of a social interaction system*. New York: McGraw-Hill.

Waltz, K. N. (1979). *Theory of international politics*. Reading: Addison-Wesley.

Wang, J. J., Ng, A. K.-Y., & Olivier, D. (2004). Port governance in China: A review of policies in an era of internationalizing port management practices. *Transport Policy, 11*, 237–250.

Weingast, B. R. (1995). The economic role of political institutions: Market-preserving federalism and economic development. *The Journal of Law, Economics and Organization, 7*(1), 1–31.

White, H. C. (1992). *Identity and control: A structural theory of social action*. Princeton, NJ: Princeton University Press.

Whittesley, D. (1935). The impress of effective central authority upon the landscape. *Annals of the Association of American Geographers, 25*(2), 85–97.

Williams, D. (1987). *The specialized agencies and the United Nations*. London: C. Hurst and Company.

Williamson, O. E. (1985). *The economic institutions of capitalism: Firms, markets, relational contracting*. New York: Free Press.

Wirth, C. (2012). Ocean governance, maritime security and the consequences of modernity in north-east Asia. *The Pacific Review, 25*(2), 223–245.

Wood, E. M. (2001). Global capital. National states. In M. Rupert & H. Smith (Eds.), *Historical materialism and globalization* (pp. 17–39). New York: Routledge.

Wright, H. T. (1977). Recent research on the origin of the state. *Annual Review of Anthropology, 6*, 379–397.

Wright-Mills, C. (1959). *The sociological imagination*. New York: Oxford University Press.

Wuisan, L., Van Leeuwen, J., & Van Koppen, C. S. A. (2012). Greening international shipping through private governance: A case study of the clean shipping project. *Marine Policy, 36*, 165–173.

Yeung, H. W. (1998). Capital, state and space: Contesting the borderless world. *Transactions of the Institute of British Geographers NS, 23*, 291–309.

Zafonte, M., & Sabatier, P. (1997). *Shared beliefs and functional interdependence as determinants of ally networks in overlapping subsystems*. Unpublished Manuscript, University of California.

Zucker, L. G. (1977). The role of institutionalization in cultural resistance. *American Economic Review, 42*, 726–743.

Zucker, L. G. (1983). Organizations as institutions. *Research in the Sociology of Organizations, 2*(1), 1–47.

Zucker, L. G. (1988). Where do institutional patterns come from? Organisations as actors in social systems. In L. G. Zucker (Ed.), *Institutional patterns and organizations: Culture and environment* (pp. 23–52). Cambridge, MA: Ballinger.

Zurn, M. (2003). Globalization and global governance: From societal to political denationalization. *European Review, 11*(3), 31–364.

Chapter 2
Form

Abstract The problems that characterise the maritime sector are constantly developing, and if there is one feature of the shipping and port industry that remains consistent, it is that it is always changing. Apart from the obvious nature of the maritime industry in that it centres upon the movement of goods and people (and consequently ships) around the world, it also displays many other features of constant change. Thus, major maritime problems such as environmental degradation, low safety standards, security violations and issues of commercial efficiency do not stand still but either by nature constantly reflect differing failures or are part of a moving programme of events deliberately manipulated by those central to the industry. The structure and development of maritime governance has failed to reflect this and instead has been characterised by institutional stasis and regulations, and rules and policy directives that are designed for a single point in time. Maritime governance is incessantly chasing maritime problems, failures and inadequacies generated by the shipping industry as it operates to its own strict commercial principles, taking advantage of any anachronisms in policy that have developed since their last revision. Maritime governance needs to address a requirement to be flexible in its institutional structures, in the vehicles it uses to face these problems, in the agencies that deliver the policies that emerge and in the nature of the measures actually taken. There is also a need to understand the difference between the current static governance and the dynamic governance that could meet these needs. This requires at the outset to address the issue of form—a central feature of the static approach to policy that its formal position currently takes. This chapter looks at the static nature of maritime governance and its focus upon form rather than process. It concludes with a discussion of the related concepts of path dependency and lock-in and their relationship to policy-making for shipping.

> Thus, whether presented as elements of a spatial distribution, as unique assemblages of physical facts and human artefacts, or as localized spatial forms, places and regions have been portrayed as little more than frozen scenes of human activity. Pred (1984: 279).

> The tendency is apparently involuntary and immediate to protect oneself against the shock of change by continuing in the presence of altered situations the familiar habits, however incongruous, of the past. Morrison (1966: 9, quoted in Gould 1973: 253).

M. Roe, *Maritime Governance*, DOI 10.1007/978-3-319-21747-5_2

The problems that characterise the maritime sector are constantly developing, and if there is one feature of the shipping and port industry that remains consistent, it is that it is always changing. Apart from the obvious nature of the maritime industry in that it centres upon the movement of goods and people (and consequently ships) around the world, it also displays many other features of constant change. Thus, major maritime problems such as environmental degradation, low safety standards, security violations and issues of commercial efficiency do not stand still but either by nature constantly reflect differing failures or are part of a moving programme of events deliberately manipulated by those central to the industry. Shipping is in constant flux and is not best represented by a static profile but by one that reflects the movement of different actors, operating at different jurisdictions where change occurs all the time.

Meanwhile, the structure and development of maritime governance has failed to reflect this and instead has been characterised by institutional stasis and regulations, and rules and policy directives that are designed for a single point in time. Undoubtedly, new policies do emerge over a period of time (e.g. the innumerable attempts at a new maritime framework for the European Union in the early years of the twenty-first century), but in many ways, this just reflects the inadequacies of the current design in that new policies are repeatedly needed to replace those outdated—and instead, it might be preferable if a single but adaptable policy process could be designed which moved with the changes that take place constantly throughout the sector. Maritime governance is incessantly chasing maritime problems, failures and inadequacies generated by the shipping industry as it operates to its own strict commercial principles, taking advantage of any anachronisms in policy that have developed since their last revision.

Given that it is clear that maritime governance needs to address the need to be flexible in its institutional structures, in the vehicles it uses to face these problems, in the agencies that deliver the policies that emerge and in the nature of the measures actually taken, there is also a need to understand the difference between the current static governance and the dynamic governance that could meet these needs. This requires at the outset to address the issue of form—a central feature of the static approach to policy that its formal position currently takes—before moving on to look at change and how this might be incorporated into the governance of the maritime sector. How process can be made part of policy-making; how flexibility can be an inherent feature; how flux, metamorphosis and entropy can be woven into governance and moves made away from the institutionalised paralysis that the current jurisdictional structure represents and the maritime failures that are generated as a consequence.

To understand change, we need first to understand what characterises a situation where there is none. As we shall see, there has been considerable debate in a number of disciplines about how to understand the features that characterise these two situations and which have centred on differences of opinion about the importance of form against that of process and, in other words, of stasis against change or a static approach against that of one that is dynamic. Consequently, we must begin with form and attempt to understand how it relates to contemporary maritime governance.

A Photographic Form of Governance

> Form poses a problem which appeals to the utmost resources of our intelligence, and it affords the means which charm our sensibility and even entice us to the verge of frenzy. Form is never trivial or indifferent: it is the magic of the works. Dalcq (1939).

Despite the miracles of modern digital photography and the flexibility intrinsic to the application of software such as Adobe Photoshop, photographs ultimately remain a snapshot in time—albeit sometimes a false one. Maritime policy is the same. Much—if not all—is just that and applicable at best only to the time it was agreed and regularly out of date before it is even applied. As we saw in the previous chapter, this formal approach to policy-making is one of the major inadequacies of maritime governance and along with other problems stemming from the relationships between globalisation and the nation-state, institutional rigidity and the inadequacies of stakeholder representation and needs to be addressed if there is to be any progress in resolving maritime failure. In fact, it can be seen in many ways as the most important of all these issues as without recognition of the failure of static policy-making, little else can be achieved in improving the other areas of deficiency. So it is to form we must first turn.

Whyte (1954: 23–27, 229–237) provides an extensive historical background to form including a detailed chronology from 2500 BC. His interpretation of the term is wide reflecting the scope of the concept as clearly extensive.

Earliest consideration of form was through the Egyptian flatland to be succeeded by the Ancient Greek concepts of proportion, balance and symmetry. Aristotle's organic forms followed. With Christianity came the desire to continue Plato and Pythagoras's search for a universal form which would transcend the individual as something 'nobler and more lasting' (Whyte 1954: 23). By the time of the Middle Ages, form was reinterpreted as not just a visually perceived shape but as a 'divinely ordered hierarchy of forms'. Figure 2.1 gives some idea of the development of its definition. What is remarkable is the change in attitude to form between 1600 and 1650, from Francis Bacon's declaration that 'the form of a thing is its very essence', to triviality defined as a 'mere formality', merely a matter of form.

Whyte (1951: 2) was the first to note how form had been neglected and that no scientific philosophy existed to act as a guide although as we shall see later the debate about atomism dates back to Democritus and the Ancient Greeks. Gibson (1951: 403–404) examined form from a psychological perspective and considered that there were at least three general meanings of the term, something he termed the 'substantial shape of an object in three dimensions'. Secondly, the 'projection of such an object on a flat surface, either by light from the object or by the human act of drawing creating images, drawings, etc.' Thirdly, there is the 'abstract geometrical form composed of imaginary lines, planes or families of them'.

Whyte (1944: 173–174, 1954: 10–11) further appreciates form and outlines Plato's conception of eternal ideas or universal intelligible *Forms*. Plato suggested that the 'Demiurge' or skilled workman was engaged in making the world and

Pythagoras	Number
Aristotle	Realized form
Euclid	Quantitative relations of space
Aquinas	Determining principle of everything. The mind of God
Leonardo da Vinci Francis Bacon	The arrangement of the spatial parts that make up the whole
Kepler Gallileo Descartes	Scientific measurement. Exactness. The world is built of minute parts and so form becomes less important
Vitalists Gestalt School	A return to the value of the whole (form) rather than the parts
Structure	As a new idea of pattern of relationships including both the minute parts and forms of the world
Atomism	Controls the form an object takes. Simultaneous consideration of both the internal and external configuration
Holism	As a response to atomism, looking at the universe as a system of systems

Fig. 2.1 The meaning of form over time. Derived from Whyte (1954: 23–27)

created all things to be like Himself as much as possible. He thus took from his own real world the forms 'representing the generic idea of everything and used them to make copies in the world of appearances'. This was what Whyte called the 'transient world of phenomena'. Taleb (2007: xxv) was highly critical of this 'Platonicity' with its overemphasis on well-defined forms whether 'objects, like triangles, or social notions, like utopias, even nationalities'. These objects tend to become 'privileged over other less elegant objects, those with messier and less tractable structures'. These include the less tangible and more difficult to grasp dynamic models of governance.

Thus, within every changing appearance, there is an unchanging form that dictates appearance. Whyte goes on to indicate that there are only 20 primary ideas (those that help us to understand the universe) that have ever been produced in the lifetime of mankind of which 12 are dominant (Fig. 2.2). These ideas are the main instruments of intellectual understanding and are all that is necessary to understand virtually everything else. Form is one of them.

Whyte (1954: 14) continues by suggesting that form is one of the more unclear ideas meaning almost anything to anyone, ambiguous and fertile with possibilities. 'Form is the dark horse'. And this is despite its clear significance. The Greek *Eidos*, *Schema* and *Morphe* along with the Latin *forma* all can be translated as form but really just mean 'the qualities that make any thing what it is'. Its significance had been emphasised by Whyte's (1951: 229–237) attempt at a chronology of form which reflects the importance of form as a concept over many millennia. He also cites the example of Leonardo da Vinci who considered all positions of

Fig. 2.2 The 12 primary ideas.
Source Whyte (1954: 11)

Number
Space Time
Atoms
Energy
Organism
Mind
Unconscious Mind
Historical process
Statistics
Form
Structure

the human body—'repose, movement, running, standing, supported, sitting, leaning, kneeling, lying down, suspended, carrying or being carried, thrusting, pulling, striking, being struck, pressing down and lifting up'. Apparently, there are ten types of noses in profile and eleven in full face. These are expressions of form but with clear relationship to process, something that Whyte re-emphasises throughout his work (see, for example, Whyte 1954: 48) and in particular that the forms of every part of the universe are harmoniously related to all the processes that characterised the holistic system.

We cannot leave the issue of form and the historical development of the concept without a few words on atomism and holism (Whyte 1951: 2–3, 1954: 51–54). The diversity in opinion between these two schools is well documented, and we shall spend relatively little time on it here. The disagreements date back to the Ancient Greeks. The atomists are represented by Leucippus, Democritus, Gassendi, Newton, Boyle, Dalton and Rutherford. Holists include Aristotle, Goethe, Bergson, Whitehead and Smuts.

As far as atomism goes, Lucretius's work of 56 BC (Lucretius 1995) provides a detailed philosophy, whilst Smuts' *Holism and Evolution* (1927) gives an opposing view. Atomism asserts that the universe is made up of ultimate particles each of which is indivisible and permanent and any changes which can be observed are simply due to their reorganisation. This idea or complex and frequent movement of indivisible particles neatly brings process alongside form. Appealing to the exact sciences, it remains short of soul and passion.

Holism suggests that the world is made up of hierarchies of unities, each part of a grand order. There is more to change in this system than the rearrangement of particles. Forms are related to the harmonious systems that characterise the hierarchies of unity.

> the 'feminine', artistic, poetic, inventive and imaginative component of human personality uses the… holistic approach, while the 'masculine' analytical, classifying component uses the… atomistic method.

In truth, we all use both masculine and feminine components, whilst both use form and process as part of their analytical framework.

Vogt (1960: 19–20) is sure that the reason why conceptualising process has been so much more difficult than the analysis of structure is because of an assumed 'premise about the nature of human society and culture that defeats us from the start'. The assumption is that cultural and social systems remain in equilibrium unless they are 'hit' by some outside force or develop an internal and unsupportable strain. The scientific model is then to establish how equilibrium is restored. Vogt suggests that this follows a long tradition characterised by Durkheim (1968) aimed at preserving social order.

He suggests we should start with the premise that change is inevitable and that as a result considering society as a focussed equilibrium is purely a convenience, undertaken to help us understand what is going on but in no way representing the reality of change (Leach 1954: 4; Henry 1955; Herskovits 1955: 443–446). Change is always present, and this is just as true in governance, policy-making and every commercial sector towards which these policies might be applied—including shipping.

Van Ginkel (1961: 57) suggests that form as applied to architecture and planning is all about impression rather than detail and that the form of a place or building is the feeling that the observer experiences, largely unaffected by specific design. Forward (1967) provides the example of the Port of St John's in Newfoundland, Canada, as an interpretation of form from a geographical perspective. Meanwhile, Eichenbaum and Gale (1971: 526) take a scientific view of form and find that to do so is confused by the commonplace usage of the word. Taking the dictionary definition as a starting point, they emphasise that form is the 'visible aspect of a thing, usually taken in the narrow sense of shape or configuration as distinguished from such properties as colour'. It often implies a 'value judgement such as orderly arrangement or regularity'.

Furster (1963: 75) looks at form in relation to structure within a geographical context and suggests that form means:

> the physical pattern of land use, population distribution and service networks while structure signifies the spatial organisation of human activities and interrelationships.

Meanwhile, Pitzl (1974: 84) considers the concept of form as intensely subtle and as a result presents difficulties in definition. Quoting Whyte (1951: 2–3), he comments on how form and matter possess a very close relationship but that:

> …it is a remarkable fact that throughout this debate, that is during eighty human generations, no one has suggested how to combine them into one simple and comprehensive way of thinking: hence much of the disorder in thought.

Johnson and Pitzl (1981: 216) comment on how loosely the term 'form' had been applied in geography and as a result taken on an elusive concept suggesting like Whyte (1954: 24) that it has become rather a question of faith. Form is an 'essence, or as a gestalt or whole, or as an idea and ideal, or in a symbolic sense, is subjective and without precise definition' and as such avoids definition. The result was indeterminate terminology with terms used such as shape, figure, structure, pattern order, arrangement, configuration, plan outline and more all used interchangeably (Gibson 1951: 403). Van Ginkel (1961) emphasises the indefinite nature of form quoting the Augustinian view that 'form is the very mode of existence, the manifestation of being'. It is rather in this context that current maritime governance designed around form can also be seen to be indeterminate and indistinct. Pitzl concludes quoting Attneave and Armoult (1966):

> Relatively few scientists have seriously applied themselves to the problems of analyzing and describing form: these problems seem to have fallen into the cracks between sciences, and no general quantitative morphonomy has ever been developed.

Cohen and Lewis (1967: 1) suggest that it is almost impossible to separate form and function. Form they define as the 'shape and structure of anything', whilst function is 'the natural, proper or characteristic action of anything'. This reflects the difference between form and process that is central to the problems of maritime governance and implies that even if we can move towards a more flexible and changeable approach to policy-making and policy design and application, the features of both form and function will need to be accommodated. Form cannot just be abandoned although its over-domination in maritime governance needs to be addressed. They go on to emphasise morphology as central to form—the 'features collectively, comprised in the form and structure of an organism or any of its parts' compared with function which was 'any quality, trait or fact so related to another that it is dependent upon and varies with that other'.

Lynch (1960: 105–107) outlines in some detail the qualities of form that are significant in terms of design. Many of these gave an indication of the nature of form and in particular its static characteristics including:

- Singularity—including sharpness of boundary, closure and contrast. The distinct opposite of blurring that comes with movement and change.
- Simplicity—limitation of parts, clarity and lacking the distortion that comes with change.
- Continuity—emphasising the continuation of what already exists and placing emphasis on repetition of what has already gone.
- Dominance—of a theme that has been identified in the past.
- Clarity of joint—reflecting the separation of the past from later events and as a result diluting the changes that occur in time.

Kaplan (1996: 143) stresses the significance of place—and indirectly therefore the form or shape of location and its historical connotations, in contrast to development, change and the future suggesting that:

> A place on the map is also a place in history. Rich (1994).

Buttimer (1976: 278) considers that many social science models are both opaque and static and need to be more dynamic to be useful. Her comments in relationship to space are relevant if only indirectly:

> For us, space cannot be reduced to geometric relationships; relations which we establish as if, reduced to the simple role of spectators or scientists, we were ourselves outside space. We live and act in space, and our personal lives, as well as the social life of

humanity, unfolds in space. Life spreads out in space without having a geometric extension in the proper sense of the word. We have need of expansion, of perspective, in order to live. Space is as indispensable as time in the development of life.

This resonates with May and Thrift's (2003: 2) interpretation of the dualism of time and space where time is seen as the domain of progress and dynamism, and space 'relegated to the realm of stasis and thus excavated of any meaningful politics', with reference to Harvey (1993), Massey (1993) and Hetherington (2003). Pred (1984: 279) concurs suggesting that 'spatial forms, places and regions have been portrayed as little more than frozen scenes for human activity'. Place is an 'inert, experienced scene', and form follows on alongside.

In similar fashion, Hagerstrand (1973) hints at the inadequacies of constraining consideration of issues in human geography to what is effectively form—and neglecting other considerations:

In a way it is an ironic circumstance that most other quantitative techniques so far applied in human geography seem to be best fitted to deal with the real-world situation of an old-fashioned, stable rural environment where friction of distance is immensely high and the projects related to human action are on the whole strongly repetitive and restricted to compact space-time 'bubbles' which are elongated in time but very narrow in space.

It is the elongation of this space that the move from form to process in maritime governance needs to address.

Not everyone agrees with a focus that is directed towards process and change. In particular, Durkheim (1968: 432–434) cited by Urry (2000: 26) suggests that the important concepts of life lie behind a 'perpetual, sensuous surface flux'. Concepts are considered outside of time and change and cannot move by themselves. 'They are fixed and immutable, and it is the task of science to reveal them, and not to be seduced by the endlessly changing sensations, perceptions and images'. However perhaps such immutable scientific concepts are not the central concern of policy-makers who have to deal with the daily periphery that does change, mutate and metamorphose on a near continuous basis.

Form and Process

Time is like a river made up of the events which happen, and its current is strong; no sooner does anything appear than it is swept away, and another comes in its place, and will be swept away too. *Marcus Aurelius Antoninus, Meditations IV,* 43.

There is a perpetual tension between form and process, between object and subject, between activity and thing. Harvey (1989: 6).

Whilst it is easy to treat form and process as rivals, in fact they have been considered closely linked for a very long time. Harrison's contribution (2005: 86) typifies this debate emphasising the intensity of the disagreement that has raged for decades. However, despite the mutual distaste that each camp appears to have for the other, there remains much to tie the two concepts together. In terms of maritime governance, we have to see whether this link can be resurrected effectively in the way that policies are derived and implemented.

Not everyone was immediately convinced of the importance that process might take over a traditional concentration upon form in science, the social sciences and elsewhere. Wooldridge (1958: 31) for example regarded it as 'quite fundamental that geomorphology is primarily concerned with the interpretation of forms, not the study of processes' although this has to be read in the light of a long-running academic argument with Strahler (1952: 924) who concluded that geomorphology needed to turn to the 'physical and engineering sciences and mathematics' for vitality and that a major part of the study needed to be of processes and their modelling.

Hartshorne (1939: 352) was an early commentator on the relationship between form and process and the need to introduce an analytical and dynamic rather than static approach. Although focussing on geographical studies, his comments are highly relevant across all types of scientific discipline. In particular, he refers to Spethmann's *Dynamische Erdkunde* (1928) in which he suggests that a dynamic view of geography was necessary to replace the static view of most others. In fact, Spethmann was actually proposing nothing new as dynamism was beginning to be accepted already by the German geographical community and that all he was doing was taking a fashionable position. He returned to these issues in later years (Hartshorne 1958: 106) where he stressed the need to incorporate the static, dynamic and chorological throughout scientific study, a view originally derived from the work of Vedova (1881).

Dodge (1935: 335) had long ago emphasised the importance of process to form considering geographical regions where much more than 'static' or 'being' and their 'becoming' to be much more important. Cohen and Lewis (1967: 1) declares that form and function (change, process etc.) were so interlinked that each should be considered a product of or direct response to the other.

Whyte (1944: 51) has much to say on the relationship between form and process claiming that the characteristics of any process are always its forms—the latter being the 'the recognisable continuity of any process'. Some forms may appear to be static, but in fact, this is simply an illusion and they are always characterised by processes at some level.

Whilst Van Ginkel (1961: 57–58) provides extensive examples of the relationship between form and function (process) from urban planning, Eichenbaum and Gale (1971: 525) link form, function and process together and provide an analysis of the metaphysical positions and methodologies that embrace each of these concepts. Form is seen as providing a measurable, geometric description of any phenomenon at a given instant in time. However, when combined with function, a cross-sectional interpretation becomes possible using the characteristics of time as a way of understanding the issues under review. Thus, form requires function to gain any true value, and function similarly requires form. The two become mutually dependent making any governance structure inadequate which is reliant on just the one. They stress that the classical form-oriented tradition may have produced a variety of epistemologies, cosmologies and theologies, as the foundation has always been the presumption of 'being'. The dominant Greek philosophy was to 'deprecate becoming and exalt being', independence and absoluteness (Eichenbaum and Gale 1971: 528–529).

Toulmin and Goodfield (1962: 47–48) continue emphasising the Greek philosophical approach to the relationship between flux and the unchanging entities that characterised much of their world. This was the problem of 'change' which exhibited three parts.

- For a theory of the natural world to be meaningful, it had to accommodate both the unchanging parts and the flux which could be observed. In other words, stability and instability at the same time.
- How can these theories be designed to be universally applied to all issues, concepts, objects and events? This is particularly difficult when trying to establish detailed and fixed solutions and explanations to objects and events which display flux.
- Physical compared with psychological change. Do we see objects or imagine that we see them? Can we observe change or do we just perceive it?

The Greeks never solved the problems of relating change and form although consumed much energy in debating them.

The principles of the form-oriented tradition have been further outlined by Beckner (1964) and Alexander (1964: 15–46) and are founded upon the 'axioms of substance and causality and the presumption of being' (Eichenbaum and Gale 1971: 528). Following the Ancient Greeks again, 'becoming' was deprecated and 'being' exalted. In Aristotle's view:

> what was altogether immutable and hence immune to influence from others was superior to that which in any way changed or depended on other things. (Eichenbaum and Gale 1971: 529).

Harvey (1969: 423) discusses the role of process in explaining events in a geographical context although his interpretation is equally as applicable across all disciplines:

> All the varied forms of the lands are dependent upon – or, as the mathematician would say, are functions of three variable quantities which may be called structure, process and time.

Using Davis's cycle of geomorphological erosion, he emphasises the role of process in determining form and how the two are interrelated but also questions the use of time as an independent variable separate from this. He cites a number of other examples of geographical models (e.g. Whittesley 1929; Broeck 1932; Taylor 1937) where process is fundamental and time arbitrarily defined as a sequence of events, difficult to measure appropriately. These crude models do, however, remain closely linked to process and change, and although their relationship to timescales is difficult to assess precisely, the drift is clear.

Berry (1973: 3) is quick to emphasise that whatever pre-eminent position form has taken up in analysis, a 'static pattern analysis is incapable of indicating which if a variety of equally plausible but fundamentally different causal processes had given rise to the patterns… studied'. However, he also stresses that the 'search for some absolutes of form in some geometric sense is understandable' (Berry 1973: 8). The consideration of form alone, however, is not enough and what was needed was a 'continuous intellectual process' that recognises everything lies within a more complex (and changeable) system.

Pred (1977: 210–211) looks at the contribution Hagerstrand (1974, 1976) made through his work on time geography and how he relates the issues of form and process. One major aim was to develop a *contextual* rather than a *compositional* model of human activity. The compositional approach was widely applied across many disciplines and considers how a set of phenomena is divided into a hierarchy of component parts and how they are then combined to form the whole. It focuses upon structure and form and eerily resembles the jurisdictional hierarchy of current maritime governance. The contextual model looks at the situation in which an object or individual is found and the connections between that individual or object and their behaviour. The emphasis is then on structure and process (Hagerstrand 1975). Both models accept that structure is important, but Hagerstrand's belief is that process was much more significant than form.

Kennedy (1979: 552) refers to the American palaeontologist George Gaylord Simpson (1963) who suggests that there are two separate but complimentary parts to scientific explanation—configurational, relating to and/or determined by unique conditions in time and space (and hence in many ways static); and immanent, representing 'unchanging properties of matter and energy and the likewise unchanging processes and principles arising therefrom' (Simpson 1963: 24) (which by definition accepts a role for process and change).

Pred (1985: 338) links form and process when commenting upon place and its determination by the 'unbroken flow of what takes place locally'. Hence, all static forms are in fact just a representation of local processes. Gertler (1988: 152) notes how in geography there had been a tendency for many years for an 'infatuation with form over process, with pattern instead of change and even with description over explanation (Pred 1977; Gertler 1987)'. This had changed substantially since the 1970s so that process and change had now become the dominant ways of interpreting geographical phenomena and had taken the central position within analysis. Problems remained with the inadequate use of 'dynamics' compared with 'statics' but the need to incorporate flexibility and flux was fully recognised. Maritime governance can take much from this.

Gertler (1988: 157) continues by looking at how form and process need to work together as analytical models and uses the capital market and the role of the firm as examples. What he calls the dialectic of fixity and fluidity has a temporal dimension as well. Firms have to commit considerable capital in fixed form (take ships and port facilities for example) in order to have the means of production (ship operations, storage, etc.). At the same time, they need resources and powers to adjust to changes in the marketplace and they never possess full information about the future conditions in which they must operate. Policies need to accommodate this dialectic as well with the ability to accept fixed form within the industry and incorporating mechanisms to change over time as circumstances change around them. At present, only the former is allowed for to any real degree and the shipping industry finds ways of manipulating maritime policies by focussing upon the fixed elements and ensuring that the more dynamic features are either minimised or excluded altogether.

In more modern times, Goodchild (2004: 709) identifies in geographical studies an historic tension between form and process and suggests that the focus on scientific methods had led to this being intensified. Rhoads (2005: 137) is more positive viewing process and form as always interactive and not in adversity. The operations of process will always cause some forms to change (if only temporarily), and in return, the changed form will affect the operation of the process. This may not necessarily lead to a changed form or process but may just maintain a situation as it stands. This does not lessen the importance of either feature (form or process), and in some ways, their interdependence becomes that much more significant.

Neo and Chen (2007: 1) emphasise how even if the principles of policy-making are good, static governance would lead to 'stagnation and decay' and the effectiveness of organisations would be challenged if there was insufficient innovation and change designed within institutional capacity.

Path Dependency and Lock in

> But you must bind me hard and fast, so that I cannot stir from the spot where you will stand me… and if I beg you to release me, you must tighten and add to my bonds. *The Odyssey* quoted in Strotz (1955/1956: 165).

> Every action of theirs, that seems to them an act of their own free will, is in historical sense not free at all, but in bondage to the whole course of previous history. Leo Tolstoy, *War and Peace*.

> Spatial form as 'outcome' (the happenstance juxtapositions and so forth) has emergent powers which can have effects on subsequent events. Spatial form can alter the future course of the very histories that have produced it… One way of thinking about all this is to say that the spatial is integral to the production of history… just as the temporal is to geography. Another way is to insist on the inseparability of time and space, on their joint constitution through the interrelationships between phenomena… Massey (1992: 84) quoted in Martin and Sunley (2006: 409–410).

Finally, we turn to a consideration of a concept that has gained in popularity over the years and which has a close relationship to the form and the domination that it has within governance as applied to the maritime sector. Path dependency has a wide range of literature associated with it (see, for example, North 1994: 365; Mueller 1997: 843; Goldstone 1998: 834; and Ramanath 2009: 67, 70), as has the linked theme of lock-in to which we turn a little later.

Definitions abound. Antonelli (1997: 643–644) suggests that:

> path dependence defines the set of dynamic processes where small events have long-lasting consequences that economic action at each moment can modify yet only to a limited extent. The trajectory of a path-dependent process however cannot be fully anticipated

> on the basis of the original events. Path dependence is different from past dependence because the former is able to accommodate the consequences of actions at each point in time. Path-dependence analysis is systemic and dynamic because it focuses attention on the process of change that is generated by the interaction of a plurality and variety of agents whose behavior is constrained by their localization in time.

Antonelli's comments are interesting as they focus upon the dynamic capabilities of path dependency and consequently its application to understanding the static nature of maritime governance, and its need to become more dynamic is clear. Altman (2000: 128) spells it out with a little more detail whereby:

> the free market typically generates suboptimal long-run equilibrium solutions to a variety of economic problems and the probability of suboptimal equilibrium outcomes increases where increasing returns (positive feedbacks) prevail.

Thus, in the free market shipping industry, suboptimal solutions commonly prevail because to those involved (operators and policy-makers), there exists considerable positive feedback in retaining the status quo. Those affected by this suboptimality in policy-making and implementation—clients, the workforce, the economy in general, the environment, are less fortunate. Dopfer (1991: 540) had earlier put it rather more simply:

> The probability that individual agents adopt an idea, such as an invention, transaction, or other behavioral pattern (or maritime policy), will increase as that idea is adopted. There is a determinate relationship between past and present actions. An initial idea is seen to constitute some sort of 'seed' that grows by the self-generating dynamics of the process. Individual behaviors receive increasingly the status of a norm… that determines individual behavior. (comments in parentheses added).

Hence, both maritime policy-making and their governance have become institutionalised, difficult to amend and find it increasingly impossible to accommodate dynamic features. Path dependency does not suggest that the choice of decisions can be predicted precisely, and thus, the future selection of maritime policies is not fully determined; however, they are heavily constrained by what has gone before and what is convenient to those in the established positions of governance (Bruggerman 2002: 415).

Other definitions abound. Liebowitz and Margolis (1995: 205) emphasise how an initial decision that provides an advantage to those taking it can have significant implications for their position in the future. Thus, change is difficult. Mahoney (2001: 111) and Scott (2001: 367) along with Webster (2008: 61) and Foray (1997: 735) provide further definitions, whilst Martin and Sunley (2006: 399) suggest it reflects an inability of organizations to 'shake free of their history'. Meanwhile, a more general consideration of path dependency has been extensive including early work by Veblen (1915: 130), Frankel (1955) and Kindleberger (1964), and more recent contributions by Atkinson and Stiglitz (1969), Arthur (1989, 1990), David (1975, 1985, 1987, 1990, 1994), Farrell and Saloner (1986), Alexander (2001), Arrow (2000), Berman (1998), Cowan and Gunby (1996), Garud and Karnoe (2001), Greener (2002a, b), Hansen (2002), Hedlund (2000), Holzinger and Knill (2002),

Mahoney (2000: 507), O'Brien (1996), Pierson (2000), Putnam (1993), Sterman and Wittenberg (1999), Torfing (1999, 2001) and Wilsford (1994).

Path dependency has a long history, and along with the sizeable array of reading, this implies that it has some substantial credibility. Martin and Sunley (2006: 397), for example, trace the origins of path dependence back to Carl Menger and Veblen's 'cumulative causation' and the rise of 'hysteresis' as a related concept through the work of Elster (1976), Franz (1990), Cross (1993) and Katzner (1993). Those who emphasise its significance include Arrow (2000: 178), Bridges (2000: 109) and Pierson (2000: 251). However, there are also those who play down its significance as a concept. Levi (1997: 27) for example suggests that:

> Path dependence does not simply mean that 'history matters'. This is both true and trivial. Path dependence has to mean, if it is to mean anything, that once a country or a region has started down a track, the costs of reversal are very high. There will be other choice points, but the entrenchments of certain institutional arrangements obstruct an easy reversal of the initial choice. Perhaps the better metaphor is a tree, rather than a path. From the same trunk there are many different branches and smaller branches. Although it is possible to turn around or to clamber from one to the other - and essential if the chosen branch dies - the branch on which a climber begins is the one she tends to follow.

Others with a critical opinion of path dependency include Dopfer (1991: 541) who was uncertain because 'the shortcoming of the present model is that it leaves undefined the process components that constitute the *seed* from which a macroscopic process may emerge'; and Altman (2000: 130–131) who cites Leibowitz and Margolis's (1990, 1994) suggestion that the related assumptions of 'inefficient equilibria' that are implied by path dependency (e.g. the inertia apparent in the IMO which whilst recognised as inefficient is seemingly unchangeable) are untenable. Much has been written about inefficient equilibria, self-reinforcement and increasing returns which we will pass over here (see, for example, the work of Altman 2000: 127, 129, Arrow 2000: 175, Pierson 2000: 252–253, Mahoney 2001: 114, Chen 2004: 434, 437, Levin et al. 2012: 135).

Meanwhile, further criticism comes from Bridges (2000: 111):

> It is a mistake to understand history as a series of cycles in which social and political processes 'are prone to consolidation or institutionalization' in arrangements which then 'reproduce' themselves until new conditions 'disrupt' or 'overwhelm' them. For one thing this reading leaves us content to recognize stability, but condemned to surprise at change.

Mahoney (2000: 507) describes it as a 'vague concept', and consequently, it is unclear why it demands special attention, founded on the idea that 'history matters' and that explanation can be based on tracing events back to 'temporally remote causes'. Hansen (2002: 270) notes others' suggestion that path dependency is little more than a 'rich metaphor for policy continuity' and Scott (2001: 367) continues the themes we noted earlier introduced by Liebowitz and Margolis (1990: 4), whilst Martin and Sunley (2006: 404–408) raise a series of questions including the uncertainty that surrounds the causes of path dependency, the lack of discussion of path creation that follows from this, the lack of theory to support the concept and whether it is an evolutionary process? They note the debate that has raged through the work of Bassanini and Dosi (2001), Hirsch and Gillespie (2001) and Castaldi and Dosi (2003), amongst others.

Examples abound of the application of path dependency from a multitude of disciplines. The application of path dependency to maritime governance would be based on a good tradition of attempting to model the difficulties of change and the tendency for institutions, organisations and policies to be dogged by inertia. Arthur (1994: 82, 93) uses video cassette recorders and clocks to illustrate the applicability of the model, Antonelli (1997) focuses upon industrial organisation and Hedlund (2000) upon Russian economic policy-making, whilst Pierson (2000: 251) notes its use for comparative politics in Europe (e.g. Lipset and Rokkan 1967), labour incorporation in Latin America (Collier and Collier 1991), state building (Ertman 1996), and health care (Hacker 1998). Hansen (2002) meanwhile suggests its applicability to immigration studies, Greener (2005: 63) cites Alexander's (2001) work on democracy, Dimitrakopoulos's (2001) examination of European integration and Berman's (1998) study of political action in times of economic depression. Martin and Sunley (2006: 398) provide an extensive review of path dependency generally and in doing so refer to examples from social behaviour (Anderlini and Ianni 1996; Goldstone 1998), power generation (Cowan 1990; Cowan and Hulten 1996), industrial technology (Ruttan 1997; Araujo and Harrison 2002), corporate governance (Bebchuk and Roe [no relation] 1999) and legal systems (North 1990). Webster (2008: 61) also suggests studies of tax rate policy (Kaplow and Shavell 2002), sulphur emissions capping (Ellerman et al. 2000), money supply (Dixit 1991) and resource extraction (Gerlagh and Keyzer 2004). In addition, there are also examples of application of path dependency to transport and in particular railways and canals. David (1985: 336) cites the problems of the UK's undersized railways which became set from the earliest development and never readjusted to the larger size common elsewhere across the world. Today, the UK retains small-scale rail (and canal) infrastructure which inhibits both freight and passenger movement simply because the path taken at the start is too difficult to change.

The relevance of the concept to the mobilisation of maritime governance which we have been pursuing is clear when you take the association of path dependency to a number of the concepts which are pursued later in this book. These include those relating to institutions—see, for example, the work of Mahoney (2000: 515, 2001: 114), Dimitrakopoulos (2001: 408) and how institutions are often self-perpetuating, Bruggerman (2002: 417) and institutional inertia, Hansen (2002: 270), Greener (2005: 62) and his consideration of historical institutionalism, Martin and Sunley (2006: 402) and their discussion of institutional hysteresis and Webster (2008: 61). There are others who have focused upon policy-making, governance and path dependence including Mahoney (2001: 111) who reflected on path dependency's ability to contribute to political analysis, Cashore and Howlett (2007: 532) who noted the policy implications of path dependency considered by Hacker (2004), Mahoney (2000) and Pierson (1993, 2000: 259, 268), Ng and Pallis (2010: 2150, 2151) who placed the inertia in institutions and its effect on governance in the ports sector within a path-dependent context, Robichau (2011: 117) who cites

Kjaer (2004: 204) in emphasising the close relationship that exists between governance, change and path dependency and Levin et al. (2012: 124) who look at the ability to use path dependence to ensure policy outcomes.

Meanwhile, path dependence and its relationship to process (Pierson 2000: 252; Martin and Sunley 2006: 408; Levin et al. 2012: 134), time (Pierson 2000: 251) and space (Oosterlynck 2012) have also been common themes.

And so to lock-in, Redding (2002: 1215) notes that innovation is path dependent: 'the historical pattern of technological development is thought to play a central role in determining the pace of future technological change'. Taken to its extreme, this can result in lock-in when 'agents continue to employ an existing technology even though potentially more productive technologies could be found' Redding (2002: 1215) a concept first introduced by David (1985) and Arthur (1989). Maritime governance exhibits just this sort of characteristic. Although taken from a political and governance context rather than technological, the similarities are clear. An old model continues to be applied despite recognition that circumstances have changed and a new one is needed (and potentially available).

Lock-in has been widely considered. In broader terms, it has been the focus of work by Arthur (1994: 82, 92), Kline (2001), Hansen (2002: 271), Martin and Sunley (2006: 414–415), Marechal (2007: 5187) and Levin et al. (2012: 134–135). Examples of application amongst very many others can be found in David (1985: 333–336), Liebowitz and Margolis (2012: 125) with their consideration of QWERTY and the typewriter keyboard; Arthur (1989) and Cowan (1990) on nuclear power, Cowan and Gunby (1996) and Cowan and Hulten (1996) on pest control; Visser and Boschma (2004) on regional economic clustering; Hassink and Shin (2005) and the impact of political lock-ins on industrial production citing Hamm and Wienert (1989), Unruh (2002) and carbon lock-ins; Hassink (2005) on regional economics; and Marechal and Lazaric (2010: 104, 108) on climate change.

A true appreciation of lock-in has much to offer to maritime governance in that the current institutional freeze appears to be an excellent example of how inertia and previous decisions can lead to a widely recognised but seemingly unalterable state of suboptimal affairs (Alexander 2001: 254; Hassink 2005: 523–524; Marechal and Lazaric 2010: 107; and Martin and Sunley 2006: 419). Consequently, the form that maritime governance takes remains divorced from the dynamism that would also appear to be essential if it is to be effective in forming a framework for policy-making.

Conclusions

The relationship between form and process is complex, and it is not just a simple matter of one replacing the other. Werlen (2004: 154) stresses how 'space is losing its importance' and continues to suggest that space 'does not exist at all'. Undoubtedly, process has much to offer to maritime governance, but form will remain an essential element—laws, policy-documents, recommendations, papers, electronic communications, ships, port facilities and much more—the list is almost endless. What would be different is an approach, an attitude to the application and design of policy that requires a sea change in underlying governance. The significance of moving away from static models has not been lost in many policy-making areas and this in turn makes the situation found in the maritime sector that much more unacceptable where shipping by its very nature and global context requires a flexible, mobile and adaptable approach to governance. The failure of static policy-making was not lost even on Mahatma Gandhi who refused to build any models of an economic situation because they all become static over time. His response was essentially dynamic, characterised by open-ended concepts amendable to any situation (Sethi 1985: xxiv).

References

Alexander, G. (2001). Institutions, path dependence and democratic consolidation. *Journal of Theoretical Politics, 13*(3), 249–270.

Alexander, P. (1964). Speculations and theories. In J. B. Gregg & F. T. C. Harris (Eds.), *Form and strategy in science* (pp. 30–46). Dordrecht: D. Reidel.

Altman, M. (2000). A behavioral model of path dependency: The economics of profitable inefficiency and market failure. *Journal of Socio-Economics, 29*, 127–145.

Anderlini, L., & Ianni, A. (1996). Path dependence and learning from neighbours. *Games and Economic Behaviour, 13*, 141–177.

Antonelli, C. (1997). The economics of path-dependence in industrial organization. *International Journal of Industrial Organization, 15*, 643–675.

Araujo, L., & Harrison, D. (2002). Path dependence, agency and technological evolution. *Technology Analysis & Strategic Management, 14*, 5–19.

Arrow, K. (2000). Increasing returns: Historiographic issues and path dependence. *European Journal of the History of Economic Thought, 7*(2), 171–180.

Arthur, W. B. (1989). Competing technologies, increasing returns and lock in by historical events. *Economic Journal, 99*, 116–131.

Arthur, W. B. (1990). Positive feedbacks in the economy. *Scientific American, 204*, 92–99.

Arthur, W. B. (1994). Positive feedbacks in the economy. *The McKinsey Quarterly, 1*, 81–95.

Atkinson, B., & Stiglitz, J. (1969). A new view of technical change. *The Economic Journal, LXXIX*(315), 573–578.

Attneave, F., & Armoult, M. D. (1966). The quantitative study of shape and pattern perception. In L. Urb (Ed.), *Pattern recognition: Theory, experiment, computer simulation and dynamic models of form, perception and discovery*. New York: Wiley.

Bassanini, A. P., & Dosi, G. (2001). When and how chance and human will can twist the arms of clio: An essay on path dependence in a world of irreversibilities. In R. Garud & P. Karnoe (Eds.), *Path dependence and creation* (pp. 41–68). London: Lawrence Erlbaum.

Bebchuk, L., & Roe, M. (1999). A theory of path dependence in corporate ownership and governance. *Stanford Law Review, 52*, 127–170.

Beckner, M. (1964). Metaphysical presuppositions and the description of biological systems. In J. B. Gregg & F. T. C. Harris (Eds.), *Form and strategy in science* (pp. 15–29). Dordrecht: D. Reidel.

Berman, S. (1998). Path dependency and political action; re-examining responses to the depression. *Comparative Politics, 30*, 379–400.

Berry, B. J. L. (1973). A paradigm for modern geography. In R. Chorley (Ed.), *Directions in geography* (pp. 3–22). London: Methuen.

Bridges, A. (2000). Path dependence, sequence, history, theory. *Studies in American Political Development, 14*, 109–112.

Broeck, J. O. M. (1932). *The Santa Clara Valley, California: A study in landscape changes*. Utrecht: N.V.A. Oosthoek's Uitg. Maatij.

Bruggerman, D. (2002). NASA: A path dependent organization. *Technology in Society, 24*, 415–431.

Buttimer, A. (1976). Grasping the dynamism of lifeworld. *Annals of the Association of American Geographers, 66*(2), 277–292.

Cashore, B., & Howlett, M. (2007). Punctuating which equilibrium? Understanding thermostatic policy dynamics in Pacific northwest forestry. *American Journal of Political Science, 51*(3), 532–551.

Castaldi, C., & Dosi, G. (2003). *The grip of history and the scope of novelty: Some results and open questions on path dependence in economic processes* (Working Paper 2003/02). Laboratory of Economics and Management, Sant'Anna School of Advanced Studies, University of Pisa.

Chen, H.-P. (2004). Path-dependent processes and the emergence of the rank size rule. *Annals of Regional Science, 38*, 433–449.

Cohen, S. B., & Lewis, G. K. (1967). Form and function in the geography of retailing. *Economic Geography, 43*(1), 1–42.

Collier, R. B., & Collier, D. (1991). *Shaping the political arena: Critical junctures, the labor movement and regime dynamics in Latin America*. Princeton, NJ: Princeton University Press.

Cowan, R. (1990). Nuclear power reactors: A study in technological 'lock-in'. *Journal of Economic History, 50*, 541–567.

Cowan, R., & Gunby, P. (1996). Sprayed to death: Path dependence, lock in and pest control strategies. *The Economic Journal, 106*, 521–542.

Cowan, R., & Hulten, S. (1996). Escaping lock-in: The case of the electric vehicle. *Technology Forecasting and Social Change, 53*, 61–79.

Cross, R. (1993). On the foundation of hysteresis in economic systems. *Economics and Philosophy, 9*, 53–74.

Dalcq, A. (1939). *Form and causality in early development*. Cambridge: Cambridge University Press.

David, P. A. (1975). *Technical choice, innovation and economic growth. Essays on American and British experience in the nineteenth century*. Cambridge: Cambridge University Press.

David, P. A. (1985). Clio and the economics of QWERTY. *American Economic Review, 75*, 332–337.

David, P. A. (1987). Some new standards for the economics of standardization in the information age. In P. Dasgupta & P. Stoneman (Eds.), *Economic policy and technological performance* (pp. 206–239). Cambridge: Cambridge University Press.

David, P. A. (1990). Heros, herds and hysteresis in technological history: Thomas Edison and the 'Battle of Systems' reconsidered. *Journal of Industrial and Corporate Change, 1*, 129–180.

David, P. A. (1994). Why are institutions the 'carriers of history'? Path dependence and the evolution of conventions, organizations and institutions. *Structural Change and Economic Dynamics, 2*, 205–220.

Dimitrakopoulos, D. (2001). Incrementalism and path dependence: European integration and institutional change in national parliaments. *Journal of Common Market Studies, 39*(3), 405–422.

Dixit, A. (1991). Analytical approximations of hysteresis. *Review of Economic Studies, 58*(1), 141–151.

Dodge, S. D. (1935). The chorology of the Claremont-Springfield region in the Upper Connecticut Valley in New Hampshire and Vermont. *Papers of the Michigan Academy of Science, Arts and Letters, 22*, 335–353.

Dopfer, K. (1991). Toward a theory of economic institutions: Synergy and path dependency. *Journal of Economic Issues, XXV*(2), 535–550.

Durkheim, E. (1968). *The elementary forms of the religious life*. London: George Allen and Unwin.
Eichenbaum, J., & Gale, S. (1971). Form, function and process. *Economic Geography, 47*(4), 525–544.
Ellerman, A. D., Koskow, P. L., Schmalensee, R., Montero, J.-P., & Bailey, E. M. (2000). *Markets for clean air: The US acid rain program*. Cambridge: Cambridge University Press.
Elster, J. (1976). A note on hysteresis in the social sciences. *Synthese, 33*, 371–391.
Ertman, T. (1996). *Birth of the Leviathan: Building states and regime in medieval and early modern Europe*. Cambridge: Cambridge University Press.
Farrell, J. R., & Saloner, G. (1986). Installed base and compatibility: Innovation, product preannouncements, and predation. *American Economic Review, 76*, 940–955.
Foray, D. (1997). The dynamic implications of increasing returns: Technological change and path dependent inefficiency. *International Journal of Industrial Organization, 15*, 733–752.
Forward, C. N. (1967). Recent changes in the form and function of the Port of St John's, Newfoundland. *Canadian Geographer, XI, 2*, 101–116.
Frankel, M. (1955). Obsolescence and technological change in a maturing economy. *American Economic Review, 45*, 296–319.
Franz, W. (1990). Hysteresis in economic relationships: An overview. *Empirical Economics, 15*, 109–125.
Furster, C. B. (1963). Form and structure of the future urban complex. In L. Wingo (Ed.), *Cities and space. The future use of urban land* (pp. 73–102). Baltimore, MD: The John Hopkins Press.
Garud, P., & Karnoe, P. (2001). *Path dependence and creation*. London: Lawrence Erlbaum Associates.
Gerlagh, R., & Keyzer, M. A. (2004). Path dependence in a Ramsey model with resource amenities and limited regeneration. *Journal of Economic Dynamics and Control, 28*, 1159–1184.
Gertler, M. S. (1987). Capital, technology and industry dynamics in regional development. *Urban Geography, 8*(3), 251–263.
Gertler, M. S. (1988). Some problems of time in economic geography. *Environment and Planning A, 20*(2), 151–164.
Gibson, J. J. (1951). What is form? *Psychological Review, 58*(6), 403–412.
Goldstone, J. A. (1998). Initial conditions, general laws, path dependence and explanation in historical sociology. *American Journal of Sociology, 104*, 829–845.
Goodchild, M. F. (2004). GIScience, geography, form and process. *Annals of the Association of American Geographers, 94*(4), 709–714.
Gould, P. R. (1973). The open geographic curriculum. In R. Chorley (Ed.), *Directions in geography* (pp. 253–284). London: Univeristy Paperbacks (Methuen).
Greener, I. (2002a). Theorising path dependency: How does history come to matter in organisations? *Management Decision, 40*(5/6), 614–619.
Greener, I. (2002b). Understanding NHS reform: The policy transfer, social learning and path dependency perspectives. *Governance, 15*(2), 161–184.
Greener, I. (2005). The potential of path dependence in political studies. *Politics, 25*(1), 62–72.
Hacker, J. (1998). The historical logic of national health insurance: Structure and sequence in the development of British, Canadian and US medical policy. *Studies in American Political Development, 12*, 57–130.
Hacker, J. (2004). Privatizing risk without privatizing the welfare state: The hidden politics of social policy retrenchment in the United States. *American Political Science Review, 98*(2), 243–260.
Hagerstrand, T. (1973). The domain of human geography. In R. Chorley (Ed.), *Directions in geography* (pp. 67–87). London: Methuen.
Hagerstrand, T. (1974). Tidsgeografisk beskrivning. *Syfte och postulat, Svensk Geografisk Arsbok, 50*, 86–94.

Hagerstrand, T. (1975). Ecology under one perspective. In E. Bylund, H. Linderholm, & O. Rune (Eds.), *Ecological problems of the circumpolar area* (pp. 271–276). Lulea: Norrsbottens Museum.

Hagerstrand, T. (1976). *Geography and the study of interaction between nature and society*. Paper presented at the 13th International Geographical Congress, Moscow.

Hamm, R., & Wienert, H. (1989). Strukturelle Anpassung altindustrieller Regionen im internationalen Vergleich. *Rheinisch-Westfalisches Institut fur Wirtschaftsgeographie, 43*, 76–89.

Hansen, R. (2002). Globalization, embedded realism and path dependence. *Comparative Political Studies, 35*(3), 259–283.

Harrison, S. (2005). What kind of science is physical geography? In N. Castree, A. Rogers, & D. Sherman (Eds.), *Questioning geography* (pp. 80–95). Oxford: Blackwell.

Hartshorne, R. (1939). The nature of geography. *Annals of the Association of American Geographers, 29*(3), 173–658.

Hartshorne, R. (1958). The concept of geography as a science of space. *Annals of the Association of American Geographers, 48*(2), 97–108.

Harvey, D. (1969). *Explanation in geography*. London: Edward Arnold.

Harvey, D. (1989). From managerialism to entrepreneurism. The transformation in urban governance in late capitalism. *Geografiska Annaler Series B, 71*(1), 3–17.

Harvey, D. (1993). From space to place and back again: Reflections on the condition of postmodernity. In J. Bird, B. Curtis, T. Putnam, & L. Tickner (Eds.), *Mapping the futures: Local cultures, global change* (pp. 3–29). London: Routledge.

Hassink, R. (2005). How to unlock regional economies from path dependency? From learning region to learning cluster. *European Planning Studies, 13*(4), 521–535.

Hassink, R., & Shin, D.-H. (2005). Guest editorial. *Environment and Planning A, 37*, 571–580.

Hedlund, S. (2000). Path dependence in Russian policy-making: Constraints on Putin's economic choice. *Post-Communist Economies, 12*(4), 390–407.

Henry, J. (1955). Homeostasis, society and evolution: A critique. *The Scientific Monthly, 81*, 300–309.

Herskovits, M. J. (1955). *Cultural anthropology*. New York: Alfred A. Knopf.

Hetherington, K. (2003). Moderns as ancients: Time, space and the discourse of improvement. In J. May & N. Thrift (Eds.), *Timespace* (pp. 49–72). London: Routledge.

Hirsch, P. M., & Gillespie, J. J. (2001). Unpacking path dependence: Differential valuations accorded history across disciplines. In R. Garud & P. Karnoe (Eds.), *Path dependence and creation* (pp. 69–90). London: Lawrence Erlbaum.

Holzinger, K., & Knill, C. (2002). Path dependencies in European integration: A constructive response to German Foreign Minister Joschka Fischer. *Public Administration, 80*(1), 125–152.

Johnson, H. B., & Pitzl, G. R. (1981). Viewing and perceiving the rural scene; visualization in human geography. *Progress in Human Geography, 5*(2), 211–233.

Kaplan, C. (1996). *Questions of travel*. Durham, NC: Duke University Press.

Kaplow, L., & Shavell, S. (2002). On the superiority of corrective taxes to quantity regulation. *American Law and Economics Review, 4*(1), 1–17.

Katzner, D. (1993). Some notes on the role of history and the definition of hysteresis and related concepts in economic analysis. *Journal of Post-Keynsian Economics, 15*, 323–345.

Kennedy, B. A. (1979). A naughty world. *Transactions of the Institute of British Geographers NS, 4*(4), 550–558.

Kindleberger, C. P. (1964). *Economic growth in France and Britain, 1851–1950*. Cambridge, MA: Harvard University Press.

Kjaer, A. M. (2004). *Governance*. Cambridge: Polity Press.

Kline, D. (2001). Positive feedback, lock-in and environmental policy. *Policy Sciences, 34*, 95–107.

Leach, E. R. (1954). *Political systems of highland Burma*. Cambridge, MA: Harvard University Press.

Levi, M. (1997). A model, a method and a map. Rational choice in comparative and historical analysis. In M. I. Liebach & A. S. Zuckerman (Eds.), *Comparative politics* (pp. 19–41). Cambridge: Cambridge University Press.

Levin, K., Cashore, B., Bernstein, S., & Auld, G. (2012). Overcoming the tragedy of super wicked problems: Constraining our future selves to ameliorate global climate change. *Policy Sciences, 45*, 123–152.

Liebowitz, S. J., & Margolis, S. E. (1990). The fable of the keys. *Journal of Law and Economics, 33*, 1–25.

Liebowitz, S. J., & Margolis, S. E. (1994). Network externalities: An uncommon tragedy. *Journal of Economic Perspectives, 8*, 133–150.

Liebowitz, S. J., & Margolis, S. E. (1995). Path dependence: Lock-in and history. *The Journal of Law, Economics and Organization, 11*(1), 205–226.

Liebowitz, S. J., & Margolis, S. E. (2012). The troubled path of the lock-in movement. *Journal of Competition Law and Economics, 9*(1), 125–152.

Lipset, S. M., & Rokkan, S. (1967). Cleavage structures, party systems and voter alignments: An introduction. In S. M. Lipset & S. Rokkan (Eds.), *Party systems and voter alignments* (pp. 1–64). New York: Free Press.

Lucretius, B. Y. (1995). *On the nature of things* [De rerum natura] (Anthony M. Esolen, Trans.). Baltimore, MD: The Johns Hopkins University Press.

Lynch, K. (1960). *The image of the city*. Cambridge, MA: The MIT Press.

Mahoney, J. (2000). Path dependence in historical sociology. *Theory and Society, 29*, 507–548.

Mahoney, J. (2001). Path-dependent explanations of regime change: Central America in comparative perspective. *Studies in Comparative International Development, 36*(1), 111–141.

Marechal, K. (2007). The economics of climate change and the change of climate in economics. *Energy Policy, 35*, 5181–5194.

Marechal, K., & Lazaric, N. (2010). Overcoming inertia: Insights from evolutionary economics into improved energy and climate policies. *Climate Policy, 10*, 103–119.

Martin, R., & Sunley, P. (2006). Path dependence and regional economic evolution. *Journal of Economic Geography, 6*, 395–437.

Massey, D. (1992). Politics and space/time. *New Left Review, 196*, 65–84.

Massey, D. (1993). Questions of locality. *Geography, 78*(2), 142–149.

May, J., & Thrift, N. (2003). Introduction. In J. May & N. Thrift (Eds.), *Timespace* (pp. 1–46). London: Routledge.

Morrison, E. E. (1966). *Man, machines and modern times*. Cambridge, MA: MIT Press.

Mueller, D. C. (1997). First-mover advantages and path dependence. *International Journal of Industrial Organization, 15*, 827–850.

Neo, B. S., & Chen, G. (2007). *Dynamic governance*. Singapore: World Scientific.

Ng, A. K. Y., & Pallis, A. A. (2010). Port governance reforms in diversified institutional frameworks: Generic solutions, implementation asymmetries. *Environment and Planning A, 42*, 2147–2167.

North, D. C. (1990). *Institutions, institutional change and economic performance*. Cambridge: Cambridge University Press.

North, D. C. (1994). Economic performance through time. *The American Economic Review, 84*(3), 359–368.

O'Brien, P. (1996). Path dependency, or why Britain became an industrialised and urbanised economy long before France. *Economic History Review, XLIX*(2), 213–249.

Oosterlynck, S. (2012). Path dependence: A political economy perspective. *International Journal of Urban and Regional Research, 36*(1), 158–165.

Pierson, P. (1993). When effect becomes cause: Policy feedback and political change. *World Politics, 45*(4), 595–628.

Pierson, P. (2000). Increasing returns, path dependence and the study of politics. *American Political Science Review, 94*(2), 251–267.

Pitzl, G. R. (1974). On the concept of form in geographical studies. *Journal of the Minnesota Academy of Sciences, 40*, 84–85.

Pred, A. (1977). The choreography of existence: Comments on Hagerstrand's time geography and its usefulness. *Economic Geography, 53*, 207–221.

Pred, A. (1984). Place as historically contingent process: Structuration and the time-geography of becoming places. *Annals of the Association of American Geographers, 74*(2), 279–297.

Pred, A. (1985). The social becomes the spatial, the spatial becomes the social; enclosures, social change and the becoming of places in Skane. In D. Gregg & J. Urry (Eds.), *Social relations and spatial structures* (pp. 337–365). London: Macmillan.

Putnam, R. D. (1993). *Making democracy work*. Princeton, NJ: Princeton University Press.

Ramanath, R. (2009). Limits to institutional isomorphism: Examining internal processes in NGO-government interactions. *Nonprofit and Voluntary Sector Quarterly, 28*(1), 51–76.

Redding, S. (2002). Path dependence, endogenous innovation, and growth. *International Economic Review, 43*(4), 1215–1248.

Rhoads, B. L. (2005). Process/form. In N. Castree, A. Rogers, & D. Sherman (Eds.), *Questioning geography* (pp. 131–150). Malden MA: Blackwell.

Rich, A. (1994). *Blood, bread, and poetry: Selected prose 1979–1985*. London: W.W. Norton.

Robichau, R. W. (2011). The mosaic of governance: Creating a picture with definitions, theories and debates. *Policy Studies Journal, 39*(51), 113–131.

Ruttan, V. W. (1997). Induced innovation, evolutionary theory and path dependence. *Economic Journal, 107*, 1520–1529.

Scott, P. (2001). Path dependence and Britain's 'coal wagon problem'. *Explorations in Economic History, 38*, 366–385.

Sethi, J. D. (1985). Foreword. In R. Diwan & M. Lutz (Eds.), *Essays in Gandhian economics*. Gandhi Peace Foundation: New Delhi.

Simpson, G. G. (1963). Historical science. In C. C. Albritton (Ed.), *The fabric of geology*. Stanford: Addison-Wesley.

Smuts, J. C. (1927). *Holism and evolution*. London: Macmillan.

Spethmann, H. (1928). *Dynamische Landerkunde*. Breslau: Ferdinand Hirt.

Sterman, J., & Wittenberg, J. (1999). Path dependence, competition and succession in the dynamics of scientific revolution. *Organizational Science, 10*(3), 322–341.

Strahler, A. N. (1952). Dynamic basis of geomorphology. *Bulletin of the Geological Society of America, 63*, 923–937.

Strotz, R. H. (1955/1956). Myopia and inconsistency in dynamic utility maximization. *The Review of Economic Studies*, 23(3), 165–180.

Taleb, N. N. (2007). *The Black Swan*. New York: Random House.

Taylor, G. (1937). *Environment, race and nation*. Toronto: University of Toronto Press.

Torfing, J. (1999). Towards a Schumpeterian workfare postnational regime: Path shaping and path dependency in Danish welfare state reform. *Economy and Society, 28*(3), 369–402.

Torfing, J. (2001). Path dependent Danish welfare reforms: The contribution of the new institutionalism to understanding evolutionary change. *Scandinavian Political Studies, 24*(4), 277–310.

Toulmin, S., & Goodfield, J. (1962). *The architecture of matter*. London: Hutchinson.

Unruh, G. C. (2002). Escaping carbon lock-in. *Energy Policy, 30*, 317–325.

Urry, J. (2000). *Sociology beyond cultures*. London: Routledge.

Van Ginkel, B. L. (1961). The form of the core. *Journal of the American Institute of Planners, 27*(1), 56–69.

Veblen, T. (1915). *Imperial Germany and the industrial revolution*. London: Macmillan.

Vedova, G. D. (1881). Il Concetto Popolare e il Concetto Scientifico della Geografia. *Bulletino della Societa Geografica Italiana, 18*, 5–27.

Visser, E.-J., & Boschma, R. (2004). Learning in districts: Novelty and lock-in in a regional context. *European Planning Studies, 12*(6), 794–808.

Vogt, E. Z. (1960). On the concepts of structure and process in cultural anthropology. *American Anthropologist NS, 62*(1), 18–33.

Webster, M. (2008). Incorporating path dependency into decision-analytic methods: An application to global climate-change policy. *Decision Analysis, 5*(2), 60–75.

Werlen, B. (2004). The making of globalized everyday geographies. In K. Simonsen & J. O. Baerenholdt (Eds.), *Space odysseys* (pp. 153–170). Aldershot: Ashgate.

Whittesley, D. (1929). Sequent occupance. *Annals of the Association of American Geographers, 19*, 162–165.

Whyte, L. L. (1944). *The next development in man*. London: The Cresset Press.

Whyte, L. L. (1951). Introduction. In L. L. Whyte (Ed.), *Aspects of form*. London: Lund Humphries.

Whyte, L. L. (1954). *Accent on form, world perspectives* (Vol. 2). Westport, CO: Greenwood Press.

Wilsford, D. (1994). Path dependency, or why history makes it difficult but not impossible to reform health care services in a big way. *Journal of Public Policy, 14*, 251–283.

Wooldridge, S. W. (1958). The trend of geomorphology. *Transactions of the Institute of British Geographers, 25*, 29–35.

Chapter 3
Time

Abstract To understand the movement from form to process in maritime policy-making and governance, there is a need to understand more fully the concept of change. This in turn has a close relationship to time. No revolutionary dynamic governance which accommodates the ever-changing characteristics of policy-making and the maritime industry can avoid taking account of temporal issues. Therefore, time is next on the agenda. This chapter looks at the concept of time in the past, present and future and the relationship it has to maritime governance. It continues with a discussion of time and space and the idea of the many different times that co-exist. It concludes by looking at time, form, process and governance and their inter-relationships in the maritime sector.

> Waste of time is thus the first and in principle the deadliest of sins. The span of human life is infinitely short and precious to make sure of one's own election. Loss of time through sociability idle talk, luxury, even more sleep than is necessary to health... is worthy of absolute moral condemnation. Weber (1930: 158), quoted in Urry (2000: 109).

> A maelstrom of perpetual disintegration and renewal, of struggle and contradiction, of ambiguity and anguish. Taylor (2003: 157).

> For since God does nothing without reason, and no reason can be given why he did not create the world sooner; it would follow, either that he has created nothing at all, or that he created the world before any assignable time. that is, that the world is eternal. But when once it has been shown, that the beginning, whenever it was, is always the same thing, the question, why it was not otherwise ordered, becomes needless and insignificant. Leibnitz in reply to a letter by the Rev. Samuel Clarke, in Alexander (1956: 38–39).

To understand the movement from form to process in policy-making and governance, there is a need to understand more fully the concept of change. This in turn has a close relationship to time. No revolutionary dynamic governance which accommodates the ever-changing characteristics of policy-making and the maritime industry can avoid taking account of temporal issues.

> Temporal order provides an alternative in which linkages are less consequential than temporal. Things are connected by virtue of their simultaneous presence or arrival. In a culture with a strong sense of monthly or yearly cycles or of birth cohorts, we should not be overtly surprised by temporal order. In many human situations the most easily identified property of objects or events is the time subscripts associated with them. March and Olsen (1984: 743).

M. Roe, *Maritime Governance*, DOI 10.1007/978-3-319-21747-5_3

Therefore time is next on the agenda.

Time has always been a popular subject. Van de Ven and Dooley (1999: 358) exemplify the situation placing it within a dynamic context, but we will begin with looking at time from the traditional approach of past, present and future, or as Schedler and Santiso (1998: 7) suggest:

> the future is uncertain, the past is past, and the present offers no salvation either.

Time Past

> Without change there is no history; without regularity there is no time. Time and history are related as rule and variation; time is the regular setting for the vagaries of history. Kubler (1962: 72) quoted in Ingold (1993: 157).

Marx and Engels (1968: 96) were explicit about the significance of time and history:

> Men (*sic*) make their own history, but they do not make it just as they please: they do not make it under circumstances chosen by themselves, but under circumstances directly found, given and transmitted from the past. The tradition of all the dead generations weighs like a nightmare on the brain of the living.

It may come as some surprise, but the concept of time needs some defining before we can even begin. There is a substantial literature on time written in only the past few years and substantially more prior to this. Some later contributions include Schivelbusch (1986), Gould (1987), Aveni (1989), Grosz (1995) and Galison (2003). Widely debated across a multitude of disciplines, the meaning of time remains unclear at best and commonly thoroughly confusing. We need to achieve some sort of clarification of what we are considering before we can begin to address the time related issues of dynamic policy and governance in any meaningful way.

Definitions abound, and for the sake of simplicity, we shall consider a minimal number here. As we go on, many others will emerge as varieties of the core concept. Elias (1992: 10) provides a good point from which to start:

> The expression 'time' therefore refers to this relating together of positions or segments within two or more continuously moving sequences of events. The sequences themselves are perceptible. The relation between them results from the elaboration of perceptions by human beings possessing knowledge. It finds expression in a communicable social symbol, the concept of 'time', which within a certain society can transmit from one person to another a memory picture which can be experienced, but not perceived through the senses.

OK so that is clear. To be fair, if you take it slowly it actually describes what goes on and begins to reveal just how complex and complicated the whole concept of time really is.

Harvey (1969: 412) adds to the debate commenting on Hallowell's (1955: 216) 'formalised reference points' to which all events of the past, present and future can be referred. They include the calendar, clock, seasons and life cycle. These reference points vary between societies leading us again to the idea of a variety of times. Levi-Strauss (1963: 301) illustrates these ideas with the Hopi kinship system which 'requires no less than three different models of time dimension'.

Prigogine and Stengers (1984: xviii) emphasise how time varies with culture. Some view time as cyclical, endlessly recurrent. For others, it 'is a highway stretched between past and future'; others again see human lives as stationary, and the future moves towards us. Individuals vary with their cultural interpretation of time—some looking only to the immediate, others to the far future. And individuals even vary in their time horizon depending on context, circumstances and experience.

Time expectancies are also important—bed time, work time, commercial break times, the length of sports events and songs. There are many more but all are significant in interpreting the meaning of time. More complex is the interpretation of the direction of time. With the discovery of the laws of thermodynamics, it became accepted that there is a continuous and inescapable loss of energy in the universe and as a result, the 'world machine is running down' (Prigogine and Stengers 1984: xix). From this, it follows that one moment is no longer exactly like the last and 'you cannot run the universe backwards to make up for entropy'. Therefore, time has directionality and is irreversible.

This in turn causes some scientific problems. As entropy drains from the universal system, this also reduces the differences within it resulting in increasing homogeneity, a concept that conflicts almost directly with theories of evolution that point towards increasing diversity and complexity. Whilst we cannot even begin to attempt to deal with these issues in a book that focuses on maritime governance, the substance of underlying debates on time remains significant and suggest that time is both important and fundamental.

Entropy has been widely used as a tool to analyse the social sciences (see for example Allen et al. 1985: 66; Li and Qi 2008; Sommers 2009). Meanwhile, Tilly (1994: 271–273) provides a rather more extensive discussion suggesting that time is a 'relational' concept, an invention as the 'humanly negotiated concordance of two or more sequences'. A single sequence cannot establish time, and time always has to be humanly conceived and formulated. It is centred on culture not some sort of 'superhuman reality' and changes as 'shared understandings and choices of sequences change'. It is in many ways artificial, constructed and entirely contextual.

Tilly continues to debate whether some sort of absolute time exists beyond human consciousness and whether time is actually grounded in genetic-based physiological rhythms. This debate between astronomical and seasonal time and locally defined time (for example church bells) continues to this day. This issue of multiple times we shall return to later, but for the moment we can see that there is also a 'prevailing time' which affects many features of life including shipping. One example is that of the original growth of flags of convenience as a response to prohibition laws in the USA in the 1920s and 1930s. Their development was a consequence of what was then prevailing and which has changed over time; and continues to do so.

Peuquet (2002: 11–32) discusses in depth the history of time suggesting it was regarded as one of the sources of the world in many 'ancient mythological, religious and philosophical systems, including *Chaos* and *Kronos* in ancient Greek mythology, *akasa* and *kala* in Indian philosophy and *Zurvan* in early Zoroastrianism' (Akhundov 1986).

In particular, the Greek concept of movement from Chaos to Cosmos has come to dominate Western philosophy. Chaos is the initial state of the universe, a 'boundless abyss of infinite space' (Peuquet 2002: 13). The God Gaia (Earth) gives birth to Kronos (Time), and order is gradually imposed creating the final state of Cosmos. Ramo (1999) noted that *Kronos* [more precisely, the exact quantification of passing time expressed in successive readings of rationalised and decontextualised devices and tools such as clocks and calendars (Sui 2012: 9)] was seen as complimentary to *Kairos* [human right and timely moments to act judiciously; or to be in the right place at the right time (Miller 1992; Couclelis 1998)]. Kairos was always characterised by three temporal dimensions (Smith 1969)—the right time, a time of tension that calls for a decision, and the opportunity to accomplish the purpose. Cyclical time followed from this based upon seasons, migrations, day and night, etc. The idea of linear, progressive and non-repeating came with the Hebrews and Zoroastrians emphasising either the final salvation of the world or the deliverance of Israel.

Meanwhile, Homer also began to identify an ordering of events with time continuous and open-ended and moving from the past, through the present and into the future, notions he developed through the *Odyssey*. However, Plato (428–347 BC) found it inappropriate to divide time into units of past, present and future. He considered that only 'Being' was real and that 'Becoming' was a journey towards Being. Time was a 'moving image towards eternity' (Peuquet 2002: 15) a concept that was continued through to Newton's days.

By the time of the Renaissance and the work of Copernicus, two significant advances in the understanding of time occurred largely emanating from the rise in scientific thought and a move away from ideas such as the physical distinction between Earth and Heaven. A continuous and unending time was scientifically established; and the concept of relating space and time closely together emerged. In the words of Newton (1962: 6):

> Absolute, true, and mathematical time, of itself and from its own nature always flows equably without relation to anything external and by another name is called duration.

Thus, absolute time forms a backdrop against which all physical objects and events can be measured. Time and space are separated, and the former becomes an 'abstract, universal order that exists by and in itself regardless of what happens *in* time' (Peuquet 2002: 19). However, this view of time continued only until the early twentieth century when Einstein, based on the work of Minkowski, developed his view of a combined space-time where time formed a fourth dimension of geometry in a 'hypercube multivariate coordinate space'.

The Kantian (1724–1804) view of time developed from that of Newton but used a different perspective which incorporated the idea that we are born with some pre-existing notion of time that is innate and intuitive in order to perceive motion or changes in objects (Kant 1955). Time, along with space, is the 'basis on which the human mind inevitably arranges knowledge' (Peuquet 2002: 21). Kant (1950) considered that time (and space) is as follows:

- Universal—there is a single temporal background.
- Unsuppressible—time is always there and continues regardless of all other things.
- Necessary—for sensory perception.
- Unique—there is only one time.
- Infinite—because time is not an object in itself it can have no boundaries.

The reality we know is filtered by current and previous preconceptions cast in the context of innate space and time and therefore represents Kant's 'construction of thought'. Periodically, we need to revise our basic assumptions to incorporate new knowledge and experience into 'noncontradictory alignment with previous experience' (Peuquet 2002: 23). The accommodation of the theories of Newton, Copernicus and Einstein, for example, are examples of this process of re-examination.

Kant's views were a substantial change from previous conceptions of time with the individual no longer a passive observer, now determining the shape of their own personal time (and space) and as such represents one of the most important developments in modern thought (Wallace 1974). It led in turn to the idea that there are multiple times—geological, astronomical, social, economic and many more—dependent on the context, individual, expectation, experience, etc.

In more recent times, there has continued a debate on the importance and role of time and this has some significant ramifications for any discussion on governance and policy, maritime or otherwise. For example, Davis (1899: 483) considered time in his analysis of the geographical cycle relating changes in the physical landscape which took place with the passage of time concluding that although its scale was important, the amount of change observed was never simply a function of time.

Russell (1926: 122) continued to develop the idea that time was dependent upon the observer suggesting that:

> we cannot point to a time itself, but only to some event occurring at that time. There is therefore no reason in experience to suppose that there are times as opposed to events; the events, ordered by the relations of simultaneity and succession are all that experience provides. Hence, unless we are to introduce superfluous metaphysical entities, we must, in defining what we regard as an instant, proceed by means of some construction which assumes nothing beyond events and their temporal relations.

Hartshorne (1939: 176) had entered the debate on the importance of time (in this case in geography) at an early stage and put forward a number of views about time's place in spatial studies referring in particular to Spethmann (1928), Hettner (1931), Sauer (1925, 1931) and their support for moves in this direction and in opposition to the great majority then in ascendance. He continued this debate some twenty years later (Hartshorne 1959: 81) and remained convinced that actually the role of the scientist was to study the world as it is, and in Hettner's words 'time in general steps into the background'. Hartshorne agreed that the dimension of time was always involved in any analysis but that the majority could take place in the context of the present as interactions amongst phenomena were commonly dependent on current processes.

Harvey (1969: 408) outlines the debate that followed between those who wished to emphasise the significance of time and those who took the opposite view. Hettner and Sauer's views in support of what Harvey calls the 'genetic forms of explanation' received extended support through the development of geomorphology. Not all gave unqualified support to the temporal conception and most were not as definite as Sauer (1963: 360):

> The geographer cannot study houses and towns, fields and factories, as to their where and why without asking himself about their origins. He cannot treat the localization of activities without knowing the functioning of the culture, the process of living together of the group; and he cannot do this except by historical reconstruction. If the object is to define and understand human associations as areal growths, we must find out how they… came to be what they are… The quality of understanding sought is that of analysis of origins and processes. The all-inclusive object is spatial differentiation of culture. Dealing with man and being genetic in its analysis, the subject is necessarily concerned with sequences.

The support for a temporal interpretation certainly grew through the twentieth century further evidenced by contributions from Wooldridge and East (1951: 682) and Sorre (1962: 44).

For example, Carlstein (1981: 43) is convinced of the importance of time to society and as such by implication to policies that affect that society (including those of the maritime sector):

> Temporality is central to the generation and perpetuation of social forms, not incidental to it, and temporality in turn makes no sense without concepts of spatial presence and absence.

Kasperson and Minghi (1969: 200) stress the significance of time in studies of political unification. Referring specifically to the stage of the process of unification that had been reached in addition to the period of history of writing or data collection, they emphasise that these were key elements in understanding what was happening and as such should not be underplayed.

Massey (1999: 267) suggests that history (and by definition time) is critical to the development of all philosophical thought, something discussed in detail by Prigogine and Stengers (1984). Meanwhile, Hagerstrand (1970: 1) emphasises how time had been neglected in scholarly work outside of astronomy—largely a consequence of the general opinion that time was fixed, defined and unquestionable—inevitable. Discussion was therefore pointless. 'As long as the Millennium and Judgement Day stood out as the ultimate goals, already planned, there was no cause to indulge in fancies about the future'. However, after a period of focussing entirely on the present and the immediate, sociologists of all sorts were beginning to recognise the significance of time to an understanding of the large majority of issues and problems. He goes on to suggest that the earlier in temporal forms had been a mistake and particularly in social terms were simply wrong.

Hagerstrand (1974a: 73) later emphasises that there was much to be learned from plant communities as much as economic and urban/industrial communities about the significance of time. Without proper temporal consideration then any analysis of social systems (including policy-making) would be shallow and insubstantial.

> Organisms, machines and buildings form populations in which generations follow each other as parcels in time. Territories of all sizes are frequently bounded not only in space but also in time.

Gertler (1988: 151) in considering geography broadly suggests that time has been neglected as a concept and deserves considerably more attention and particularly in the analysis of economic issues. Citing support from economics more generally (for example Shackle 1968; Robinson 1974, 1980; Arrow 1978; Bausor 1983), he suggests that the use of time needs to be reconstructed and more closely integrated into geographical studies of this sort. Meanwhile, Adam (1990: 24) comments on Bergson's (1924) contribution:

> To him the future is becoming in a way that can never be a mere rearrangement of what has been.

Time Present

> 'Time moves very fast these days', says Venturi and Co. But even on the bullet train of post-modernity we're in for a bumpy ride... (O'Connor 1981, in Wark 1988).

Thrift (1977: 69–70) following Parkes and Thrift (1977) provides a discussion on the relationship between temporal level and time suggesting that a hierarchical model of time could be developed which could be applied to any social situation. Four types of time were identified each related to a societal level. Each level includes all the elements of the levels below it, and each is constantly active in trying to subsume that below it. The levels are termed as follows:

- Superstructure,
- Built environment,
- Activity system, and
- Attitude and perception.

With some interpretation, this temporal model can be applied to the maritime sector by marrying up its structure to that of the jurisdictional model identified by Roe (2013). Thrift suggests that each level exudes its own time signal. Superstructure associates with the global level and has a longer-term, overarching characteristic. The built environment can be seen operating at the supranational level with a shorter time focus, but it still is extensive. The activity system is the national level, whilst the attitude and perception level is much more of an individual concept which can be associated with local action and even with the seafarer or port worker. Thrift continues to suggest that three other times—biological, psychological and socio-ecological—cut across the hierarchy and operate at all levels.

In the most modern of terms, time has been frequently considered as synonymous with money and this in turn reflects its importance to everyday life. Thrift (1996: 178) quotes Lakoff and Johnson (1980: 8):

> In our culture TIME IS MONEY (*emphasis original*) in many ways: telephone message units, hourly wages, hotel room rates, yearly budgets, interest on loans, and paying your debt to society by 'serving time'. They have arisen in modern industrialised societies and structure our basic everyday activities in a very profound way, corresponding to the fact that we act as if time is a valuable commodity – a limited resource, even money – we conceive of time that way. Thus we understand and experience time as the kind of thing that can be spent, wasted, budgeted, invested wisely or poorly, saved or squandered.

They go on to suggest a multitude of metaphors that emphasise further the place that time holds within society:

- You're wasting my time.
- This gadget will save you hours.
- I don't have the time to give to you.
- How do you spend your time these days?
- The flat tyre cost me an hour.
- I've invested a lot of time in her.
- I don't have enough time to spare for that.
- You're running out of time.
- You need to budget your time.
- Put aside some time for ping-pong.
- Is that worth your while?
- Do you have much time left?
- He's living on borrowed time.
- You don't use your time profitably.
- I lost a lot of time when I was sick.
- Thank you for your time.

Low and Barnett (2000) provide an all-encompassing interpretation of the role of time in globalisation suggesting that there is no one scale of time and that globalisation is characterised by multitemporality. Far from diluting the significance of either time or globalisation, this implies an even greater importance to analysing the relationship between the two. They go on to discuss the tendency for globalisation to overwhelm and to the detriment of other significant issues in academic debate such as historicism and other ways that exist of accessing interdisciplinary concepts.

Time Future

Arrow (1978: 157) considers the relevance of studies in economics of the present compared with looking at those of the future and concludes like Alchian (1950) that the present should be considered relatively 'small and unimportant'. To concentrate on holdings rather than future flows, on the perishable rather than the durable was understandable but short-sighted. Whilst to emphasise the importance of time in economics was hardly new, the need to reconfigure this explicitly rather than implicitly was clear.

Hagerstrand (1970: 1) stresses much the same, seeing the world as moving away from what had been viewed as automatic progress towards increasing chaos. 'If mankind shall have a future at all, we need to be able not only to forecast coming events but consciously and purposely to invent this very future'. To achieve this, the need was (and is) to understand much more about the complex systems of concern to society.

Tilly (1994: 291–293) takes it further, contemplating time in the future and suggesting the emergence of three types. The first was an unfashionable possibility associated with the time of nation-states whereby the process of globalisation does not eradicate the state's potential to control the flows of materials, people, finance and information over controlled time periods. Instead the state retains effective control of citizens' time and much more so than is anticipated.

Secondly, the more fashionable conclusion that the supranational authority will extend its power so that timekeeping will become one more string to its economic, social and political bows. Alternatively, he envisages a final temporal future with citizens living in multiple times for 'protection, production, consumption, procreation, recreation, friendship, worship and other zones of activity, each individual and group knotting them together in their own distinctive times' producing what he sees as a period of unparalleled diversity.

Baumann (2000: 113) sees future time as speeding up and associates this with the consideration of space as well. Time is to become 'processual, mutable and dynamic, not preordained or stagnant'. It will be the continuation of what has been happening for some time—perhaps forever—but with increasing acceleration. Space would be conquered by ever faster machines creating larger usable space but in turn demanding ever greater space and thus faster machines—mirroring the dilemmas of the capitalist society and its chase for ever more spatial fixes. 'Space was the value, time was the tool'. Weber suggests it was necessary to sharpen the tools of society to achieve the future time scenario that was desired. This 'instrumental rationality' focuses on designing ways to perform tasks faster 'while eliminating unproductive, idle, empty and so wasted time'. Baumann (2000: 117–118) takes this further with consideration of the importance of increasing instantaneity as we move into the future. Georg Simmel (1900) provides an interesting view on the value of instantaneity suggesting that values are valuable as far as they are to be gained by foregoing other values. Instant (or effectively instant) anything suggests the expenditure of no time at all, and hence, its value can then be questioned. This in turn devalues the space that the instantaneity has revealed.

Time and Space

> …social theory must acknowledge, as it has not done previously, time-space intersections are essentially involved in social existence… Giddens (1979: 54).

> An adequate account of human agency must, first, be connected to a theory of the acting subject/the human individual; and second, must situate action in time and space as a continuous flow of conduct, rather than treating purposes, reasons, etc., as somehow aggregated together. Giddens (1979: 2).

Pop (2006) analyses in some depth the contribution of James T. Mangan and his creation, the Nation of Celestial Space—over which he acted from 1948 to 1970 as the first representative. This new nation incorporated the whole of outer space which Mangan claimed as sovereign territory, filing the appropriate documents and making all the traditional claims of sovereignty and inviolability. The story is both fascinating and informative in its dealings with globalisation and the state's role, and although this argument has never formally been recognised, it provides a multitude of lessons about international, global and extraterrestrial relationships and also the relationship of space (in its areal sense) to time.

To quote Pop (2006: 212–213):

> One year after having founded Celestia, Mangan declared that he had willed his claim to the territory to his children, aware that he might not live to see the day when the Nation of Celestial Space would be considered anything but 'fantastic'(Statesville Record 1949). Mangan's death on 14 July 1970 left his son, James C. Mangan, in control of Celestia, in what the inheritor calls 'the biggest inheritance in history' (Suburbanite Economist 1970). While his reign may have reached the bounds of the universe, Mangan's life lasted a mere 73-year-long moment. Perhaps this is because as, suggested by an editorialist in 1949, under the space-time conception, 'the claimant to all space seems to have missed a bet when he failed to stake out a claim to all time too'.

Time has a long history of a close relationship with spatial issues (see for example Wilson 1955; Van Fraassen 1970; Sack 1974), and there has been considerable comment about the significance of this relationship. Maritime governance clearly has close relationships with the spatial characteristics of the industry manifested in the distribution of seafarer origins, ship registration, flag registries, port facilities, the location of financial and insurance services, and so on. These in turn have a temporal dimension which we explore further in this section and which is increasingly seen as inseparable from the spatial.

Crang (2003: 190) cites Kofman and Lebas (1995: 16), quoting Lefebvre (1970: 224):

> Space is nothing but the inscription of time in the world, spaces are the realizations, inscriptions in the simultaneity of the external world of a series of times, the rhythms of the city, the rhythms of urban population… the city will only be rethought and reconstructed on its current ruins when we have properly understood that the city is the deployment of time.

Russell (1926: 121) suggests that the question of time is actually 'rather less complicated than space'. Without wishing to argue with such an authority, this seems a little doubtful, whilst Schumm and Lichty (1965: 110) notc thc importance of time in determining form when looking at both physical and human geography where both reflect the impact of history upon their characteristics. Time and space are thus essential to the study of systems.

Harvey (1968: 71) examines the spatial features of objects and events and attempts that had been made to explain the patterns that emerge from the causal mechanisms that generated them. The implication is that spatial patterns could be deduced from temporal processes and although simple sequential mapping could achieve some of this, rather more sophisticated and formal models would be needed. These would include the space-time languages of Carnap (1958) and of course the work of Hagerstrand on time geography, which at the time of Harvey's writing was only in embryonic form.

Hagerstrand (1974a: 80–81) was strongly influenced by the relationship between time and space through his utilisation and adaption of time-space budgeting ultimately to manifest itself in time geography—much more of which in a later chapter. He suggests that 'one accepts spatial location to be strongly determined by the sequence of events in time'. The 'space-time division of power' was unlikely to turn out to be easy but one that would become increasingly necessary. Hagerstrand's views are supported by Pred (1977: 209) who sees a need to 'get away from the overly *strong emphasis upon the spatial cross-sectional view* (Hagerstrand 1974a) of human phenomena and to focus a great deal more on time'.

Sack (1974: 1) quotes Blaut (1961: 3) in expressing doubts about whether the traditional approach of geographers to treat space separately—the *spatial separatist theme*—'all that seems to be required is a belief that withdrawing the temporal dimensions from a section of reality, along with all objects, somehow leaves something spatial behind for the geographers to study'. Lukerman (1965) reaffirmed this with his assumption of 'absolute space'.

Sack continues (1974: 3) citing support for using a space-time system for the identification of facts (rather than a separatist spatial approach) from Russell (1948), Quine (1950), Wilson (1955), Carnap (1958), Strawson (1963) and Harvey (1969). Lukerman (1958: 5) provides further back-up: 'the basic concepts of time and distance relate to all facts' and 'temporal/spatial concepts are used in all theories'.

Van Paasen (1976) looks at space-time relationships in terms of anthropology, whilst Gregory (1978: 119) comments that it is important to assign levels of development their own temporality, following on from the comments of Althusser and Balibar (1970: 99) on the decision by Marx not to follow Hegel in reducing history to a single essence that unfolded through time. Taking this approach, the assumption is that the capitalist mode of production contains 'different rhythms which punctuate the different operations of production, circulation and distribution'. Each has its own temporal agenda and spatial consequences (Vilar 1973: 188), and these may well not coincide. Castells (1977: 442–444) has his own view and that analysis needs to be directed towards 'an historically defined space-time, a space constructed, worked, practised by social relations', and organised 'into specific articulated units according to the arrangements and rhythms of the means of production'.

Seamon (1980: 159–160) analyses the relationship between time and space in terms of 'routines' which encompass both aspects of daily life. Most activities then do not need detailed consideration because they are 'routine' and they are considered (if at all), together. It is meaningless to consider them separately. Their inherent routineness releases time for consideration of those events that need core attention. Seaman also goes on to describe *place-ballets* which are combinations of time-space routines and *body-ballets*. Such notions provide for the combination and analysis of people, space, place and time.

Soja (1980: 210) considers both space and time and within a social context. Quoting Lefebvre (1976: 31):

> Space is not a scientific object removed from ideology and politics; it has always been political and strategic. If space has an air of neutrality and indifference with regard to its contents and thus seems to be 'purely' formal, the epitome of rational abstraction, it is precisely because it has been occupied and used, and has already been the focus of past processes whose traces are not always evident on the landscape. Space has been shaped and moulded from historical and natural elements, but this has been a political process. Space is political and ideological. It is a product literally filled with ideologies.

Much the same could be said of time. Both space and time therefore cannot be separated from social product or practice, and activities and objects must then be considered in both time and space and not one or the other.

Bird (1981: 131) suggests that without a consideration of time, space will always be static and any spatial analysis inadequate. However, space and time possess very different properties despite the energetic attempts of many to integrate the two (for example Thornes and Brunsden 1977; Carlstein et al. 1978; Parkes and Thrift 1980). Bird cites Ullman (1974) who characterises time as the more active and mental construct, while space is the more passive and concrete dimension.

Urry (1985: 21) debates whether time and space can be considered as 'absolute entities, possessing their own natures or particularities'. Is either of them 'causally productive' and possessing their own structure or merely relative and a way of characterising the relations between the constituents of the physical world? This latter view is that of Liebniz (1898) who suggests that space is something merely relative—it is an order of coexistences as time is an order of sequences.

Urry thus sees events as distributed in time-space, structured with specific relationships. These relationships might change spatially, temporally or more commonly on both dimensions but not necessarily to the same extent.

Dear (1986: 374) is convinced of the close proximity of space and time which had taken on a new significance with the onset of postmodernism. Using concepts explored by Jameson (1984: 83–84), the old systems of organisation and perception had been 'destroyed and replaced by a postmodern *hyperspace*'. The boundaries of space and time had been stretched to accommodate the new multinational global space and both were necessary to achieve this. Dear and Flusty (1998: 50) remain convinced some years later suggesting that new geographies of a postmodern era had been created alongside a new time-space fabric.

Gertler (1988: 154) emphasises the close relationship between capital and time which in turn has significant repercussions for the analysis of space and place. Robinson (1953) had stressed much earlier how temporally sensitive capital was; Gertler agreed and went on to link this with a need to add a spatial dimension as well. However, he was not as convinced of Solow's (1956) neoclassical view of the firm and capital as essentially malleable with an almost unrestricted potential for use which is variable in both space and over time in response to changing price signals. This he described as 'vulgar' and 'radical'.

Gertler (1988: 160) continues by emphasising the need to accommodate a temporal dimension into economic studies but is aware of the theoretical failings which remain prominent. He notes one genuine attempt by Massey (1978) to use a geological analogy in interpreting how 'successive rounds of accumulation wash one over another depositing layers of industrial relations and social apparatus which interact dialectically with the prior people, production systems and political characteristics they find'. Massey's approach (but not the principle of temporality) is criticised by Warde (1985) for its failure to specify how one round of accumulation is distinguished from another, something that Gertler also noted.

Harvey (1990: 205) is convinced of the significance of time over space—but without dismissing the importance of the latter. He suggests that social theory has always concentrated on 'social change, modernisation and revolution (technical, social, political). Progress is its object and historical time its primary dimension'. Progress was commonly seen as the conquest of space, removing spatial barriers and the 'annihilation of space through time'. Space reduction is inherent in progress, and the latter is characterised by similar reduction. Thus, computers get smaller, communications get quicker. One facilitates the other. In broad terms, modernity has focussed on *becoming* rather than *being*, something noted earlier. Harvey used Foucault (1984: 70) to summarise wondering why and when it happened that:

> "space was treated as the dead, the fixed, the undialectical, the immobile" (while) "time, on the contrary, was richness, fecundity, life, dialectic."

The inadequacies of space to act as a dominant means of interpretation and the desire to increasingly represent flux and change was a feature of the Futurist art movement desirous of representing speed and motion on a two dimensional canvas, an example of which is shown in Fig. 3.1.

Harvey noted that they took their art beyond canvas to be part of a revolution thus introducing mobility and change. In similar fashion, Walter Pater (see Schoen 1942) argued that art aspires to music, providing a transition to a dynamic medium; the next step was to film and it is interesting to consider that increasingly art forms have become progressively more mobile, less static.

Not everyone agrees and Harvey points out reaction to increasing mobility, speed and flux in the early twentieth century (when the Futurists and other such genre were emerging). In particular, he cites Heidegger (1927) who had

Fig. 3.1 Natalia Goncharova, cyclist (1913)

proclaimed the permanence of being over the temporariness of becoming. His opposition to 'velocity, instantaneousness and simultaneity' led him to suggest:

> All this implies that this nation, as a historical nation, must move itself and thereby the history of the West beyond the centre of their future 'happening' and into the primordial realm of the powers of being.

In turn, this led him to the 'inner truth and greatness of the National Socialist movement of Germany'. He commented on its withdrawal from the League of Nations and that this would inevitably bind people into the great will of the German state and the people might:

> … grow in its unity as a work people, finding again its simple worth and genuine power, and procuring its duration and greatness as a work state. To the man of this unheard of will, our Fuhrer Adolf Hitler, a three-fold Sieg-Heil! (Taken from Blitz 1981: 217).

I do not think we need to say much more here except to note that the points made by Heidegger concerning *being* over *becoming* may be retrogressive but continue to have some support (for example the recent debate in the UK over EU membership which focuses on the supposed unification of individual will into the British nation-state).

Luke (1991: 320) continues the debate on the relationship between time and space discussing the growth of importance of the 'informational' society and chronopolitics which he saw 'grounded in the pace of exchange; how rapidly the flows can travel, expand and unfold without meeting resistant barriers or closed borders'. Luke sees society moving at that stage from the Westphalian system of autonomous nation-states to a new system of global networks, transnational flows and informational communities set within a chronopolitical program. The division between the two would remain blurred—the original Westphalian definitions are themselves a product of technological 'velocity', itself reflecting temporarily. The conflict between the chronopolitical pace of today and the geopolitical space of the past which remains dominant both reflects the significance of time and space and the contradiction between their definitions.

Virilio (1995: 2) dismisses space as being superseded by time as 'distances and surfaces become irrelevant in favour of time span'. Meanwhile, Forsberg (1996: 355) reminds us through the work of Walker (1993: 131) that conceptions of space and time cannot be treated as 'some uniform background noise, as abstract ontological conditions to be acknowledged and then ignored'. Forsberg (1996: 365–366) goes on to indicate that territorial identity appears to be in decline because of a number of reasons including the erosion of state sovereignty and more significantly because concepts of time have become more important than those of space. However, this is based on a fallacious argument itself centred on a non-existent dichotomy between time and space. Progress is represented by increasing control over space, and this is achieved more by the 'annihilation of time' than anything else.

Massey (1999: 262) continues the debate quoting Raper and Livingstone (1995: 363):

> Space and time must be considered relative concepts, i.e., they are not determined by the nature and behaviour of the entities that 'inhabit' them (the concept of 'relative space'). This is the inverse of the situation where space and time themselves form a rigid framework which has an existence independent of the entities (the concept of 'absolute' space).

They conclude that time and space cannot be thought of separately but as a combined and fourth dimension—space-time. This is important for maritime policy as much as anything else for as Raper and Livingstone suggest (1995: 262), 'the way that spatio-temporal processes are studied is strongly influenced by the model of space and time that is adopted'. Grossberg comments (1996: 178) 'the bifurcation of time and space was perhaps the founding moment of modern philosophy'; meanwhile, Unwin (1993) suggests that a reconceptualisation of time-space was needed centred upon its reunification. Massey (1999: 263) blames it all on Kant and his emphasis on a debilitating separation of the two concepts. She goes on to contemplate the need for change in representation of space so that time is also given its fair place, so that 'representation is no longer stasis, but an element in a continuous production; a part of it all, and constantly becoming'. Historically, representation has been associated only with space but in truth, it also represents a fixed point in time. Not only is it important that this relationship is recognised but the need to represent space-time as a unity needs to be accommodated.

Dacin et al. (1999: 340) place the consideration of time in an organisational context seeing it important to view sources, mechanisms and outcomes of organisational embeddedness in a 'broad historical and comparative perspective' and stressing the value of historical and longitudinal studies. Others note temporality in situated interactions (Barley 1988), organisational control, group dynamics and interaction (Gersick 1988, 1994), entrainment (Ancona and Chong 1992; Lacey et al. 1998) and within organisations in general (Hassard 1996). Meanwhile, Abbott (1988) looks at the relationship between the sequence of processes in key events and the variation in pattern.

Castells (2000: 13–14) considers the issue of time and space extensively, and we shall return in a later chapter to consider more of his contribution. He sees the emergence of new social structures occurring with the advent of globalisation focussing particularly on the relationship between time and space. He cites Giddens (1984), Thrift (1990), Harvey (1990), Lash and Urry (1994) and Adam (2000) as evidence. Castells suggests that two emergent social forms of time and space were characterising what he termed the new network society. Timeless time and space of flows.

Timeless time he defines by the 'use of new information/communication technologies in a relentless effort to eliminate time'. Time is compressed (for example electronic communications) and desequenced (for example the blurring of life cycle patterns).

Space of flows refers to 'the technical and organisational possibilities of organising the simultaneity of social practices without geographical contiguity'. He suggests that the majority of social functions are now organised with these characteristics drastically changing social networks, physical locations and the relationship that exists between time and place. Place is not irrelevant, but it is the location of networks that matters far more than what he termed the 'spaces of places' which used to be dominant.

The debate over the relationship of space and time has continued into the twenty-first century. Nielsen and Jespersen (2001) extensively discuss their relationship to freight transport. May and Thrift (2003a, b: 2) are convinced that the two are inseparable quoting Massey (1994: 260–261). He stresses that instead of prioritising either space or time we need to:

> Overcome… the very formulation of space/time in terms of this kind of dichotomy…[and to recognise instead] that space and time are inextricably interwoven.

This space-time was 'multidimensional', capable of accommodating multiplicity (Rodowick 1997; Assad 1999).

May and Thrift continue to examine the four prime spatial characteristics of time. Time is shaped by our responses to *rhythms and timetables* which themselves are defined by the relationship between time and space in the natural world. They cite examples such as the diurnal cycle, the tides, the seasons and body rhythms (Parkes and Thrift 1980: Young 1988). These rhythms vary spatially in terms of the impact they have on lives (for example comparing urban and rural life; developed and developing countries); and also according to social circumstances (for example the impact of the menstrual cycle; the regime of the car assembly worker; and the daily routine of a child going to bed) (Valentine 1997).

Social discipline emerges from this discussion both secularly and religiously defined. The former can be seen in the design of workplaces to ensure that employees are adequately monitored during work time (Stein 1995). During family time at home, work may often intervene thus placing a different time-consuming activity in the 'wrong' place—and vice versa in the context of family demands on work time (for example generated by a new baby) (Schivelbusch 1988; Schlor 1998).

The importance of *instruments and devices* also needs to be noted. These range from the more obvious sundial and digital clock to the DVD recorder and iPod. These can play two differing roles—either to flag up the time and its progress (for example the use by a bus driver of their watch) or to disguise these very characteristics (the use of an iPod on a long-haul flight). The extended use of recording devices for visual and audio images has had substantial spatial effects both directly—in freeing up locational constraints that previously existed, requiring listeners/viewers to be in a place at a specific time to experience the occasion—or indirectly in allowing the consumption of information at different times and places and the consequential impacts on life this has. Initially, the land-line and subsequently the mobile phone are significant examples (Kern 1983; Urry 1995).

Finally, May and Thrift (2003a, b: 4–5) suggest that the relationship between time/space and *texts* is important—a rather vaguer concept than the earlier ones referring to 'vehicles of translation (attempts to render social meaning from new conceptualisations of time itself)'.

Rather than privileging space over time or vice versa, this four part social structure of the relationship between the two attempts to provide a balanced account where one characteristic is not necessarily more important than another. May and Thrift suggest that this has commonly been the case citing Thompson on labour control (1967), Kern (1983), Young on the natural universe (1988), Harvey (1990), and Urry on technology (1995). Instead, we should consider various and uneven networks of time stretching over a variable social space. They describe this as 'a multiple, heterogeneous and uneven time-space'.

Amin (2002: 386) considers the 'historicity of spatiality', a concept derived from Agnew (1999: 504) which combines the notions of space and time with respect for the geographical embeddedness of power relationships. In this way, the temporal dimension of spatial characteristics is emphasised with layers of spatial power jostling over time. Herbert and Matthews (2004: 164) comment on Sack (1972) and his suggestion that 'space has limited independent meaning and is in effect a relational concept'. It has to be qualified by time, context and a range of economic, social and political factors'. Meanwhile, Dale and Burrell (2007: 5), for example, note the continued exercise of a dichotomised relationship between space and time.

Many Times

> It was then I began to understand that everything in the room had stopped, like the watch and the clock, exactly, a long time ago. I noticed that Miss Havisham put down the jewellery exactly on the spot from which she had taken it up. As Estella dealt the cards, I glanced at the dressing table again, and saw that the shoe upon it, once white, now yellow, had never been worn. I glanced down at the foot from which the shoe was absent, and saw that the silk stocking on it, once white, now yellow, had been trodden ragged. Without this arrest of everything, this standing still of all the pale decayed objects, not even the withered bridal dress on the collapsed form could have looked so like grave-clothes, or the veil so like a shroud. *Great Expectations*, Charles Dickens (1861).

It is clear from the discussion so far that the concept of time is not quite as simple as might first be thought. In particular, a variety of times has been identified (Sorokin and Merton 1937: 615–616), and this theme continues to reappear over the decades since then. Reichenbach (1958: 117) reaffirms that there are a number of ways of measuring the differing times that exist, whilst Schedler and Santiso (1998) continue to note how this is related closely to time in its context of past, present and future.

Thrift (1977: 69) suggests that 'just as there are many spaces within spaces, so there are many times within times'. This essential multidimensionality means that any research into life and the world has to 'dynamize' its curriculum. Arrow (1978: 158) notes Hick's (1946) consideration of the variation that exists in time when contemplating commodities and regarding them as different things dependent on the dates they are considered are reflected as much as anything in the price which can be associated with them.

Harvey (1990: 224–225) provides an extensive discussion of the variety of times and in particular refers to Gurvitch's (1964) social tines (Fig. 3.2) each of which derives from different social formations in turn generating their own temporality. Some of these can run concurrently depending on the social formation—he cites the contrast between academic time and revolutionary time in France in 1968. This was reaffirmed some time ago by Sorokin and Merton (1937: 615) who saw time as a 'necessary variable in social change' and also by Graham (1998: 179) citing Thrift (1996: 2)—'time is a multiple phenomenon; many times are working themselves out simultaneously in resonant interaction with each other'.

Urry (2000: 105–106) cites Adam (1995) in commenting upon the variety of times that exists, supported by Hawking (1988: 33): 'there is no unique absolute time, but instead each individual has his own personal measure of time that depends on where he is and how he is moving'. This personal view of time (*eigenzeit*) is stressed by Nowotny (1994) and goes hand in hand with the idea of seeing time as related to its measurement.

Urry continues with a discussion of natural and social times (Urry 2000: 118–119) suggesting that the difference between the two is largely imaginary as social time itself is 'generalised through nature' and thus characteristic of the physical sciences. Consequently, the characteristics of social time such as past, present and future, the

Type	Level	Form	Social Formation
Enduring time	Ecological	Continuous time in which past is projected in the present and future; easily quantifiable	Kinships and locality groupings (particularly rural peasant societies and patriarchal structures)
Deceptive time	Organized society	Long and slowed down duration masking sudden and unexpected crises and ruptures between past and present	Large cities and political 'publics'; charismatic and theocratic societies
Erratic time	Social roles, collective attitudes (fashion) and technical mixes	Time of uncertainty and accentuated contingency in which present prevails over past and future	Non-political 'publics' (social movements and fashion followers); classes in process of formation
Cyclical time	Mystical unions	Past, present and future projected into each other accentuating continuity within change; diminution of contingency	Astrology followers; archaic societies in which mythological, mystical and magical beliefs prevail
Retarded time	Social symbols	Future becomes present so late as to be outmoded as soon as it is crystallized	Community and its social symbols; guilds, professions etc, feudalism
Alternating time	Rules, signals, signs and collective conduct	Past and future compete in the present; discontinuity without contingency	Dynamic economic groups; transition epochs (inception of capitalism)
Time in advance of itself	Collective transformative action and innovation	Discontinuity, contingency; qualitative change triumphant; the future becomes present	Competitive capitalism; speculation
Explosive time	Revolutionary foment and collective creation	Present and past dissolved into a transcendent future	Revolutions and radical transformations of global structures

Fig. 3.2 Typology of social times. *Source* Gurvitch (1964)

qualitative experience of time, are actually all integral to the natural sciences (Adam 1990: 150). The only exception was clock time. However, this has been taken as the determining feature of natural time and in turn is inappropriate because clock time is Newtonian and Cartesian—based on the notion of absolute time that is not liable to change. This absolute time is 'invariant, infinitely divisible into space-like units, measurable in length, expressible as a number and reversible' (Urry 2000: 119). It is Cartesian space because 'it is premised upon the dualisms of mind and body, repetition and process, quantity and quality, form and content, subject and object' (Urry 2000: 119).

The modern conception of time has rejected this Newtonian and Cartesian interpretation:

> Space and time are now dynamic qualities; where a body moves or a force acts, it affects the curvature of space and time – and in turn the structure of space-time affects the way in which bodies move and forces act. (Hawking 1988: 33).

Despite much discussion, this integration of natural and social time into one variable and flexible concept remains elusive, and therefore, it has proved difficult to comprehend how 'nature, society and individuals are embedded in each other and are interdependent' (Elias 1992: 16).

Urry (2000: 127–129) also stresses the significance of instantaneous time, something which has particular relevance to changes in daily life in recent years. This has seen events which are occurring 'now' (or instantaneously) as more important in social consideration than those with some sort of theme. Despite the fact that they share nothing in common, they are all happening instantaneously. This 'collage' effect is accompanied by the intrusion of distant events into everyday life. This space-time compression reflects the presence of instantaneity which in turn diminishes the impact of space. Thus, technology allows multiple and (almost) instantaneous events to be absorbed together and in one location. Time (and space) loses one of its most significant characteristics. Virilio (1986) suggests that this 'violence of speed' transcends and destroys place. Modern youth culture sees the day consisting of 24 h (and not 16 plus sleep during conventional hours) which can be divided up at will through the use of technology and mind-inducing substances.

Crang (2003: 189) takes Lefebvre's use of discrimination between types of cities, using an 'assemblage of different beats' as a basis for distinguishing different times and tempos. Using the city as his example, he suggests that it is the location where 'multiple temporalities collide', quoting Mehrotra (1999: 65–66) and his example of Bombay where there is an:

> inter-twining of times, of attitudes, of the coming together and moving apart of the past and present(which) has historically created Bombay's urban kaleidoscope. It is an urban phenomenon that does not lend itself to simplistic readings of its form, which is pluralistic in nature and does not make explicit its origins, intention or rationale.

Crang goes on to suggest that we think of everyday rhythms in multiple forms including some which are speeding up continuously (for example Bombay), and others which are slowing (daily commuting) and others which are regular (the school run; the annual holiday; shop opening hours, and so on). In Felski's (2000: 18) terms, 'everyday life is above all a temporal norm' and Lefebrvre's vision has been taken up successively by Quick (1998) and retrospectively by de Certeau (1984), Harvey (1985).

Stalder (2006: 156) is emphatic in how many different conceptions of time exist. He notes the work of Urry (1985) who comments on 'computime' and its contrast in temporality with 'glacial time' and that of 'clock time'. Castells' 'timeless time' we shall return to later where he considers how all temporalities come together creating differing interactions. No particular time is dominant overall as it depends entirely on context. Time is constructed rather than natural, something widely accepted in the social sciences for many years but a much newer idea in the scientific community.

The concept of 'anti-time' provides an alternative perspective to the conventional interpretation of time. The concept is far from new and reflects the attitude that time has only a limited place in the understanding of society. Hartshorne (1939: 176) within a geographical context suggests that to consider the spatial nature of the discipline in its proper context, any consideration of time relations must be 'secondary and merely supplementary'. Others suggest that time is important as it places the phenomena under study into their appropriate point of development—emphasised in particular at that time by Spethman (1928). Hartshorne goes on to discuss the example of Sauer and Meigs' (1927) work on the site and culture of San Fernando de Velicata which he considers 'history rather than geography'.

> The question is whether geographers who wish to study the present geography of a region are required first to produce works for which few of us are technically prepared and which can hardly be distinguished from other fields.

Hartshorne continues to criticise the significance that time is accorded by some geographers and thus remains firmly in the 'anti-time' school of thought. Of course since then, the attitude towards space and time and their interrelationship has changed substantially. Sauer (1974: 190) notes with some surprise how the physical geographer W.M. Davis promoted geomorphology as a discipline 'free of concern with chronology of time and change'. Whilst his cycle of erosion was characterised by temporal phases (youth, maturity and old age), the interpretation was of these phases topically rather than over time and there was no consideration of a phase within time or whether a phase was long or short. Thus, time was relegated to something that was the concern of geologists.

Guelke (1977: 3) identifies theoretical difficulties in incorporating time into any studies that focus upon a real or spatial relationships (and there is an argument that this might also cover the maritime and logistics sectors). He quotes Hartshorne (1939: 184–188), who suggests that time 'steps into the background' and that in geography, the only historical study with validity was one that presented a 'cross section or period picture'. Although dated, specifically focussing upon geography and widely condemned, Hartshorne's comments remained strongly influential.

Massey (1993: 141, 147) spends some considerable effort in assessing the role of time and its significance particularly in contrast to the pre-eminence placed upon space. Evidence for this trend comes from many places. Massey quotes Berger (1974): 'it is space, not time, that hides consequences from us'; 'the difference that space makes' (Sayer 1985); 'the new spatiality implicit in the postmodern' (Jameson 1984); 'it is space rather than time which is the distinctively significant dimension of contemporary capitalism' (Urry 2000); 'the anxiety of our era has to do fundamentally with space, no doubt a great deal more than time' (Foucault 1986). To these clearly spatial opinions, she adds Laclau (1990: 41) who comments in the context of a definition that 'temporality must be conceived as the exact opposite of space'. This perpetuates the idea of a dichotomy, largely irreconcilable and both thoroughly unhelpful and widely criticised. Massey (1993: 147) continues:

> All the strings of these kinds of opposition with which we are so accustomed to work (mind-body; nature-culture; reason-emotion; and so forth) have been argued to be at heart problematical and a hindrance to either understanding or changing the world.

Such dichotomies are considered to work to the advantage of certain, always dominant, social groups—and this includes all such dichotomies not just temporal/spatial ones, and therefore forms part of a conspiracy typified by the black/white and male/female scenarios.

Forsberg (1996: 371) reaffirms the significance of space over time suggesting that it occupies a stronger mental category:

> I think that it is at least empirically arguable that our daily life, our psychic experience, our cultural languages, are today dominated by categories of space rather than categories of time, as in the preceding of high modernism. Jamieson in Keith and Pile (1993: 2).

Forsberg (1996: 370) continues by emphasising the false dichotomy between time and space. The process of deterritorialisation emphasised by Harvey, amongst many others, reflects the significance of both dimensions and reveals how an attitude of anti-time is both negative and unhelpful.

Timeless time is a concept widely discussed by Castells (for example 2000: 16). He suggests that the characteristics of societal, structural transformations have come from the extensive introduction of information networks as the main organisational form. This stems from the 'simultaneous availability of new, flexible information technologies and a set of historical events which came together by accident around the late 1960s and 1970s.' These include the emphasis upon deregulation and liberalisation of capitalism; the difficulties faced by nation states with respect to intensified globalisation; the rise of 1960s counter-culturalism; and the development of new media adopting 'global hypertext'. Together, these favoured the adoption of information networks reflecting a move towards what Castells terms 'timeless time' where temporality becomes both intensely important (and continuously pursued to annihilation) and yet meaningless (in that by being annihilated it becomes of less significance). Virilio (1995: 1) concurs suggesting that 'real time now prevails above both real space and the geosphere. The primacy of real time, of immediacy, over and above space and surface is a *fait accompli* and has inaugural value'. He focusses far more on the significance of time that stems from its elimination, rather than its belittling. The result is 'global time', a result of increasing instantaneity which is overcoming the local and the spatially bound (1995: 2).

Webster (2002: 108) comments upon Castells' ideas on timeless time and his consideration of well-trodden arguments about space-time compression. Castells' contribution is to envisage a 'network society' within which the significance of time takes on the dichotomy identified above of simultaneous significance and insignificance. Castells' examples include the growth of flexitime to maximise the effective use of time and the emergence of 'electronically managed global capital markets' (Castells 1996: 417). Other vignettes of timeless time include the blurring of lifestyles as individuals start to ignore or overcome traditionally, temporally defined functions such as childbearing (at an extended age); defying death (through cryogenics); and almost endless other possibilities afforded by genetic engineering. A permanent present is the result, reflected in news stories available anywhere and immediately. What is generated is 'systemic perturbation in the sequential order of phenomena' (Castells 1996: 464; Stalder 2006: 156); constant instantaneity; lack of continuity; and spontaneity.

We shall return to this in discussion of form and process in a later chapter, but the extended debate in social theory between the *contextual* and *compositional* (and its implications for science as a whole) and briefly noted earlier has some relevance here. Thrift (1983: 27–28) provides a detailed consideration of the issues, whilst Subramanian et al. (2003) look at neighbourhoods and social capital, Veenstra (2005) considers health and social capital in British Columbia, Kamphuis et al. (2008) provide an example of application to recreational cycling in Melbourne and Kulu and Boyle (2009) consider fertility in city suburbs. Stemming from a number of writers but perhaps most significantly the work of Hagerstrand (1974b), with major contributions from Simpson (1963) and Kennedy (1979) (the latter distinguishing between the *immanent* and *configurational*), the compositional approach finds its highest point in the 'structural-genetics' of Marx. The activity of humans is divided into a set of structural categories which have homogeneity and are derived through a process of abstraction. These categories can then be combined to form an explanation of society. Meanwhile, Thrift (1983: 28) sees elements of the contextual approach in the work of Schutz (1967) on phenomenology, in Berger and Luckmann's phenomenological–dialectical approach (Berger and Luckmann 1991), in Goffman's (1986)frame analysis, and in Hagerstrand's (for example 1970) time geography. Human activity is considered a series of situated, social events within their immediate spatial and temporal settings. Kennedy explains the immanent (unchanging) properties of matter and energy and the likewise unchanging processes and principles arising therefrom (Simpson 1963: 24) as his equivalent of the compositional. Meanwhile, the configurational (relating to and or determined by unique conditions of time and space) is a concept which enables physicists to separate their scientific concerns from the complexities of change over time and the influence of history upon the present and future. The consideration of time is never dull.

Time, Form and Process

> ...without time we cannot study change. Thus process is a word liberally introduced... Thrift (1977: 65).

> Truly temporal processes are continuous or invisible in the sense that, the very process of differentiating them into phases of before and after serves, not to separate them into a 'patchwork of disjointed parts' as Dewey puts it, but on the contrary, to relate their phases as aspects of the same dynamic unity. Shotter (1983: 21).

Le Poidevin (2003: 14–15) examines Aristotle's view (384–322 BC) of time which on the one hand dismissed it as not existing and on the other saw it represented by change in the form of things. Time and change were considered one and the same, and without change in form, there was no time to observe. In some ways, Plato (429–347 BC) was the source of this view as he accounted for the birth of time when the celestial bodies started to move for the first time. Thus, change in form represented time and without those changes, time did not exist.

Eichenbaum and Gale (1971: 531) comment on how Schaefer (1953: 243) is dismissive of the importance of time as 'purely geographical laws contain no reference to time or change' an opinion backed by Bunge (1966: 199) referring to the science of geography as one of locations. Schaefer continues (1953: 243–244): we cannot 'deny that the spatial structures we explore are, like all structures anywhere, the result of processes', but 'the geographer, for the most part, deals with them as he finds them, ready made'. Despite its focus on geography, these comments are of relevance in suggesting a failure to appreciate the significance of time in the study of form (and by assertion, other issues such as governance and policy). The implication is that the temporal dimension can be discarded leaving the true elements of the system behind to be analysed.

Schumm and Lichty (1965: 110) are some of the earliest commentators on the relationship between form and time referring back to the debates by Strahler (1950, 1952), Von Bertalanffy (1952: 109). Although directed specifically towards the debate within the field of geomorphology, their comments are both interesting and relevant. Von Bertalanffy was significantly opposed to the idea that time was particularly relevant to the study of form:

> In physical systems events are, in general, determined by the momentary conditions only. For example, for a falling body, it does not matter how it has arrived at its momentary position, for a chemical reaction it does not matter in what way the reacting compounds were produced. The past is, so to speak, effaced in physical systems. In contrast to this, organisms appear to be historical beings.

Schumm and Lichty (1965: 110) disagree considering that landforms reflect systems influenced by history. Thus, the geomorphologist must try to relate causality to evolution and ultimately to form, only possible by accommodating the temporal context.

Berry (1973: 8) cites Harvey (1969) in suggesting that it would be profitable to examine 'interactions between temporal process and spatial form'. Berry goes on to consider these relationships:

> Not only is the 'reality' of any element within a system relative to the entire system of elements; it is also time-relative. To seek any fixed thing is to deal in false imagination, therefore all phenomenal existence is immediately also seen to be transitory when the dimension of time is added. No particular thing is 'real' in any absolute sense; it is passing into something else at every moment. Every individual, for example, is a progressively ageing, temporarily-organized 'bundle' of energy flows faced with ultimate disintegration.

He considers that to search for absolute, geometric form is understandable as it stems from a society that tends to understand things through a process of codifying and classifying reality. However, to advance science is conditional upon recognising the 'relativity of existence and the relative truth of perceptions'. Hinting at what we shall discuss in some depth in a later chapter, Berry suggests that what we need is 'a more continuous intellectual process… that recognises that every system and every interpretation needs assessment in the light of a more complete system'. Or to put it more succinctly—the temporal dimension is as essential as the spatial.

Hagerstand (1974a: 76) considers the potential for developing a new framework for analysis that incorporates both spatial and temporal concepts. The majority of geographers, planners and even statisticians always evaluate their findings in spatial terms including location and distribution. Even Harvey's early writings (1969: 410) considering space-time issues focus upon the redistribution of real income in an urban system and revolve around issues of accessibility and proximity, both essentially spatial. Hagerstrand goes on to consider how to accommodate the temporal dimension to the same extent that the spatial dimension has been making 'form and process not so different as they seem'. Combining the two was realistic—both represent resources to be consumed and this is doneso normally together. The significance of time must be emphasised.

The tendency to neglect the issue of time when compared with the consideration of space was maintained over a considerable period. Gertler (1988: 152) considers how wide discussions on inter-regional convergence and divergence tended to focus upon 'putative outcomes, results or spatial distributions' and as a result largely ignore issues of process or how such divergence or convergence evolve through time (Borts and Stein 1964; Romans 1965; Lande and Gordon 1977; Smith 1979). However, Sabatier (1988: 102), in his discussion of the application of the 'advocacy coalition framework', was more positive in consideration of the relationship of process to time rather than space suggesting that policy interpretation requires an appreciation of 'hundreds of actors from dozens of organisations seeking to influence the overall policy process over periods of a decade or more in situations where relatively technical information concerning problem severity and causes cannot be ignored'. Meanwhile, Virilio (1986, 1999) provides support for Sabatier arguing that the 'acceleration of communication has led to a replacing of geographical space with time' (Elden 2005: 8).

These arguments were never satisfactorily resolved and in fact remain rumbling on today at least to a certain extent. The view that change and time are synonymous can only work if there is a universal and constant time against which to measure change as the latter is only observable in relative terms. In that case, time must exist but rather than this helping to cement time's position *vis a vis* space, it seems to have if anything done the reverse. Consequently, space and form have dominated discussion. The whole situation is summarised through what is known as the 'experience argument'.

1. During a period of time without change, there would be no experience at all—since experience itself is a form of change—and so there can be no experience of the period of time without change.
2. A period of time by itself changes nothing, and so makes no difference to what we could experience after that period.

3. We can establish that some contingently true statement is true only if its being true could make some difference to what we experience, either now or at some later stage.

Therefore,

1. We cannot possibly establish that a period of time without change has occurred.
2. If it is impossible to establish whether or not some statement is true (or likely to be true) than that statement has no meaning.

Therefore,

1. Any statement to the effect that a period of time without change has occurred would have no meaning.

At which point we shall leave this argument. Suffice to say that there remains substantial disagreement about time and change and their relationship to form and space.

Rhoads (2005: 133), in contrast to others in their consideration of Davis's 'Cycle of Erosion', considers it as developmental change within a physical system which equates process with time:

> Time, thus became, at least for many of those concerned with adapting the evolutionary notion to wider fields, almost synonymous with 'development' and 'change' such that it was viewed not merely as a temporal framework within which events occur but as a process itself. It was in this sense that Davis employed the concept of evolution as the basis for the cycle of erosion. (Chorley et al. 1973: 193).

Cram (2011: 637–638) is clear about the importance of time and its relationship to process highlighting a number of scholars who had criticised analyses of current developments that were based upon 'snapshots' and focussing on historical institutionalist approaches (Bulmer 1994; Pierson 1993, 2004; Thelen and Steinmo 1992). She stresses that the temporal dimension is central with the present only understood in the context of the past. However, the past itself can change as new narratives emerge and these in turn have an effect on the future.

Time and Governance

> Understanding the process of policy change – and the role of technical information therein – requires a time perspective of a decade or more. Such a time-span is also necessary to get a reasonable assessment of policy impacts. Sabatier (1988: 99).

Soja (1980: 210) provides commentary on the notion of space as a social (and therefore policy) construct, but his contribution to this debate could equally be applied to time. Quoting Lefebvre (1976: 31) and adapting what he says for our purposes (noted in parentheses) the similarity between the two concepts clearly can be seen:

> Space (time) is not a scientific object removed from ideology and politics: it has always been political and strategic. If space (time) has an air of neutrality and indifference with regard to its contents and thus seems to be 'purely' formal, the epitome of rational abstraction, it is precisely because it has been occupied and used, and has already been the focus of past processes whose traces are not always evident on the landscape. Space (time) has been shaped and molded (*sic*) from historical and natural elements, but this has been a political process. Space (time) is political and ideological. It is a product literally filled with ideologies.

Lefebvre's comments on space clearly are heavily interrelated with temporal issues as well with the frequent reference to process and history, so that even without the interpretation made here, time rears its head as a serious political dimension of policy-making.

Harvey (1990: 202) emphasises the significance of the choice of time horizon to policy decision-making and effectiveness suggesting that differing time horizons will generate different policy decisions. The result is that efficient governance requires the tangible inclusion of a temporal dimension if it is to be meaningful.

Tilly (1994: 271) is interested in the relationship of time to the state and particularly the state's role as policy-maker. He saw three dimensions—the time of eras within which a state was located over an extended period of time characterised in recent time as 'powerful, sharply bounded, relatively centralised coercion-wielding organisations; the medium of time within which states existed which are characterised by the 'temporal organisations of other actors with which agents of states had to contend'; and the influence that states have in influencing their subjects through the organisation of time.

Tilly (1994: 273) also suggests that the features of life vary considerably depending upon the time in which they take place. Shipping is a prime example of a time influenced activity with maritime policies changing as circumstances around them change. Thus, flags of convenience emerged as a response to policies towards alcohol prohibition in the USA between 1919 and 1923. Meanwhile, environmental and security policies since 2001 have had an immense effect upon the industry. Tilly goes on to cite Aminzade (1992) who suggests four temporal features that can affect the social meaning of processes (including policy-making)—pace, duration, cycles and trajectory.

The importance of the state in shaping prevailing time was also considered by Tilly (1994: 275). He noted three different effects:

- By pre-empting and ordering citizens' time directly, as in government employment, conscription or obligatory voting.
- By absorbing portions of citizens' times indirectly in such activities as earning to pay taxes, answering official inquiries or attending political meetings.
- By establishing their own inescapable temporal references; clock times, calendar times, schedules of school and work, cycles of military service, and so on.

Marsh and Smith (2000: 11) focus upon an analysis of agricultural policy in the UK since the 1930s and to do this identify a number of dialectical relationships. These are identifiable only because a temporal perspective is taken—without this the policies would have remained obscure and hidden. Snapshots of policy networks would provide little to analyse, and a true understanding of policies could only be obtained once how they were formed was understood and how they have changed over time.

Stoker (1998: 26) considers that traditional governance is time and place specific. This is far from adequate in that almost as soon as defined, the policies emerging from this governance process are out of date. He urges policy-makers to accommodate 'an evolutionary way to capture the processes of adaptation, learning and experiment'. This dynamic dimension to governance mirrors the discussion in Chap. 1 reflecting the need for a policy-making process that accommodates change.

LaGro (2007: 4) suggests that the issue of temporality and its place in governance is not a new one but that discussion of temporal issues in political analysis is so far inadequate. The limited number of academic works in this area is almost all specifically aimed at the EU dimension (Schedler and Santiso 1998; Pierson 2000; Tilly 1994, 1995; Schmitter and Santiso 1998; Ekengren 1997, 2002; Jerneck 2000; Goetz 2006; and Meyer-Sahling 2007). LaGro goes onto note the example of time inconsistency that is apparent in EU policy-making and governance (noted also by Tocci 2005: 78), whereby reforms are completed in the relatively short-term, whilst membership is long-term. Tocci comments: 'the process is front-loaded with obligations and back-loaded on the delivery of the benefits'.

In a similar vein to Marsh and Smith's earlier (2000) work, Cram (2011: 636–637) emphasises the importance of the temporal dimension in interpreting and understanding new modes of governance. This is founded upon the problem of 'snapshot' governance noted in earlier work by Cram.

These policy relationships with time are clearly emphasised through transport in general and shipping in particular suggesting that a 'snapshot' governance can never be adequate. There are many examples that could be taken. Bird (1981: 137) considers decision-making in port policy finding that decision-makers in European ports react to decisions forced upon them commonly by ship owners who in turn have to react to international competition. Decisions by each are made with appropriate consideration of the time when they are made but with little consideration of the long-term implications before being projected onto space in the form of port location and structure in a hypothetico-deductive sequence.

Meanwhile, Shaw (2006: 237) stresses the relatively minor attention given to time in transport research. He notes examples of where it has been important including using travel time as an impedance measure in spatial interaction models and studies of time in activity-based modelling. In addition, the work of Janelle (1969, 1975) on time-space convergence and that of Knowles (2005) on the differential collapse of time-space relationships can be noted. Shaw goes on to suggest that modern communications changes and the growth of instantaneity have lessened the value of conventional time-space models such as the gravity model, facility locations models and spatial choice models. More focus is needed on time in general and real time in particular, and their relationship to classical modelling. Brooks and Pallis (2008: 414) re-emphasise this in relation to the ports sector how policies take time to have an effect—sometimes lengthy, and governance changes can be even more protracted.

Conclusions

Time is clearly a big issue. A big issue in policy and governance as much as anywhere and the changes in communications with their impact upon globalisation have made the incorporation of a temporal dimension essential.

Maritime governance is no different from many other governance frameworks worldwide in neglecting time as a dimension, producing a series of static frameworks and policies that fail to recognise the significance that time can play. However, the intensely global nature of the shipping industry makes the absence of a temporal framework that much more significant.

Rather curiously, the increasing time compression that has characterised the moves towards globalisation has meant that the time and the changes in its significance have become more important rather than less. This is the case when considering the increased importance of process and change over place and form and is essential if governance in the maritime sector is to be improved and reflect better the policy issues towards which it is directed. Time is a formidable element in attempts to improve maritime governance and reflect the dynamic nature of the sector. The near disappearance of time with the growth of virtually instantaneous communications has made its presence even more important in governance (Urry 2000: 125).

Time is also highly related to the concept of reducibility. It is traditional in the maritime sector to attempt to reduce time to a minimum. Sometimes, there are conflicting trends (for example in the maritime sector, cruise liners and also slow steaming), but the broader trend is clear. Progress is quicker, not slower. In contrast, there is considerable emphasis on reducing time to its smallest elements at which point change is measurable and time itself becomes more important. Both trends—reducing time and raising its status by reducing its size—are two sides of the same coin. Both reflect the importance of time.

Time is also rearranged to reduce its significance. It is sometimes considered in extremely small units—and thus, its impact in consideration of a context can be minimised. It can be enclosed (and thus ignored) and events can be explained entirely by their causal antecedents which have resulted in 'now'. Meanwhile, we shall return to look at complexity and chaos in a later chapter where these concepts view time as irreducible, ever present and uni-directional (Turner 1999).

Change and time have also been identified as serious partners that need to be considered together. Change can be of varying speed from the infinitesimally slow to the almost instantaneous. Some would say that there is always change and that a static situation is a mere reflection of the human inability to measure the change

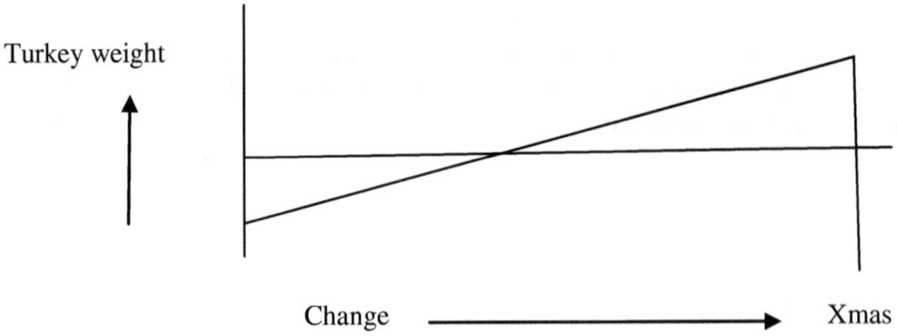

Fig. 3.3 Turkeys at Christmas

going on. Others would suggest that if you cannot identify the change occurring then effectively there is none.

We conclude this discussion with a thought for turkeys at Christmas who provide a widely used example of variations in the interpretation of changes in time (Fig. 3.3). The turkey is lulled into believing that humans are kind, providing accommodation, food and good company over a number of months through the year… until one day just before Christmas, chop. And the moral—even if you do not believe change is coming, it always is albeit often slowly (see for example Taleb 2007).

References

Abbott, A. D. (1988). *The system of professions. An essay on the division of expert labor*. Chicago, IL: University of Chicago Press.

Adam, B. (1990). *Time and social theory*. Oxford: Polity Press.

Adam, B. (1995). *Timewatch*. Cambridge: Polity Press.

Adam, B. (2000). The temporal gaze: The challenge for social theory in the context of GM food. *The British Journal of Sociology, 51*(1), 125–142.

Agnew, J. (1999). Mapping political power beyond state boundaries: Territory, identity and movement in world politics. *Millennium, 28*, 499–521.

Akhundov, M. D. (1986). *Conceptions of space and time: Sources, evolution, directions*. Cambridge, MA: MIT Press.

Alchian, A. (1950). Uncertainty, evolution and economic theory. *Journal of Political Economy, 58*, 211–221.

Alexander, H. G. (Ed.). (1956). *The Leibnitz-Clarke Correspondence*. Manchester: Manchester University Press.

Allen, P. M., Sanglier, M., Engelen, G., & Boon, F. (1985). Towards a new synthesis in the modelling of evolving complex systems. *Environment and Planning B, 12*, 65–84.

Althusser, L., & Balibar, E. (1970). *Reading capital*. London: New Left Books.

Amin, A. (2002). Spatialities of globalisation. *Environment and Planning A, 34*, 385–399.

Aminzade, R. (1992). Historical sociology and time. *Sociological Methods and Research, 20*(4), 456–480.

Ancona, D., & Chong, C.L. (1992). *Entrainment. Cycles and synergy in organizational behavior* (Working Paper). Cambridge, MA: Sloan School of Management.

Arrow, K. J. (1978). The future and present in economic life. *Economic Inquiry, 16*, 157–169.

Assad, M. L. (1999). *Reading with Michel Serres: An Encounter with time*. Albany, NY: State University of New York Press.

Aveni, A. (1989). *Empires of time. Calendars, clocks and cultures*. Boulder, CO: University Press of Colorado.

Barley, S. R. (1988). On technology time and social order: technically induced change in the temporal organization of radiological work. In F. A. Dubinskas (Ed.), *Making time: Ethnographies of high technology organisations* (pp. 123–169). Philadelphia, PA: Temple University Press.

Baumann, Z. (2000). *Liquid modernity*. Cambridge: Polity Press.

Bausor, R. (1983). The rational-expectations hypothesis and the epistemics of time. *Cambridge Journal of Economics, 7*, 1–10.

Berger, J. (1974). *The look of things*. New York: Viking Press.

Berger, P. L., & Luckmann, T. L. (1991). *The social construction of reality*. London: Penguin.

Bergson, H. (1924). Bergson à E. Peillaube. *Revue de Philosophie*, July 24, 440.

Berry, B. J. L. (1973). A paradigm for modern geography. In R. J. Chorley (Ed.), *Directions in geography*. London: Methuen.

Bird, J. (1981). The target of space and the arrow of time. *Transactions of the Institute of British Geographers NS, 6*, 129–151.

Blaut, J. (1961). Space and process. *The Professional Geographer, 13*, 1–7.

Blitz, M. (1981). *Heidegger's being and time: And the possibility of political philosophy*. Ithaca, NY: Cornell University Press.

Borts, G. H., & Stein, J. L. (1964). *Economic growth in a free market*. New York: Columbia University Press.

Brooks, M., & Pallis, A. (2008). Assessing port governance models: Process and performance components. *Maritime Policy and Management, 35*(4), 411–432.

Bulmer, S. (1994). The governance of the European Union: A new institutionalist approach. *Journal of Public Policy, 13*(4), 351–380.

Bunge, W. (1966). *Theoretical geography*. Lund studies in geography (Series C, No. 1, 2nd ed.).

Carlstein, T. (1981). The sociology of structuration in time and space: A time-geographic assessment of Gidden's theory. *Svensk Geografisk Arsbok, 57*, 41–57.

Carlstein, T., Parkes, D., & Thrift, N. (Eds.). (1978). *Timing space and spacing time (3 Vols)*. London: Arnold.

Carnap, R. (1958). *Introduction to symbolic logic*. New York: Dover Publishing.

Castells, M. (1977). *The urban question: A Marxist approach*. London: Edward Arnold.

Castells, M. (1996). *The information age*. Cambridge, MA: Blackwell.

Castells, M. (2000). Materials for an exploratory theory of the network society. *British Journal of Sociology, 51*(1), 5–24.

Chorley, R. J., Beckinsale, R. P., & Dunn, A. J. (1973). *The history of the study of landforms. Volume 2: The life and work of William Morris Davis*. London: Methuen.

Couclelis, H. (1998). Aristotelian spatial dynamics in the age of geographic information systems. In M. Egenhofer & R. G. Golledge (Eds.), *Spatial and temporal reasoning in geographic information systems* (pp. 109–118). New York: Oxford University Press.

Cram, L. (2011). The importance of the temporal dimension: New modes of governance as a tool of government. *Journal of European Public Policy, 18*(5), 636–653.

Crang, M. (2003). Rhythms of the city. In J. May & N. Thrift (Eds.), *Timespace* (pp. 187–207). Geographies of Temporality, London: Routledge.

Dacin, M. T., Ventresca, M. J., & Beal, B. D. (1999). The embeddedness of organizations: Dialogue and directions. *Journal of Management, 25*(3), 317–356.

Dale, K., & Burrell, G. (2007). *The spaces of organization and the organization of space*. London: Palgrave Macmillan.

Davis, W. M. (1899). The geographical cycle. *The Geographical Journal, 14*(5), 481–504.
Dear, M. J. (1986). Postmodernism and planning. *Environment and Planning D, 4*, 367–384.
Dear, M. J., & Flusty, S. (1998). Postmodern urbanism. *Annals of the Association of American Geographers, 88*(1), 50–72.
De Certeau, M. (1984). *The practice of everyday life*. Berkeley, CA: California University Press.
Dickens, C. (1861). *Great expectations*. London: Penguin.
Eichenbaum, J., & Gale, S. (1971). Form, function and process: A methodological inquiry. *Economic Geography, 47*(4), 525–544.
Ekengren, M. (1997). The temporality of European Governance. In K. E. Jorgensen (Ed.), *Reflective approaches to European Governance*. London: Macmillan.
Ekengren, M. (2002). *The time of European Governance*. Manchester: Manchester University Press.
Elden, S. (2005). Missing the point: Globalization, deterritorialization and the space of the world. *Transactions of the Institute of British Geographers NS, 30*, 8–19.
Elias, N. (1992). *Time: An essay*. Oxford: Blackwell.
Felski, R. (2000). The invention of everyday life. *New Formations, 39*, 13–32.
Forsberg, T. (1996). Beyond sovereignty, within territoriality. Mapping the space of late-modern (geo)politics. *Cooperation and Conflict, 31*(4), 355–386.
Foucault, M. (1984). *The history of sexuality*. Paris: Galimard.
Foucault, M. (1986). Of other spaces. *Diacritics, 16*, 22–27.
Galison, P. (2003). *Einstein's clocks, poincare's maps: Empires of time*. New York: W.W. Norton.
Gersick, C. J. G. (1988). Time and transition in work teams. *Toward a new model of group development, Academy of Management Journal, 31*, 9–41.
Gersick, C. J. G. (1994). Pacing strategic change. *The case of a new venture, Academy of Management Journal, 37*, 9–45.
Gertler, M. S. (1988). Some problems of time in economic geography. *Environment and Planning A, 20*, 151–164.
Giddens, A. (1979). *Ideology, science and human geography*. London: Hutchinson University Library.
Giddens, A. (1984). *The constitution of society: Outline of a theory of structuration*. Cambridge: Polity Press.
Goetz, K. H. (2006). Territory, temporality and clustered Europeanization. *Reihe Politikwissenschaft, 109*.
Goffmann, E. (1986). *Frame analysis: An essay on the organization of experience*. Boston, MA: Northeastern University Press.
Gould, S. (1987). *Time's arrow. Time's cycle*. Cambridge, MA: Harvard University Press.
Graham, S. (1998). The end of geography or the explosion of place? Conceptualizing space, place and information technology. *Progress in Human Geography, 22*(2), 165–185.
Gregory, D. (1978). *Ideology, science and human geography*. London: Hutchinson.
Grossberg, L. (1996). The space of culture, the power of space. In I. Chambers & L. Curti (Eds.), *The post-colonial question* (pp. 169–188). London: Routledge.
Grosz, E. (1995). *Space, time and perversion: Essays on the politics of bodies*. London: Routledge.
Guelke, L. (1977). Regional geography. *The Professional Geographer, XXIX, 1*, 1–7.
Gurvitch, G. (1964). *The spectrum of social time*. Ann Arbor, MI: University of Michigan Press.
Hagerstrand, T. (1970). Regional forecasting and social engineering. In M. Chisholm, A. B. Frey, & P. Haggett (Eds.), *Regional forecasting* (pp. 1–7). London: Butterworths.
Hagerstrand, T. (1974a). The domain of human geography. In R. J. Chorley (Ed.), *New directions in geography* (pp. 67–87). New York: Cambridge University Press.
Hagerstrand, T. (1974b). Tidgeografisk beskrivning – syfte och postulat. *Svensk Geografisk Arsbok, 50*, 86–94.
Hallowell, A. I. (1955). *Culture and experience*. Philadelphia, PA: University of Pennsylvania Press.

Hartshorne, R. (1939). The nature of geography: A critical survey of current thought in the light of the past. *Annals of the Association of American Geographers, 29*(3), 173–658.

Hartshorne, R. (1959). *Perspective on the nature of geography*. London: John Murray.

Harvey, D. (1968). Pattern, process and the scale problem in geographical research. *Transactions of the Institute of British Geographers NS, 45*, 71–78.

Harvey, D. (1969). *Explanation in geography*. London: Arnold.

Harvey, D. (1985). *Urbanisation of consciousness*. Oxford: Blackwell.

Harvey, D. (1990). *The condition of postmodernity*. Oxford: Blackwell.

Hassard, J. (1996). Images of time in work and organizations. In S. R. Clegg, C. Hardy, & W. R. Nord (Eds.), *Handbook of organization studies* (pp. 581–598). Thousand Oaks, CA: Sage.

Hawking, S. (1988). *A brief history of time*. London: Bantam.

Heidegger, M. (1927). *Being and time*. Tubingen: Max Niemeyer Verlag.

Herbert, D. T., & Matthews, J. A. (2004). The significance of place: Introduction. In D. T. Herbert & J. A. Matthews (Eds.), *Unifying geography. Common heritage, shared future* (pp. 162–170). London: Routledge.

Hettner, A. (1931). Die Geographie als Wissenschaft und als Lehrfach. *Geogr. Anz., 32*, 107–117.

Hicks, J. R. (1946). *Value and capital: An inquiry into some fundamental principles of economic theory*. Oxford: Clarendon Press.

Ingold, T. (1993). The temporality of the landscape. *World Archaeology, 25*(2), 152–174.

Jameson, F. (1984). Postmodernism or the cultural logic of late capitalism. *New Left Review, 146*, 53–92.

Janelle, D. G. (1969). Spatial reorganization; a model and concept. *Annals of the Association of American Geographers, 59*(2), 348–364.

Janelle, D. G. (1975). Time space convergence and urban transportation issues. In C. K. Blong (Ed.), *Systems thinking and the quality of life* (pp. 594–600). Washington, DC: The Society for General Systems Research.

Jerneck, M. (2000). Europeanization, territoriality and political time. *Yearbook of European Studies, 14*, 27–49.

Kamphuis, C. B. M., Giskes, K., Kavanagh, A. M., Thornton, L. E., Thomas, L. R., Van Lenthe, J., et al. (2008). Area variation in recreational cycling in Melbourne: A compositional or contextual effect? *Journal of Epidemiology and Community Health, 62*, 890–898.

Kant, I. (1950). *Critique of pure reason*. New York: Humanities Press.

Kant, I. (1955). *Kant's Prolegomona to any future metaphysics*. La Salle IL: Open Court.

Kasperson, R. E., & Minghi, J. V. (1969). Process. Introduction. In R. E. Kasperson & J. V. Minghi (Eds.), *The structure of political geography* (pp. 195–210). Chicago, IL: Aldine Publishing.

Keith, M., & Pile, S. (1993). Space and politics. In S. Pile & M. Keith (Eds.), *Place and the politics of identity*. London: Routledge.

Kennedy, B. A. (1979). A naughty world. *Transactions of the Institute of British Geographers NS, 4*, 550–558.

Kern, S. (1983). *The culture of time and space 1880–1918*. Cambridge, MA: Harvard University Press.

Knowles, R. D. (2005). *Transport shaping space*. Annual Meeting of the Association of American Geographers, Denver, CO.

Kofman, E., & Lebas, E. (1995). Lost in transposition—Time, space and the city. In E. Kofman & E. Lebas (Eds.), *Henri Lefebvre: Writings on cities* (pp. 3–60). Oxford: Blackwell.

Kubler, G. (1962). *The shape of time: Remarks on the history of things*. New Haven, CO: Yale University Press.

Kulu, H., & Boyle, P. J. (2009). High fertility in city suburbs: Compositional or contextual effects? *European Journal of Population, 25*, 157–174.

Lacey, R., Gruenfeld, D., & Ventresca, M. J. (1998). *Work group sources of organizational innovation* (Working Paper 98-14). Evanston, IL: Kellogg Teams and Groups Center, North Western University.

Laclau, E. (1990). *New reflections on the revolution of our time*. London: Verso.
LaGro, E. (2007). *The temporalist of enlargement: Comparing East Central Europe and Turkey*. In Biennial Conference of the European Studies Association, Montreal, Canada.
Lakoff, G., & Johnson, N. (1980). *Metaphors we live by*. Chicago, IL: University of Chicago Press.
Lande, P. S., & Gordon, P. (1977). Regional growth in the United States: A re-examination of the neoclassical model. *Journal of Regional Science, 17*, 61–69.
Lash, S., & Urry, J. (1994). *Economics of signs and space*. London: Sage.
Lefebvre, H. (1970) *From rural to urban*. Paris: Economica (in French).
Lefebvre, H. (1976). Reflections on the politics of space. *Antipode, 8*(2), 30–37.
Le Poidevin, R. (2003). *Travels in four dimensions*. Oxford: Oxford University Press.
Levi-Strauss, C. (1963). *Structural anthropology*. New York: Doubleday Anchor Books.
Li, H., & Qi, A. (2008). Impact of corporate governance on voluntary disclosure in Chinese listed companies. *Corporate Ownership and Control, 5*(2), 360–366.
Liebniz, G. W. (1898). *The monadology and other philosophical writings*. London: Oxford University Press.
Low, M., & Barnett, C. (2000). After globalisation. *Environment and Planning D, 18*, 53–61.
Luke, T. W. (1991). The discipline of security studies and the codes of learning from Kuwait. *Alternatives, 16*, 315–344.
Lukerman, F. (1958). Toward a more geographic economic geography. *Professional Geographer, 10*, 2–10.
Lukerman, F. (1965). Geography: de facto or de jure. *Journal of the Minnesota Academy of Science, 32*, 189–196.
March, J. G., & Olsen, J. P. (1984). The new institutionalism: Organizational factors in political life. *The American Political Science Review, 78*(3), 734–749.
Marsh, D., & Smith, M. (2000). Understanding policy networks: Towards a dialectical approach. *Political Studies, 48*, 4–21.
Marx, K., & Engels, F. (1968). *Selected works*. London: Lawrence and Wishart.
Massey, D. (1978). Regionalism: Some current issues. *Capital and Class, 6*, 106–125.
Massey, D. (1993). Politics and space/time. In M. Keith & S. Pile (Eds.), *Place and the politics of identity* (pp. 141–161). London: Routledge.
Massey, D. (1994). *Space, place and gender*. Cambridge: Polity Press.
Massey, D. (1999). Space-time, 'science' and the relationship between physical geography and human geography. *Transactions of the Institute of British Geographers NS, 24*, 261–276.
May, J., & Thrift, N. (Eds.). (2003a). *Timespace, geographies of temporality*. London: Routledge.
May, J., & Thrift, N. (2003b). Introduction. In J. May & N. Thrift (Eds.), *Timespace, geographies of temporality* (pp. 1–46). London: Routledge.
Mehrotra, R. (1999). Working in Bombay: The city as generator of practice. In C. Davidson (Ed.), *Anytime* (pp. 64–69). Cambridge, MA: MIT Press.
Meyer-Sahling, J.-H. (2007). *Time and European Governance: An inventory*. Workshop on Temporality of Europeanization and Enlargement, 15–16 February, Potsdam.
Miller, C. R. (1992). Kairos in the rhetoric of science. In S. P. Witte, N. Nakadate, & R. D. Cherry (Eds.), *A rhetoric of doing: Essays on written discourse in honor of James I. Kinneavy* (pp. 310–327). Carbondale, IL: Southern Illinois University Press.
Newton, S. L. (1962). *Mathematical principles of natural philosophy and his system of the world*. New York: Greenwood Press.
Nielsen, L. D., & Jespersen, P. H. (2001). Time and space in freight transport. In L. D. Nielsen & H. H. Oldrup (Eds.), *Mobility and transport* (pp. 63–72). Aalborg: Transportradet, Aalborg University, Department of Development and Planning.
Nowotny, H. (1994). *Time*. Cambridge: Polity Press.
O'Connor, J. (1981). The meaning of crisis. *International Journal of Urban and Regional Research, 5*(3), 301–329.

Parkes, D., & Thrift, N. (1977). Putting time in its place. In T. Carlstein, D. Parkes, & N. Thrift (Eds.), *Timing space and spacing time in socio-economic systems* (pp. 119–129). London: Edward Arnold.
Parkes, D., & Thrift, N. (1980). *Times, places and space: A chronogeographic perspective*. London: Wiley.
Peuquet, D. J. (2002). *Representations of space and time*. New York: The Guilford Press.
Pierson, P. (1993). When effect becomes cause: Policy feedback and political change. *World Politics, 45*(4), 595–628.
Pierson, P. (2000). Not just what but when: Timing and sequence in political processes. *Studies in American Political Development, 14*, 72–92 (Spring).
Pierson, P. (2004). *Politics in time*. Princeton, NJ: Princeton University Press.
Pop, V. (2006). The nation of celestial space. *Space Policy, 22*, 205–213.
Pred, A. (1977). The choreography of existence: Comments on Hagerstrand's time-geography and its usefulness. *Economic Geography, 53*(2), 207–221.
Prigogine, I., & Stengers, I. (1984). *Order out of chaos*. London: Heinemann.
Quick, A. (1998). Time and the event. In S. Lash, A. Quick, & R. Roberts (Eds.), *Time and value* (pp. 103–112). London: Sage.
Quine, W. (1950). Identity, ostension and hypostasis. *The Journal of Philosophy, 47*, 621–633.
Ramo, H. (1999). An Aristotelian human space-time manifold: From chronochoros to kairostopos. *Time and Society, 8*(2), 309–328.
Raper, J., & Livingstone, D. (1995). Development of a geomorphological spatial model using object-oriented design. *International Journal of Geographical Information Systems, 9*, 359–383.
Reichenbach, H. (1958). *Space and time*. London: Constable.
Rhoads, B. L. (2005). Process/form. In N. Castree, A. Rogers, & D. Sherman (Eds.), *Questioning geography* (pp. 131–150). London: Wiley Blackwell.
Robinson, J. (1953). The production function and the theory of capital. *Review of Economic Studies, 21*, 81–106.
Robinson, J. (1974). *History versus equilibrium*. Thames Papers in Political Economy. Woolwich, London: Thames Polytechnic.
Robinson, J. (1980). Time in economic theory. *Kyklos, 33*, 219–229.
Rodowick, D. N. (1997). *Gilles Deleuze's time machine*. Durham, NC: Duke University Press.
Roe, M. S. (2013). *Maritime Governance and policy-making*. London: Springer.
Romans, J. T. (1965). *Capital exports and growth among US regions*. Middletown, CT: Wesleyan University Press.
Russell, B. (1926). *Our knowledge of the external world*. London: George Allen and Unwin.
Russell, B. (1948). *Human knowledge: Its scope and limits*. New York: Simon and Schuster.
Sabatier, P. (1988). The advocacy coalition framework: Revisions and relevance for Europe. *Journal of European Public Policy, 5*(1), 98–130.
Sack, R. D. (1972). Geography, geometry and explanation. *Annals of the Association of American Geographers, 62*, 61–78.
Sack, R. D. (1974). The spatial separatist theme in geography. *Economic Geography, 50*(1), 1–19.
Sauer, C. O. (1925). The Morphology of Landscape. *University of California Publications in Geography, 2*, 19–53.
Sauer, C. O. (1931). Cultural geography. *Encyclopaedias of social sciences* (pp. 621–623). New York: Macmillan.
Sauer, C. O. (1963). *Land and Life*, Berkeley, CA: University of California Press.
Sauer, C. O. (1974). The fourth dimension of geography. *Annals of the Association of American Geographers, 64*(2), 189–192.
Sauer, C., & Meigs, P. (1927). Site and culture at San Fernando de Velicata. *University of California Publications in Geography, 2*, 271–302.
Sayer, A. (1985). The difference that space makes. In D. Gregory & J. Urry (Eds.), *Social relations and spatial structures* (pp. 49–66). London: Macmillan.

Schaefer, F. K. (1953). Exceptionalism in geography: A methodological examination. *Annals of the Association of American Geographers, 43*, 226–249.

Schedler, A., & Santiso, J. (1998). Democracy and time: An invitation. *International Political Science Review, 19*(1), 5–18.

Schivelbusch, W. (1986). *The railway journey: The industrialization of time and space in the 19th century*. Berkeley, CA: University of California Press.

Schivelbusch, W. (1988). *Disenchanted night: The industrialisation of light in the nineteenth century*. Berkeley, CA: University of California Press.

Schlor, J. (1998). *Nights in the Big City: Paris, Berlin, London, 1840–1930*. London: Reaktion.

Schmitter, P. C., & Santiso, J. (1998). Three temporal dimensions to the consolidation of democracy. *International Political Science Review, 18*(1), 69–92.

Schoen, M. (1942). Walter Pater on the place of music among the arts. *The Journal of Aesthetics and Art Criticism, 2*(6), 12–23.

Schumm, S. A., & Lichty, R. W. (1965). The space and causality in geomorphology. *American Journal of Science, 263*, 110–119.

Schutz, A. (1967). *The phenomenology of the social world*. Evanston, IL: Northwestern University Press.

Seamon, D. (1980). Body-subject, time-space routines and place-ballets. In A. Buttimer & D. Seamon (Eds.), *The human experience of space and place* (pp. 148–165). London: Croom Helm.

Shackle, G. L. S. (1968). *Uncertainty in economics and other reflections*. Cambridge: Cambridge University Press.

Shaw, S.-L. (2006). What about 'time' in transportation geography. *Journal of Transport Geography, 14*, 237–240.

Shotter, J. (1983). 'Duality of structure' and 'intentionality' in an ecological psychology. *Journal for the Theory of Social Behavior, 13*, 19–43.

Simmel, G. (1900). A chapter on the philosophy of value. *American Journal of Sociology, 5*, 577–603.

Simpson, G. G. (1963). Historical science. In C. C. Albritton (Ed.), *The fabric of geology* (pp. 24–48). Stanford, CA: Stanford University Press.

Smith, D. M. (1979). Neoclassical growth models and regional growth in the US. *Journal of Regional Science, 15*, 165–181.

Smith, J. F. (1969). Time, times and the right time, chronos and kairos. *The Monist, 53*(1), 1–13.

Soja, E. W. (1980). The socio-spatial dialectic. *Annals of the Association of American Geographers, 70*(2), 207–225.

Solow, R. M. (1956). The production function and the theory of capital. *Review of Economic Studies, 23*, 101–108.

Sommers, J. (2009). The Anglo-American model of economic organization and governance: Entropy and the fragmentation of social solidarity in twenty-first century Latvia. *Debatte. Review of Contemporary German Affairs, 17*(02), 127–142.

Sorre, M. (1962). The role of historical explanation in human geography. In: P. L. Wagner & M. W. Mikesell (Eds.), *Readings in cultural geography* (pp. 44–47). Chicago, IL: University of Chicago Press.

Sorokin, P. A., & Merton, R. K. (1937). Social time: A methodological and functional analysis. *The American Journal of Sociology, XLII, 5*, 615–629.

Spethmann, H. (1928). *Dynamische Landerkunde*. Breslau: Zwölf Jahre Ruhrbergbau.

Stalder, F. (2006). *Manuel Castells. The theory of a network society*. Cambridge: Polity Press.

Statesville Record. (1949). Celestia bans trips to the Moon. In *Statesville Record*, July 2.

Stein, J. (1995). Time, space and social discipline; factory life in Cornwall, Ontario, 1867–93. *Journal of Historical Geography, 21*(3), 278–299.

Stoker, G. (1998). Governance as theory: Five propositions. *International Social Science Journal, 50*(155), 17–28.

Strahler, A. N. (1950). Equilibrium theory of erosional slopes approached by frequency distribution analysis. *American Journal of Science, 248*, 673–696, 800–814.

Strahler, A. N. (1952). Dynamic basis of geomorphology. *Geological Society American Bulletin, 63*, 923–938.
Strawson, P. (1963). *Individuals: An essay in description metaphysics*. New York: Doubleday.
Subramanian, S. V., Lochner, K. A., & Kawachi, I. (2003). Neighborhood differences in social capital: A compositional artefact or a contextual construct? *Health and Place, 9*, 33–44.
Suburbanite Economist. (1970). Celestia loses Prime Minister. *Suburbanite Economist*, July 19th.
Sui, D. (2012). Looking through Hagerstrand's dual vistas: Towards a unifying framework for time geography. *Journal of Transport Geography, 23*, 5–12.
Taleb, N. N. (2007). *The Black Swan*. London: Penguin.
Taylor, P. J. (2003). Time: From hegemonic change to everyday life. In S. L. Holloway, S. P. Rice, & G. Valentine (Eds.), *Key concepts in geography* (pp. 151–164). London: Sage.
Thelen, K., & Steinmo, S. (1992). Historical institutionalism in comparative politics. In K. Thelen, S. Steinmo, & F. Longstreth (Eds.), *Structuring politics: Historical institutionalism in comparative analysis*. Cambridge: Cambridge University Press.
Thompson, E. P. (1967). Time, work discipline and industrial capitalism. *Past and Present, 38*, 56–97.
Thornes, J. B., & Brunsden, D. (1977). *Geomorphology and time*. London: Taylor and Francis.
Thrift, N. (1977). Time and theory in human geography. *Progress in Human Geography, 1*(1), 54–101.
Thrift, N. (1983). On the determination of social action in space and time. *Environment and Planning D, 1*, 23–57.
Thrift, N. (1990). The making of capitalism in time consciousness. In J. Hassard (Ed.), *The sociology of time* (pp. 105–129). London: Macmillan.
Thrift, N. (1996). *Spatial formations*. London: Sage.
Tilly, C. (1994). The time of states. *Social Research, 61*(2), 269–295.
Tilly, C. (1995). To explain political processes. *The American Journal of Sociology, 100*(6), 1594–1610.
Tocci, N. (2005). Europeanization in Turkey: Trigger or anchor for reform? *Southern European Society and Politics, 16*(1), 73–83.
Turner, F. (1999). Foreword. Chaos and social science. In R. A. Eve, S. Horsfall, & M. E. Lee (Eds.), *Chaos, complexity and sociology: Myths, models and theories*. London: Sage.
Ullmann, E. L. (1974). Space and/or time: Opportunity for substitution and prediction. *Transactions of the Institute of British Geographers NS, 63*, 125–139.
Unwin, T. (1993). *The place of geography*. Harlow: Longman.
Urry, J. (1985). Social relations, space and time. In D. Gregory & J. Urry (Eds.), *Social relations and spatial structures* (pp. 20–48). London: Macmillan.
Urry, J. (1995). *Consuming places*. London: Routledge.
Urry, J. (2000). *Sociology beyond societies*. London: Routledge.
Valentine, M. (1997). 'My Son's a Bit Dizzy'. 'My Wife's a Bit Soft': Gender and cultures of parenting. *Gender, Place and Culture, 4*, 37–62.
Van de Ven, A. H., & Dooley, K. J. (1999). Explaining complex organisational dynamics. *Organization Science, 10*(3), 358–372.
Von Bertalanffy, L. (1952). *Problems of life*. London: Watts and Co.
Van Fraassen, B. (1970). *An introduction to the philosophy of time and space*. New York: Random House.
Van Paasen, C. (1976). Human geography in terms of existential anthropology. *Tijdschrift voor economische en sociale geografie, 6*, 324–341.
Veenstra, G. (2005). Location, location, location: Contextual and compositional health effects of social capital in British Columbia, Canada. *Social Science and Medicine, 60*, 2059–2071.
Vilar, P. (1973). Histoire. *Marxiste, histoire en construction: essai de dialogue avec Althusser, Annales ESC, 28*, 165–198.
Virilio, P. (1986). *Speed and politics*. New York: Semiotext(e).

Virilio, P. (1995). *Speed and information: Cyberspace alarm.* www.ctheory.net/articles.aspx?id=72.
Virilio, P. (1999). *Polar inertia*. London: Sage.
Walker, R. B. J. (1993). *Inside/outside: International relations as political theory*. Cambridge: Cambridge University Press.
Wallace, K. D. (1974). *Epistemological basis for Ernst Cassirer's Philosophy of science and its application to spatial theory*. Ph.D. dissertation, Fordham University, New York.
Warde, A. (1985). Spatial change, politics and the division of labour. In D. Gregory & J. Urry (Eds.), *Social relations and spatial structure* (pp. 190–212). London: Macmillan.
Wark, M. (1988). On technological time: Cruising Virilio's over-exposed city. *Arena, 83*, 82–100.
Weber, M. (1930). *The Protestant ethic and the spirit of capitalism.* London: Unwin.
Webster, F. (2002). *Theories of the information society*. London: Routledge.
Wilson, N. (1955). Space, time and individuals. *The Journal of Philosophy, 52*, 589–598.
Wooldridge, S. W., & East, W. G. (1951). *The spirit and purpose of geography*. London: Hutchinson.
Young, M. (1988). *The Metronomic Society*. London: Thames and Hudson.

Chapter 4
Process

Abstract Starting from the premise that one of the causes of failure in maritime governance comes from its static nature, our examination of the significance of form in policy-making and governance has led us to consider the role of time and how its incorporation might be a significant ingredient in overcoming the lack of movement that characterises governance of the sector, something that as far back as Galileo might have seemed unacceptable. This chapter focuses upon process and the need for a more fluid and flexible approach to maritime governance. After discussing the origins, significance and context for process as a central part of maritime governance, it goes on to compare this with the current 'snapshot' approach upon which maritime policy-makers tend to rely. The issues of form, object and flow are then introduced along with change and governance and the potential offered by a process approach. The chapter concludes with a consideration of the variety of process models that might be applied to the redefinition of maritime governance.

> The great adventure of my scholarly life has been to transcend the map. I see, almost liberally, the opulence of the world as a moiré of processes in conversation. Torsten Hagerstrand, in Gren (2003: 210).

> 'In nature there is certainly nothing older than motion', emphasised Galileo. It was also he who initially proved that falling motion could be described mathematically. This was not just of theoretical significance. It was not beyond Galileo to place his discoveries at the disposal of practical application. He lectured at the Arsenal of Venice on the parabolic strategies of cannon balls (Elias 1984: 89), and he further aimed his telescope at the planets and the moon and made observations which could clean up among previous speculations about the celestial bodies and their rotation. But even in this connection there were those interested in application, this time amongst the merchants. The telescope could help them to identify distant vessels heading for the harbour. Such prior knowledge had high commercial value. Hagerstrand (2004: 315–316).

Starting from the premise that one of the causes of failure in maritime governance comes from its static nature, our examination of the significance of form in policy-making and governance led us to consider the role of time and how its incorporation might be a significant ingredient in overcoming the lack of

M. Roe, *Maritime Governance*, DOI 10.1007/978-3-319-21747-5_4

movement that characterises governance of the sector, something that as far back as Galileo might have seen unacceptable. He comments: 'in nature there is nothing older than motion'. Williams (1951: 457) also emphasises the importance of the individual in this moving space:

> at every moment, each of us finds himself the apparent centre of the world, enjoying a little bit of foreground of the here and now, while around him there looms, thing beyond thing, event beyond event, the plethora of the universe.

The importance of movement, forces, change and travel as processes is widely recognised. For example, take Augé quoted in Virilio (2007: 56):

> We have never been as close as we are today to the real, technological, possibility of ubiquity… This time, we will be able to handle immobility, but will we still be travellers? It's certainly not for nothing that the metaphor of travel is so often associated with cybernetic activity: you surf, you travel on the Internet.

Or Bigelow et al. (1988: 614):

> Tradition branded the force of gravity in particular as occult. It seemed that it could not be observed, that it acted at a distance across empty space, and that it was propagated at infinite speed. Magnetism too, seemed queer and disquieting.

The analysis of process has been well documented, and there are many examples over many years where it has raised its head particularly in its spatial context—take for example Sauer (1941) who considers its relationship to historical geography; Harvey (1969: 419–421) who considers its relationship to geographical analysis; Shotter (1983: 22) who reflects upon the significance of processes to ecological psychology and the broader social sciences; Doolittle (1984) examining agricultural change as a process; Engelen (1988) who considers the world as a continuous process; Morris (1988: 7) who considers where 'fixed and mobile meet' using the example of a motel; Bennett (1991: 219, 230) who examines the relationship between dynamism, convergence and policy-making; Dopfer (1991: 539) and his consideration of institutions, structure and process; Massey (1993) in relation to politics, space and time; Lee (1997: 20) in relation to the complexity of process and change; Hesse (1999: 33) who considers the importance of processes to society; Goodchild (2004) who attempts to place processes within the context of GIS; and O'Riordan (2004) and Urry (2010: 348) who stress the increasing mobility of society.

Mead (1932: 641), quoted in Mihata (1997: 30) comments:

> When things get together, there then arises something that was not there before, and that character is something that cannot be stated in terms of the elements which go to make up the combination. It remains to be seen in what sense we can now characterize that which has so emerged.

Despite this body of support, Markusen (2003: 703–4) urges some caution suggesting that 'an increasing emphasis on process rather than structure, agency and performance accompanies the proliferation of fuzzy concepts' and that they are 'often not well-defined and, worse, are abstracted from actors and institutions in ways that impoverish the quality and impact of the resulting work'.

Process is a big issue—and will take two chapters to deal with appropriately. In the first, we look at the main issues that underlie the concept and its relationship to flow, time and change. In the second, we move on to more indirect, but no less important issues including process philosophy, reductionism, atomism, policy transfer and in particular the role of metaphors. From there, we can begin to consider the attempts to make governance more dynamic.

Definition, Origins, Significance and Context

> Human consciousness is dominated by a pervasive sense of process... existence was viewed as process of dissolution and creation, evidenced by death and reproduction, mortality and fertility, decay and growth. A perpetual process of transformation as recognised in human life, in organic nature, in the alternation of the seasons, in the waxing and waning of the moon and the rhythms of the sky, and this transformation was sometimes traced to its source in the myths of the gods that die and are reborn, Osiris and Adonis. Whyte (1954: 15).

Definitions of process abound. James and Jones (1954: 5) define a process as 'a sequence of change systematically related as in a chain of cause and effect. The phenomena that can be observed at any one moment of time result from the operation of these sequences of change'.

Mintzberg et al. (1976: 1) consider a 'decision process' as a set of actions and dynamic factors that begins with the identification of a stimulus for action and ends with the specific commitment to action'. The emphasis is on change.

Van de Ven (1992: 170) gives three definitions of process of which one is particularly relevant here. This refers to a 'sequence of events or activities that describes how things change over time, or that represents an underlying pattern of cognitive transitions by an entity in dealing with an issue'. This is an historical perspective looking at process as occurring over the life of the issue involved—in our case a maritime policy or the broader concept of maritime governance. Many process models are based on this concept. We shall return to process modelling at a later stage, but for the moment relevant models include those derived by Scott (1971), Greiner (1972), Cohen et al. (1972), Mintzberg et al. (1976), Quinn (1980), Gluck et al. (1980) and Lorange (1980).

Rescher (1999: 37) has published widely on process philosophy. His definition of a process is brief: 'a sequentially structured sequence of successive stages or phases'. This produces three characteristics:

1. Processes are complex consisting of a 'unity of distinct stages or phases'.
2. Process has a temporal coherence and integrity which cannot be removed.
3. It has a structure, what he terms a 'formal generic patterning' generating a fixed shape or format.

McKelvey (1999: 8) cites Mackenzie (1986: 45) who suggests a process is 'a time dependent sequence of elements governed by a rule called a process law'. He identifies five components:

1. The entities involved in performing the process;
2. The elements used to describe the steps in a process;
3. The relationships between every pair of these elements;
4. The links to other processes; and
5. The resource characteristics of these elements.

Process laws define the structure of the elements, their interrelationships and the links to other processes. Processes are always linked to others and activated by events.

Johnston et al. (2000: 639), in their *Dictionary of Human Geography*, define process as 'a flow of events or actions which produces, reproduces or transforms a system or structure', with clear reference to the significance of change and movement. They note that the idea of process, although widely recognised, is seen as only a simple idea and its inherent complexity did not emerge until Blaut's (1961) work was published which emphasised that the traditional Kantian split between spatial structure and temporal process was discredited by relativism.

Braganza and Lambert (2000: 179) spend some time defining what process means in the context of business. Citing a variety of sources they suggest that business processes are 'similar to vertical functions' (Ettlie and Reza 1992; Atkinson et al. 1997), whilst others see them as activities that are performed across different functions of an organisation (Davies 1991; Euske and Player 1996). Meanwhile, Davenport (1993) and Kettinger et al. (1996) see them as operating in either individual or across functions. However, Craig and Yetton (1993) suggest that all these definitions are unhelpful as they imply that all business activities are processes.

Consequently, Hammer and Champy (1993), Edwards and Peppard (1994) and Braganza and Lambert (2000: 179), citing Crowston (1997), consider business processes as meaning the 'coordination and integration of activities performed in different functions to create outputs that are of value to one or more stakeholder'. Three characteristics emerged from this:

1. Processes add value to stakeholders.
2. Processes coordinate activities that should be undertaken to address stakeholder expectations. Anything else is superfluous.
3. Processes normally cross functional boundaries. Intra-functional activities form part of a process rather than being one in their own right.

Matthews and Herbert (2008: 29) note a process revolution that has taken place specifically in physical geography from the mid-twentieth century. This has resulted in a much better understanding of the processes that underlie the variable characteristics of the Earth's surface and reflects a general move in the social sciences towards the flexible, changing and indefinable.

Meanwhile, Whyte (1944: 7) says it all:

> To be alive is to undergo ceaseless change. Man fears change and seeks to dent it by imposing on it a principle of permanence. That may offer the illusion of escape but it cannot bring understanding. To understand nature, and himself, Man must accept change and identify the universal form of process which underlies the variety of particular processes.

Von Bertalanffy (1952: 134–5) is one of the earlier commentators to interpret the 'antithesis between structure and function, morphology and physiology (as) based upon a static conception of the organism'. In truth, everything is undergoing an orderly process even though this process is supported by an underlying structure and organised form. 'What is described in morphology as organic forms and structures is in reality a momentary cross-section through a spatio-temporal pattern'. Structures and forms are actually slow processes of (sometimes very) long duration. Benninghoff (1938) provides a useful commentary:

> Thus, while within the body the components are in a state of flux, the body itself seems to persist. But also the individual represents a series of events that starts with fertilisation and ends with death… What is in a slow flow, and is relatively persistent and quasi-stationery, is impressive as an organic form, the quicker flow of events is the function maintaining that form… If I look from the lower levels to the higher, then forms are apparent. The higher system acts as the form into which all subordinate events are integrated. Looking the other way, travelling down through the various levels, the forms are resolved one after another into processes whose speed increases with decreasing size of system.

Schaefer (1953: 248) soon followed this up with: 'mature social science looks for process laws'. Some time later, Kasperson and Minghi (1969: 196) analyse the role of process in the study of politics and identify three essential contributions: towards an understanding of political integration and disintegration; towards a better understanding of the transfer of sovereignty; and towards fuller comprehension of the political dimension associated with growth and development.

Buttimer (1976: 277) considers the significance of processes in what she terms the 'lifeworld'. By questioning the use of the word 'dwelling' as a noun (artefact) or verb (process), she introduces the idea that science tends to consider life as an artefact, thing, noun and that in fact the humanist view of the world as a process is far more illuminative. Man's complex relationship with Earth was essentially one of processes with dwelling far more than occupation and much more to do with living in tune with rhythms of nature. Dwelling occurs both in space and time and as such is a process. The fundamental significance of processes as a result can hardly be over-stressed.

Thrift (1977: 65) looks at the role of time in spatial analysis and concludes that without time there can be no study of change; and a study of change requires an understanding of process something subsequently reiterated by Gregory (1982b: 191–194). This however can be very difficult to achieve effectively. Gertler (1988: 152) considers that:

> most popular conceptions of change within the mainstream traditions are primitive and inadequate. They are typically characterised by either or both of two closely related temporal abstractions; the use of comparative statics as a substitute for dynamics, and the assumption of inevitability.

Pred (1984a: 279) focuses upon place-centred studies and as such process takes a back seat. However, he does suggest that place is not what Giddens (1979: 206–207, 1981: 39, 45) sees as 'fleetingly observed on the landscape, a locale, or setting for activity and social interaction'. Place is in effect a process that 'takes place ceaselessly, that contributes to history in a specific context through the creation and utilization of a physical setting' (Pred 1984a: 279). Meentemeyer (1989: 165) fundamentally agrees: 'the value of a phenomenon at a particular place is usually driven by causal processes which operate at differing scales'.

Lash and Urry (1994: 323) interpret what they see as 'disorganised capitalism' as processes and flows including, for example, the movement of capital and technologies between the 170 or so independent countries about which each has little choice or control; time-space compression in financial markets; the globalisation of culture; huge increases in personal mobility; and the declining effectiveness of the nation-state. Processes and flows are becoming evermore important.

Rojek and Urry (1997: 1) look at the importance of change, process and movement in tourism. 'Peoples, cultures and objects migrate… It is now clear that people tour cultures; and that cultures and objects themselves travel' (Said 1983; Clifford 1992). Tourism and travel are seen as not on the margins of society but central to it and 'symptomatic of an increasingly mobile society and almost come to occupy centre-stage rather than being a marginal, peripheral activity' (Rojek and Urry 1997: 10).

Gregory (2000: 125) notes a change in attitude towards process through the twentieth century in particular with reference to geological and geomorphological analysis, with a 'realist dynamic basis' replacing the 'rather static functional treatment previously used'. Urry (2010: 348) concurs describing not only people but also objects, images, information and waste as mobile and thus subject to processes virtually at all times. This mobility manifests itself through imaginative travel, movements of information, virtual travel, object travel and corporeal travel amongst many others.

Radaelli (2000: 25) cites Caporaso (1996: 30) in suggesting that there is a distinct move by EU scholars towards examining integration through its process rather than through its characteristics. This needed development of a series of conceptual models appropriate to the task.

Matthews and Herbert (2004: 379) feel that an understanding of process would allow insights into ways in which patterning in society has emerged and will go on to change. Rhoads (2005: 144) describes processes as pivotal to inquiry in his consideration of physical and human geography—the 'key to developing adequate comprehension of how physical and human systems change in form or are sustained in dynamic, yet enduring configurations'. Processes are seen to enrich the theoretical content of any discipline, and their relationship to theory is always going to be close because they are difficult to observe directly consequently relying upon theoretical interpretation.

Noe and Rebello (2007: 2) suggest that the relationship between governance and business performance 'evolves dynamically in a non-stationary world', but the development of formal dynamic models of this relationship remains awaited.

Snapshots

> Gandhi's starting point was that he refused to build any systematic model because all models become static after having their full run. But he systematically went about initiating, developing, and perfecting open-ended concepts which were amenable and adequate to produce many models, that is, models for a given context, situation and society. Sethi (1985: xxiv).

> The political, legal, social, economic and cultural context is also always fluid. Interconnections occur at all levels and change is continuous. Huggett and Perkins (2004: 230).

The very nature of process is that it evolves over time and even if this involves seemingly very little change it suggests that given long enough and sufficient sensitivity, change will always occur. Although there are those who argue that effectively stasis remains a significant component [e.g. Thietart and Forgues (1995: 359) in organisational studies], the support for change remains very much wider (Koput 1992; Cheng and Van de Ven 1996; Jayanthi and Sinha 1998) and typified perhaps by Brenner's (1998: 461) enthusiasm for the relationship that exists between capitalism and the contradictions of fixity and motion also stressed by Harvey (1982) and Rittel and Webber's (1973: 156–157) discussion of systems, form and process. Maritime governance has failed to recognise this continuous process of change, and almost always is designed around single event policies which represent and are appropriate for only snapshots in time. Much of what this book centres around is this very problem—to recognise the need for a dynamic, process-based governance that accommodates change and then to move away from the snapshot focus that has come to dominate policy-making in the sector.

Snapshot policy-making and the problems that this brings have been well recognised as detrimental to the governance of any sector. In the maritime sector, it remains a mystery. Dodge (1936: 335), in considering geography as a whole, suggests that regions are treated as 'static, as being, whereas it is the becoming which is important. What processes have shaped regions, what processes are continuing the elusive transformations, and what are the trends for the future?' Hartshorne (1939: 352) follows this up by noting the debate within German geography current at the time between the static and dynamic approach. He views them as much the same, but the need for an appreciation of the dynamism in society and the inadequacies of taking snapshots of issues that are in continuous change is clear. Sack (1972) contemplates the relationship between static and dynamic laws, whilst Berry (1973: 3) suggests that Dacey (1964) was showing clearly enough in his consideration of the point pattern case that:

> static pattern analysis is incapable of indicating which of a variety of equally plausible but fundamentally different causal processes had given rise to the patterns he was studying.

At the same time, Hagerstrand (1973: 81–82) was also commenting in a geographical context how human geography models seemed to be increasingly designed for stable environments where the friction of distance (and therefore movement) was high—ones 'strongly repetitive and restricted to compact space-time *bubbles* which are elongated in time'. Buttimer (1976: 278) considers the whole of social science and despite some criticism of her over wide-ranging perspective feels the need in modern times to accommodate 'our ways of being in the world'. In this context, she is disappointed with what exists. 'Elaborate descriptions of (overt) behaviour, *explained* in terms of disciplinary models (or philosophical dictum), remain opaque and static (recording) facets of experience as emanating from a past, but shed little light on direction or meaning'. Much the same could be applied to maritime policy-making with its static approach to problems that have emerged from a history that becomes ever-increasingly irrelevant as time moves on and processes make their changes to the sector.

Her desire to incorporate more than just 'now' into scientific and social scientific philosophy and thought was clear in her use of the work of Minkowski (1933: 400):

> For us space cannot be reduced to geometric relations; relations which we establish as if, reduced to the simple role of spectators or scientists, we were ourselves outside space. We live and act in space, and our personal lives, as well as the social life of humanity, unfolds in space. Life spreads out in space without having a geometric extension in the proper sense of the world. We have need of expansion, or perspective, in order to live. Space is as indispensible as time in the development of life.

Gregory (1978: 30–31) emphasises that recognising the significance of process in many disciplines was constrained in the 1930s and for some time thereafter as the whole idea was uncomfortably close to the 'discredited thesis of environmental determinism'. Schaefer (in Bunge 1968: 19) also considers the significance of morphological laws (or in his terms patterns) over those of process, but the debate got lost in a bitter dispute between Schaefer and Hartshorne that raged for years (and in some ways remains unresolved). Writing in 1939 (176, 179), Hartshorne had stressed a number of arguments centring on geography which promoted the static perspective. Some years later (1959: 81), he was still defending his essentially static view against the dynamicists; that his job was to study the world as it is and as Hettner (1905: 184) puts it, 'time in general steps into the background'. However, the year before he had recognised that static, dynamic and chorological approaches to the discipline were all valid—although the latter was most appropriate (Hartshorne 1958: 106). Dodge (1936) came in for some considerable criticism and in particular his assertion as Hartshorne saw it that to study 'what a region is like, is no problem worthy of a scientist' (Hartshorne 1939: 183).

Shotter (1983: 20) comments on the differences between what he terms static configurations and evolving totalities in his consideration of ecological psychology citing the work of Prigogine (1980). Meanwhile, Pred (1984a: 279) clearly does not wish to dismiss the study of place and form and replace it with process and change but does hint that there is a growing awareness of the limitations inherent in the frozen scenes for human activity that places and regions represent. Studies that concentrate upon form rather than process could be conceived as inert and 'experienced' (rather than 'experiencing').

Dyck (1990: 461) is also clear in treating rules and resources as never static but 'the media of production and reproduction of practices' and as such 'human agency has a transformative capacity involving active negotiation among actors'.

Clifford (1997: 2) in his consideration of travel and translation in the twentieth century assumes the importance of movement to society and argues that travel and contact are essential for modernity. Human location (and by definition the social activities of humans including governance and policy-making for maritime trade) is 'constituted by displacement as much as by stasis'.

May and Thrift (2003: 2) are certain about the inadequacy of snapshot policy-making and analysis citing Harvey (1993), Massey (1993) and Hetherington (2003) in suggesting that spatial is suggestive of stasis and temporal of dynamism:

> Time is understood as the domain of dynamism and progress, the spatial is relegated to the realm of stasis and thus excavated of any meaningful politics.

Urry (2005: 238) is enthusiastic about the application of complexity sciences over a wide field, citing the earlier work of Prigogine (1997: 189) noting how it 'repudiates the dichotomies of determinism and chance, as well as stasis and change'. He sees that 'order and chaos are often in a kind of balance where the components are neither fully locked into place but yet do not dissolve into anarchy'. Policies that attempt to address a stable situation in any sector will be inadequate.

Medd and Marvin (2005: 46) consider the tendency for inertia to become an important feature of governance, thus ensuring that stasis rather than dynamism becomes dominant. This resilience they suggest could take many forms, some more suggestive of flexibility and process than others—resistance, stability, continuity, innovations, adaptation, transformation, immunity and recovery. Whilst sounding progressive, many of these forms actually represent attempts to avoid specific changes by adapting in retrogressive ways. Evidence of the interest in resilience—and hence indirectly recognition of the significance of change and process in policy-making—has risen since 2000 including in particular the work of Adgar (2000) and Berkes et al. (2003). Issues remain to be understood. These include the relationship between different forms of resistance, the impact of technological developments, institutional design and resilience and also the geopolitical contexts for different forms of resilience and their impact upon governance.

Sheller and Urry (2006: 207–210) recognise that globalisation has changed society so that movement has become a universal and essential feature. Although much of what they consider relates to travel and communication, they identify a more generic paradigm of change, movement and dynamism that encompasses almost all aspects of society and the international shipping industry and its associated governance is no exception. Contributions from the literature abound: Mol and Law (1994), Serres (1995), Appadurai (1996), Kaplan (1996), Clifford (1997), Virilio (1997), Urry (2000), Cresswell (2001), Degen and Hetherington (2001), Pascoe (2001), Riles (2001), Ginsburg et al. (2002), Kaufmann (2002), Coleman and Crang (2002), Verstraete and Cresswell (2002), Amin and Thrift (2002), Ahmed et al. (2003), Crouch and Lubbern (2003) and Sheller (2003) amongst others.

They continue to show how the majority of the social sciences (and this includes generic governance and policy-making) fails to understand what the spatialities of life presupposes—in other words the continuous need for change and movement and to adapt what they say to our purposes. Dynamism in policy-making and governance is seen as a 'black box, a neutral set of technologies and processes predominantly permitting forms of economic, social and political life' and which do not need to be accommodated in governance in any direct or formal way. Snapshot policies will suffice. Heidegger (1985) actually recommends this as 'the way in which humans should inhabit the Earth'. Also interpreted as *sedentarianism*, this treats as normal 'stability, meaning and place and treats as abnormal distance, change and placelessness' (Sheller and Urry 2006: 208). Heavily associated with national identity, it rejects any concept of flux or dynamism and associates in most cases what exists—the present situation—as the optimum. Consequently, snapshot governance is wholly appropriate. Although there are exceptions [see, e.g. Lynd and Lynd (1937); Hawkins (1986); Lynch (1993)], the social sciences have tended to be static.

In more recent years, Neo and Chen (2007: 1) suggest that 'even if the initial chosen set of principles, policies and practices are (*sic*) good, static efficiency and governance would eventually lead to stagnation and decay', and consequently policies that focus on a single point in time must be suboptimal. Edelenbos et al. (2009: 176–177) outline the differences between project and process management and consequently between dynamic and static approaches to problems and policy-making (Fig. 4.1). The clear association of the current model of maritime

Dimension	Project Management	Process Management
Main Focus	A well thought-out substantive solution to the problem	The involvement of stakeholders and their interests
Dealing With Dynamics	Thorough decisiveness and control: dynamics are approached as dysfunctional because they lead away from the original designed solution	Through resilience, responsiveness and being open to other options: dynamics around a project must be taken into account and can lead to changes in the initial solution
Self-Organization	Autopoietic self-organization: changing circumstances must not affect the planned course of action	Dissipative self-organization: the initiative must be and remain open and attractive for actors
Coevolution	Almost separately from the environment. A singular process system is seen as desirable to stay in control	In interaction with the environment. A composite process system is seen as necessary to realise consensus
Most Important Problems	Acceptance of results	Time consuming

Fig. 4.1 Overview of process and project management. *Sources* Gage and Mandell (1990), Kickert et al. (1997), Mandell (2001), Agranoff and McGuire (2003), Meredith and Mantel (2000), Mantel (2005), De Bruijn et al. (2004), Susskind and Cruikshank (1987)

governance with the project management approach rather than process is clear. Solutions are sought without consideration of change that might be occurring; the words 'decisiveness' and 'control' feature prominently rather than 'resilience' and 'responsiveness'. The idea that dynamism represents dysfunction rings horribly true for a maritime sector where moves away from an agreed and institutionalised policy are seemingly abhorred and solutions emerge from lengthy and stale discussions that have lost touch with changes in the market place. Just take the IMO deliberations on climate control and air emissions as an example. Process management would look for dynamism leading to changes from initial plans to accommodate the flexible and dynamic nature of the sector. Whereas in maritime governance changing circumstances do not affect planned courses of action, what is needed is for all initiatives to remain permanently open to amendment and attractive to all stakeholders to become involved. No more snapshot 'Blue Papers' but a continuously moving policy target.

Teisman et al. (2009a, b: 3–5) look at the emergence of complex systems from a developmental perspective rather 'than in their characteristics in a certain place and time' and in particular the use of such systems within governance rather than within an institutional system—an approach that addresses some of the fundamental issues troubling maritime governance we identified earlier. Their approach is rooted in evolutionary biology (Odim 1971) but can also be found in economics, psychology and sociology. Here, development is looked at in terms of 'selection out of variety' which ultimately 'enhances a species' fitness'. In economics, the application of an evolutionary approach rejects the conventional approaches that dwell on equilibrium and rationality and as such might be highly applicable to the sometimes irrational and commonly unbalanced situation in the maritime sector (Hodgson 1993). Thus, the assumption that policies can be designed for some sort of ultimate maritime utopia where every player has achieved the best that can be achieved and that policy needs little if any adjustment is clearly unrealistic—but policy-making and its governance at present looks towards that kind of final solution. Instead, an acceptance of continuous change, moving policy targets and the need for governance to keep up would certainly be a more positive move. Maritime institutional and technical change (the former rare but needed and the latter common) should no longer be considered external influences to be accommodated where possible but instead 'endogenous to the systems being researched'. Teisman et al. (2009a, b: 4) go on:

> Rather than focusing on the establishment of supposedly stable states, evolutionary economics focuses on the ongoing development of systems from one temporal state to another. (c.f. Nelson and Winter 1982; Norgaard 1984; Van den Bergh and Gowdy 2000).

This could provide fertile ground for maritime governance. To have any chance of transcending from a conventionally static approach to policy-making, there needs to be focus on the dynamic and temporal dimensions of change; on adaptive processes; and on recognising change as contingent and path-dependent (Teisman et al. 2009a, b: 5).

Cram (2011: 637) has been particularly excited by the problems inherent in governance that are characterised by analysis that 'treats current developments as snapshots'. Citing the historical institutionalists [e.g. Bulmer (1994), Pierson (1993, 2004) and Thelen and Steinmo (1992)], Cram goes on to stress how the 'temporal dimension is central; the present needs to be understood in the context of the past; but the past is not fixed but may be recast as new narratives emerge over time; and these narratives cast a long shadow into the future'. Historical institutionalism and its penchant for snapshot governance are inadequate.

Process, Form and Object

> Our nature lies in movement: complete calm is death.
> Blaise Pascal (Pensees 1669).

A geographical perspective is particularly relevant to a discussion of process in governance as the relationship between space, place, form and time has been a core debate in the discipline for many years. Schaefer (1953: 243–244) however is unusual in that he sees no place for processes in geography:

> Geography is essentially morphological. Purely geographical laws contain no reference to time and change. This is not to deny that the spatial structures we explore are, like all other structures anywhere, the result of processes. But the geographer, for the most part, deals with them as he finds them, ready made. (Quoted in Bunge 1979: 131).

However, there are few who today who would wholeheartedly support his view, and in considering maritime governance, there is clearly a need to move on. The contradiction that inevitably exists between process, with its inherent dynamism and movement, and form and object, with the characteristics of stasis and stability, is one that has dominated much of the literature and is significant because it reflects the inherent difference between a static and dynamic form of governance. The contention here is that the existing framework for maritime governance is typically static, providing snapshot policies for the sector which as a consequence is inadequate.

Hartshorne (1939: 372) can kick us off suggesting that form and process are always related, but despite this there remains much confusion. He firmly rests on the side of form implying that if Newton really had discovered the concept of gravity through an apple falling on his head, then 'it was not the force itself he observed by his sense of feeling but simply the moving apple' (quoted in Gregory 1978: 29–30). Vogt (1960: 19) in his consideration of cultural anthropology, placed the discussion of process, and form and object in the context of structures suggesting that the analysis of processes had been made difficult because the assumption was always that social and cultural (and quite possibly physical) processes would work naturally towards an equilibrium. Based on the structural–functional model and originally derived from Durkheim's (1895) organismic analogy which looked for social order, in fact it might be more beneficial to try to understand processes by assuming constant change and not that the ultimate is always an equilibrium defined in terms of a permanent (or virtually so) structure, form or object (Leach 1954: 4; Henry 1955; Herskovits 1955: 443–446). Change is always present; forms and structures are just profiles of events that are part of that process taken with permanence simply for convenience. Thus, the two are always related. Vogt comments: 'every class of phenomenon we observe in the natural world is characterised by describable processes of change'. This idea of constant change and perpetual processes is emphasised by Von Bertalanffy (1952: 134) who sees forms and structures as 'in reality a momentary cross-section through a spatio-temporal pattern'. As noted earlier, 'structures are slow processes of long duration'.

Blaut (1961: 4) sees no difference between structures (forms, objects) and processes viewing real world structures simply as 'slow processes of long duration'. Elements or forms that are apparently static are merely relatively slowly dynamic in comparison with human interpretation and experience. Commonly, we map or model processes reinterpreting a dynamic feature as one that is fixed in time. Thus, process and form are one and the same thing and in reality, if not perception, all are dynamic.

Blaut (1962: 4) returns to the issue of form and process in a consideration of the philosophical interpretation of relationships and objects. He denies that there is a distinction between phenomena and relations as in detail, any relationship is actually made up of a 'prosaic sequence of mundane events'. Thus, it is not unreasonable to consider phenomena and relationships as one and the same thing and by deduction process and form, something reiterated by Janelle (1969: 348). Hagerstrand (1970: 4) emphasises the need to relate process and spatial form in regional planning and analysis and goes on to attempt to do just this and develop a 'calculus of process patterns' through the modelling of time geography—more of which we shall see in a later chapter.

Eichenbaum and Gale (1971: 525–529) extensively debate the metaphysical relationship between form, function and process. In particular, they define process as a 'continuous or regular action or succession of actions, taking place or carried on in a definite manner, and leading to the accomplishment of some result; a continuous operation or series of operations.' This implies the omnipresence of change.

They claim that few social scientists have ever paid anything more than cursory attention to the philosophical ramifications of the relationship between form and process, something clearly apparent in maritime governance and policy-making and despite their mutual dependency. The conclusion is that the three concepts of form, function and process need to be considered together in a 'mutually interactive role' if governance is to be relevant and effective.

However, it is not all a one-way trend. Rather surprisingly, Berry (1973: 8) felt that 'the search for absolutes of form in some geometric sense is understandable'. We perceive the world through ideas, and the idea system is limited by a 'language oriented to classifying objects, naming things and hence codifying their *reality*'. However, we need to move on and Berry continues by qualifying his earlier comments and stressing the need for a 'continuous intellectual process' that recognises the systems that exist and that each interpretation of contexts and scenarios needs to be reinterpreted in the light or ever-increasingly complete systems.

Pred (1984a: 280) presents his theory of place as an 'historically contingent process' based upon the theory of structuration and derived from Hagerstrand's (1970) time geography. Places are seen as actually part of an unbroken series of events or processes and people, rather than atomised individuals performing unconnected tasks at unconnected places, and are actually 'objects and subjects' all at the same time. Thus, process and place, form and object become one and the same.

> Place is therefore a process whereby the reproduction of social and cultural forms, the formation of biographies and the transformation of nature ceaselessly become one another at the same time that time-space specific activities and power relations ceaselessly become one another. The components of the theory are universal in the sense that they are inextricably interwoven with one another in the becoming of every place (object, form) or region. (Pred 1984a: 282).

Pred's theory of place (or form and object) is thus historically contingent.

Castells and Portes (1989: 11) examine the informal economy identifying movable social boundaries derived from historical realities which also apply to governance and policy-making in the maritime sector. Issues relating to seafarer labour supply, environmental damage, terrorist supply and safety at sea are all essentially social in nature and are constantly in flux. They also emerge from a history that cannot be denied and should not be ignored. As a consequence, it is a process rather than an object. Meanwhile, Meentemeyer (1989: 164) sees any space as defined by both spatial and process elements. A relevant space will always have processes operating to control and affect it—migrations and commuting, watersheds, pollutant dispersal, idea diffusion and many more. These processes might well be operating at different scales for the same features and within the same spaces.

Der Derian (1990: 297) following Virilio (1986) introduces the idea of process as speed in the context of international relations seeing the emergence of chronopolitics which elevates 'chronology over geography; pace over space in their political impacts'. This leads on to an examination of simulation, surveillance and speed in international relations.

We shall return to speed at a later time but meanwhile can note Der Derian (1990: 306) who quotes Filippo Marinetti who in 1909 suggested that:

> The Futurist writer will make use of free verse, an orchestration of images and sounds in motion to express our contemporary life, intensified by the speeds made possible by steam and electricity, on land, on the seas, and in the air. (from Lista 1986: 12–14).

Whilst it is interesting to see the use of the seas as a form of transport associated with speed and despite the exhortations of the Futurists, it was only in the 1970s that the issue raised its head again in the fields of political and social theory and largely as a response to the explosion of globalisation and the work of Virilio (see, e.g. 1975, 1977, 1978, 1980, 1983, 1984a, b, c, 1986, 1988). Virilio's aim is to raise the issue of politicising speed which he contrasts with the attention given to the politicisation of wealth. This he associates with the move from geopolitics (associated with space and place) to chronopolitics (associated with time, movement and process) with the distribution of territory viewed as 'outmoded, minimal'. We return to Virilio in more depth in Chap. 7.

Brenner (1999: 52) discusses the relationships between form and process extensively in his examination of state-centrism, space, territoriality and scale. Global scale could be seen through a state-centric analysis of a globally 'stretched territorial grid' or form; or through emphasis upon the 'processes of deterritorialization which purportedly trigger the demise, erosion or contraction of state territoriality'. The former focuses upon the physical and relatively fixed form of states as vehicles of globalisation; the latter considers the world as stateless, borderless and supraterritorial and requires the erosion of the form of the state by the process of globalisation. Although Brenner rejects both as inadequate ideas to explain globalisation, he does accept elements of each. He retains the notion of a state, but instead of the erosion of its form, he sees a revision of the processes taking place and consequent reterritorialization (Brenner 1999: 53). This contrasts with the more traditional view of Robertson (1992) who sees globalisation creating a 'global unicity', a single place. This state-centric approach takes an historic view of place and its significance without recognising that history is in fact a process and will continue into the future. 'Globality' is thus a macrogeographical static form of state domination and place (Brenner 1999: 55).

MacLeod and Goodwin (1999: 504–505) take up this theme suggesting that there was an 'undeveloped notion of the state' which with globalisation was being drastically reconfigured (Goodwin et al. 1994; Goodwin and Painter 1996; Brenner 1997; Jessop 1997a, b). The spatial context for each jurisdiction (national, supranational, international, etc.) was taken as given despite the fact that each are 'perpetually redefined, contested and restructured' (Swyngedouw 1997: 141) and that to understand the issue there is a need to focus on 'theorizing and understanding process' (MacLeod and Goodwin 1999: 505).

Goodchild (2004: 709) notes the historical tension between form and process something that also manifested itself throughout the long debate on atomism that continues to this day and over a range of issues [beginning, e.g. with Russell's (1918) comprehensive assessment and continuing with Sailor (1964), Deleuze and Guattari (1988: 361), Davis (1989), Friedmann (2000: 113, 118), Peuquet (2002: 14), Snowden (2005) and Francke and Ham (2006: 6)] and also the debate between wholeness and separatism [e.g. Eichenbaum and Gale (1971: 527) and Buttimer (1976: 279, 286)]. Anaxagoras (c510–c428BC) introduced the idea of infinite divisibility ('for of the small there is no smallest, but always a smaller'), whilst atomism developed in response to this whereby everything is seen as infinitely separable and separate. These separate particles are adrift in the 'void' which acts as both a boundless box and the generator of their order.

Russell (1918: 496) devotes considerable energy to the issue of atomism opposing the concept to Hegel's monism expressing a belief that there are 'many separate things'. His support of atomism in discussing the relationship of form to process is not without accepting that there is a need for both general and particular things:

> Suppose that you had succeeded in chronicling every single particular fact throughout the universe, and that there did not exist a single particular fact of any sort anywhere that you had not chronicled, you still would not have got a complete description of the universe unless you also added: 'These that I have chronicled are all the particular facts that there are.' So you cannot hope to describe the world completely without having general facts as well as particular facts. Russell (918: 503).

Monism by contrast is focused on the belief that there is a unity with the universe as one, single complex entity (Russell 1918: 511):

> All the kinds of things to which we habitually give proper names are on the face of them complex entities: Socrates, Piccadilly, Romania, Twelfth Night or anything you might like to think of, to which you give a proper name, they are all apparently complex entities. They seem to be complex systems bound together into some kind of a unity, that sort of a unity that leads to the bestowal of a single appellation.

Russell's view is that such entities do not exist and that the world is essentially atomistic—made up of small parts that only when combined together form entities. Thus, the atomistic form of things includes processes which are the dynamic actions of these individual parts moving in some sort of unison.

Shrader-Frechette (1977: 411–2) considers that atomistic theories have dominated science for its entire history characterised by a 'search for a substance representing the foundation of observable phenomena'. Heisenberg (1976) expresses considerable dissatisfaction with this suggesting that 'good physics has sometimes been unconsciously spoiled by poor philosophy (atomism)'. This stems from the earliest Greek atomists who considered that the 'ultimate constituents of matter… were minute, hard, and indivisible units of definite size and shape'.

Bohm's (1980) process metaphysics considers that the universe should not be regarded as made up of things but 'of a complex hierarchy of smaller and larger flow patterns in which the things are invariant or self-maintaining features of the flow' (from Platt 1970: 2). This diverts considerably from the views of the Ancient Greek, Democritus (c460–c370BC), who believed that nature consisted of things (atoms) in a void. In fact, Bohm's views brought modern opinion back to those of another Ancient Greek, Heraclitus (c535–c475BC), whose view was that 'all was flow, or fire'. Bohm's view was very much a holist one with 'fields of flow extending outward indefinitely'. Everything is in relation to everything else, echoing the Tao, Huseri and Polanyi transactional psychologists and the words of Whitehead (1925: 72, 1929: 41): 'the reality is a process… an actual entity is a process'.

Disagreement between those who see the world as divisible into ever smaller parts (and then understandable as a combination of these parts put together)—reductionists or atomists—and those who have a more holist vision has continued for thousands of years. The contradiction between form (and its atomistic tendencies) and flow/process (and its holistic tendencies) remains unresolved. To confuse the issue further, there are even those of the process school who see the value in reducing process studies to their smallest constituent parts, thus creating a reductionist process philosophy (Sims 2003: 11).

Russell (1926: 109, 110) attempts to put it all in context:

> In the world of immediate data nothing is permanent; even things that we regard as fairly permanent, such as mountains, only become data when we see them and are not immediately given as existing at other moments…
>
> The belief in 'indestructible' things very early took the form of atomism. The underlying motive in atomism was not, I think, any empirical success in interpreting phenomena, but rather an instinctive belief that beneath all the changes of the sensible world there must be something permanent and unchanging. This belief was, no doubt, fostered and nourished by its practical successes, culminating in the conservation of mass: but it was not produced by those successes. On the contrary they were produced by it. Philosophical writers on physics sometimes speak though the conservation of something or other were essential to the possibility of science, but this, I believe, is an entirely erroneous opinion. If the a priori belief in permanence had not existed, the same laws which are now formulated in terms of this belief might just as well have been formulated without it. Why should we suppose that, when ice melts, the water which replaces it is the same thing in a new form? Merely because this supposition enables us to state the phenomena in a way which is consonant with our prejudices. What we really know is that, under certain conditions of temperature, the appearance we call ice is replaced by the appearance we call water. We can give laws according to which the one appearance will be succeeded but the other, but there is no reason except prejudice for regarding both as appearance of the same substance.

Quite.

Meanwhile, Pred (1986: 48) places the form/process debate into the context of the work of Hagerstrand and the development of the theory of structuration (see, e.g. Abrams 1983; Berger and Luckmann 1967; Bhaskar 1978, 1979; Bourdieu 1977a, b, 1984; Giddens 1976, 1979, 1981, 1983; Gregory 1982a; Kosik 1976; Layder 1981; Philo 1984; Pred 1983, 1984a, b, 1985a, b; Thrift 1983; Touraine 1977; Williams 1977).

Hagerstrand provides a contextual and processual view of the world which considers that the:

> flow of human practices characterizing any bounded area is connected not only to physical presences and absences but also to the control of material and other resources, (or to the power relations touching upon locally present individuals, institutions, groups and classes) and to the thoughts feelings and knowledge, (or consciousness) of human agents. Pred (1986: 48).

Consequently, the content of a person's life over a period of time is defined by the people, institutions, artefacts and natural features which are present along with their position within local networks of power relations which govern access to the phenomena. Structuration sees the significance of power relations which constrain and permit social activity and structure and also sees this happening in both ways—the shaping of practice by structure and the 'perpetuation or transformation of structure by practice', creating a dialectical condition. Thus, form (people, institutions, artefacts, etc.) and process (the workings of society) are both dialectically opposed and yet intrinsically interlinked.

This debate on the nature of process and form, dynamism and stasis, and network and structure all leads back to notions of dispersion and concentration as outlined by Francke and Ham (2006: 6) in their consideration of Castells' *Space of Flows*. They see information technology as central to the simultaneous dispersal and concentration that characterises globalisation contrasting the increasing ability to work at home with the tendency for concentrations of specialisation cities to become more centralised and significant. Linking together form and process, they comment:

> the emphasis on interactivity between places breaks up spatial patterns of behaviour into a fluid network of exchanges that underlies the emergence of a new kind of space; the space of flows.

Castells had already confirmed this (1998: 295): 'in territorial terms, the age of information is not just the age of spatial dispersal, it is the age of generalized urbanization… characterized by territorial sprawl and locational concentration'. Such arrangements are only possible because of the growth of the significance of flows—'the power of flows takes precedence over the flows of power' (Castells 1996: 449).

> The space of flows… dominates the space of places of people's cultures. Timeless time as the social tendency toward the annihilation of time by technology supersedes the clock time logic of the industrial era. Capital circulates, power rules, and electronic communication swirls through flows of exchanges between selected, distant locales, while fragmented experience remains confined to places. Technology compresses time to a few, randomized instants, thus de-sequencing society and de-historicising history. By secluding power to the space of flows, allowing capital to escape from time, and dissolving history in the culture of the ephemeral, the network society disembodies social relationships, introducing the culture of real virtuality. Castells (1998: 349).

Friedmann (2000: 118) sees the Castells' model like an apartment block. The penthouse is occupied by all powerful and near invisible financial networks, whilst the basement tenants are nation-states—irrelevant, free-floating, self-centred, populated by excluded masses and generic labourers of the global economy, 'communes of resistance' formed around the primary identities of god, nation, family and territory (Castells 1997: 356). Webster (2002: 107) further confirms Castells' view of the city (and by implication the globalised world) seeing it as a process rather than a place or thing through which information (money, people, goods, etc.) flows.

Process and Flow

> The real world is a never-ending flow of interrelated processes and absences… What we have to search for when we try to understand how the world seen in this perspective moves along is a complicated counterpoint of collateral processes. Hagerstrand (1984, quoted in Pred 1986: 47).

> The body is in constant motion. Even at rest, the body is never still. As bodies move they trace out a path from one location to another. These paths constantly intersect with those of others in a complex web of biographies. These others are not just human bodies but also all other objects that can be described as trajectories in time-space: animals, machines, trees, dwellings, and so on. Thrift (1996: 18).

Processes imply flow, and so the relationship between the two perhaps should be easy to identify and understand. Instead of a contrast (e.g. between process and form), it would seem inevitable that processes require a flow of something and thus, the formulation of dynamic governance should also require the presence of flows of materials, people, money, information and more—and all of this sounds very familiar in the context of the maritime sector.

The relationship between the two is well documented, and we shall consider only a limited sample here although the concept of flow is further considered in a later chapter. In the specific context of political geography, Kasperson and Minghi (1969: 199) are explicit in the emphasis that should be placed upon future research into flows and processes and follow this up with a considerable proportion of their edited book focussing on the value of process and flow analysis.

Bird (1981: 134) links events, processes and flows together through the medium of geomorphology something taken on further by Renwick (1992: 265). When a hydrological regime of a river is studied a state is observed becoming another through a series of processes; no observation and the river continues to change through these same processes but they manifest themselves as two events in sequence. Noting Grunbaum (1971: 214) and quoting Zwart (1973: 133):

> Events do not just have their place in time, like pieces of wood floating in a river, but events constitute time. There is no flow of time beside or beneath the flow of events, but the flow of time is nothing but the flow of events. Therefore one should not compare events in the flow of time with objects floating in a river, but with the molecules of water the river is composed of. As the passing molecules of water constitute the flowing river, so the passing events, i.e. their occurrence, constitute the flow of time.

Bird is beginning to tie together all the elements of time, flow, process and events (forms) so that we can begin to understand their relationship and how they need to be integrated to provide the basis for a maritime governance that is dynamic whilst incorporating the effective reality of form within a moving and flexible political, economic and organisational landscape.

Shotter (1983: 20) follows Prigogine's (1980) consideration of the processes of flow and the inevitability of irreversible, evolutionary processes and argues in favour of evolutionism. Citing Bohm (1975), he stresses the inadequacy of a world of functional entities with 'static geometrical structures' and instead the existence of a permanently evolving world full of agencies, 'containing everywhere, structurizing activities or formative causes' or otherwise known in Giddens' (1979) terms as *structuration*.

Heidegger (1985) suggests that 'being-in-the-world' is not just an organism or ego experiencing a stream of experiences but a 'mode of comportment' with shared readiness to deal 'appropriately with the people and things with which it encounters. Quoting Dreyfus (1991: 159): 'a way of being that is concerned about its own being, and yet must get its meaning by assigning itself to the occupations (including roles and equipment) provided by the one', the latter being cultural norms. This focus on process and flow is continued by Searle (1983: 143) using a party as an example of the centrality of movement to any situation:

> Think of what is necessary to go to the refrigerator and get a bottle of cold beer to drink. The biological and cultural resources that I must bring to bear on this task, even to form the intention to perform the task are (considered in a certain light) truly staggering. But without these resources I could not form the intention at all: standing, walking, opening and closing doors, manipulating bottles, glass, refrigerators, queuing, partying and drinking.

Slot and Grabant (1986: 548–550) interpreted the relationship between process and flow (and also place) in the context of globalisation and the changes that were occurring in the territoriality of the state. They saw jurisdiction defined always in terms of territory but not always simply. Where events take place outside the territory of a state but either involve the domestic state's nationals or have an impact upon that state a new interpretation has emerged that suggests that jurisdiction exists beyond the state border (think of shipping and the role of the nation-state flag) and is intrinsically linked to the role of process and flow of people, information, money, goods and so on. Thus, seafarers are subject to domestic jurisdiction on-board ship, those affecting a state's security by sharing and distributing information by the internet can be subject to domestic laws wherever they might do it, and cargo on-board ship is carried under the jurisdiction of the flag state, whilst money and investment is complex but certainly is commonly an extraterritorial item. The processes that characterise globalisation are almost always tied up with flows, and together, these make jurisdictional interpretation interesting and frequently complex.

Gertler (1988) sees it all as a dialectic between fixity and fluidity operating in both a temporal and spatial context. His analysis is in terms of the firm needs to retain an element of fixity (the plant, office), whilst increasing fluidity in the marketplace, in moving location with increasing rapidity as circumstances dictate and in investing in communications. The fixed element takes a second place (but certainly remains), whilst the fluid processes and flows of information, money, power, people become increasingly integrated, confused and complex—yet essential and inevitable. Shipping provides an excellent example.

Braganza and Korac-Kakabadse (2000: 46) examine the need for new policy-making practices in business and how contexts have changed rapidly in the late twentieth century with the development of sophisticated communication tools. In particular, they look at the role of information flows which act as 'patterns of connectivity' between the forms and objects that make up business. These patterns of connectivity are taken from the work of Ruggles (1998) referring to 'information being generated, shared, stored, retrieved and exploited within each process, that is within and across processes' (Braganza and Korac-Kakbadze 2000: 49).

Allmendinger (2001: 226) comments on the importance of fluid processes, structures and flows in his consideration of postmodern planning practices and notes how a more active and creative role is needed. Meanwhile, Matthews and

Herbert (2004: 381) consider the importance of process, time, flux and change to geographical studies, but their message is equally applicable to almost any discipline. Little can be understood until attention is focussed upon how patterns have developed and evolve and also the underlying processes that direct and influence the flows of information, materials, money, people and much more.

Mol and Spaargaren (2005: 97) spend some time considering the various interpretations of flows and their substantial significance in society. Starting with Sassen (1994) and Castells (1996, 1997, 1998), they emphasise the domination of flows in economic terms and their relationship to technical communications, manifested primarily in Castells' *Space of Flows* focussing on money and information. This contrasts with the view of Urry (2010, 2003) who sees flows and fluids as the key elements of social systems—'utterly crucial categories of analysis in the globalising social world' (Urry 2003: 61).

They continue in a later publication (Spaargaren and Mol 2008: 32) to stress the importance of processes to the understanding of systems generally with the 'system approach developing out of the process revolution'. To understand the interrelationships that go on within any system, there is a need to understand the functional relationships between the elements that make up that system—and this in turn requires an appreciation of the relationship of processes to the flow of information, materials, money, etc., that might be moving through it. This requires a dynamic appreciation of the system under consideration (e.g. policy-making) that in turn takes account of the processes and flows (e.g. in maritime governance of information, finances, seafarer employment, legal constraints, cargo and much more).

Process and Change

> A foolish consistency is the hobgoblin of little minds, adored by little statesmen and philosophers and divines. With consistency a great soul has simply nothing to do. He may as well concern himself with his shadow on the wall. Speak what you think now in hard words, and to-morrow speak what to-morrow thinks in hard words again, though it contradict every thing you said to-day. - 'Ah, so you shall be sure to be misunderstood.' - Is it so bad, then, to be misunderstood? Pythagoras was misunderstood, and Socrates, and Jesus, and Luther, and Copernicus, and Galileo, and Newton, and every pure and wise spirit that ever took flesh. To be great is to be misunderstood. R.W. Emerson, *Self Reliance*, 1841.

Or Bertrand Russell:

> There was a young man who said Damn!
> I learn with regret that I am
> A creature that moves
> In predestinate grooves
> In short, not a bus, but a tram.

The mere concept of process suggests change although of course it is always possible that processes can be ineffectual or in fact circular in motion and eventually lead the observer or participant back to the original situation. However in most cases, intuitively, the two go hand in hand. Much has been written about the relationship between process and change. In the context of maritime governance, the two are intimately related, and once we have decided that maritime governance needs to become more processual, it is necessary to understand how this relates to change within the maritime marketplace.

Change has been on the agenda as long as man has been interested in time and the processes that go on around him:

> Nature… creates ever new forms: what exists has never existed before, what has existed returns not again - everything is new and yet always old…. There is an eternal life, a coming into being and a movement in her; and yet she goes not forward. Georg Christoph Tobler, *Essay on Nature* (1783), although commonly attributed to Goethe.

Or even Carl Jung:

> … others who seemed to live in a most curious condition of consciousness, as if the state they had arrived at today were final, with no possibility of change, or as if the world and the psyche were static and would remain so forever. *Man and His Symbols* (1964), p. 39.

Chorley (1973: 155) cites Rousseau's view in his *Discourssur Les Science et Les Arts* (1750) of human development in 'terms of the paradox that man deteriorates as material civilisation advances'. Whilst this rather pessimistic view of change may be open to debate, clearly the issue was relevant even some 250 or so years ago.

Whyte (1944: 7) opens the discussion of the significance of process and change: 'to be alive is to undergo ceaseless change. Man fears change and seeks to deny it by imposing on it a principle of permanence… To understand nature and himself, man must accept change and identify the universal form of process which underlies the variety of particular processes'. He also saw change as universal with anything that appeared to be permanent as having no 'substance' (31). Change was not seen as arbitrary in any way but 'unfolding continuously out of the present'.

Whyte (1944: 171) also returns to Heraclitus who saw everything as 'strife and change' with harmony far from static but 'lying in a developing relation between opposites'. Whilst all pervasive, there also was order within it. Simplicius (490–560 BC) (commonly but erroneously attributed to Heraclitus) summed up the latter's philosophy on change:

> Into the same river you could not step twice, for other waters are flowing… in change is rest… craving and satiety… God is day and night, winter and summer, war and peace, plenty and want… for men to have whatever they wish, would not be well. Sickness makes health pleasant and good; hunger, satiety, weariness, rest… war is the father and king of all… it is hard to contend with passion, for whatever it craves it buys with its life. Peters (1967: 178).

Plato meanwhile spoke for the antithesis contrasting the 'process of the senses and the permanence of timeless ideas' (Whyte 1944: 174). Only what he considered permanent could be assumed to have a real existence, whilst the confusion of the world of process and change was simply an illusion.

Vogt (1960: 19) continues the debate on process and change: 'every class of phenomenon we observe in the natural world is characterized by describable processes of change' and although he does not refer to the societal, political, economic or any other type of change that relates directly to the commercially maritime, his comments remain indicative and in many ways valid. Particularly since the natural and societal worlds are highly interactive. In fact he goes on to stress that 'no society is ever static, that one of its most fundamental properties is change' (Vogt 1960: 20). However, despite agreement on the significance of process and change, there remain major difficulties in accommodating and understanding this relationship.

> Our problem becomes one of describing, conceptualising, and explaining a set of on-going processes which proceed at varying rates, move in varying directions, and are triggered off and maintained by complex interactions of technological-environmental social, cultural and psychological variables. Vogt (1960: 20).

In so commenting, Vogt brings the relationship of process and change from the physical to the social and, as such, draws the debate closer to that of governance and policy-making including those of the maritime sector.

Toulmin and Goodfield (1962: 46–48) consider early studies of change by the Ancient Greeks who whilst spending much time on individual problems such as the phases of the moon and mathematical properties of geometric shapes were primarily concerned with the 'unifying principle' of life. The Greeks saw life presenting a flux, and the need was to understand the 'enduring, unchanging entities behind' it. Then, the world would make sense. In turn, this gave rise to three issues that needed to be resolved:

1. The theory needed to explain the world had to accommodate both the unchanging nature of things and how these unchanging things can generate the flux that existed and therefore contain elements of both stability and change.
2. How can all the multitude of changes with all their many different characteristics be explained by one common vocabulary? Can one single language (e.g. mathematics) explain all the variety in smells, noises, sights, colours and much more that exist?
3. Only once these two questions had been addressed could the Ancient Greeks turn to specific issues such as earthquakes, tides, fire, vision and so on. To do this, they first had to address mechanisms (processes) as these (they thought) were confined to those visible. Forms were constant, largely unchanging and hence only the victims of such change.

Whilst it is easy today to pick holes in the Ancient Greeks' approach to form, process and change and the 'unifying principle', much remains philosophically the same. Process and change remain central to life and its understanding, and it is only when the relationship of these two to form is appropriately understood that any planning, decision-making, scientific or social process or anything else can be effective. Maritime governance cannot be excluded from this.

Hagerstrand (1970: 1) was early on the scene in understanding the need to accommodate change in policy with some of his work dating from the 1950s. He suggested that analysing the processes of change was bound to be difficult because of the need to switch between scales, or in the words of the British astrophysicist and philosopher Arthur Eddington: 'if we had two eyes of different sizes, we might have evolved a faculty for combining the points of view of the mammoth and the microbe'.

Eichenbaum and Gale (1971: 526) are confident that 'ongoingness or change, for example in the time domain, is everywhere present'. They continue by emphasising the relationship of process and change to relations and relativity citing Buddhistic 'dependent origination' (Eichenbaum and Gale 1971: 529). This hints at a discussion of process philosophy in the next chapter.

Berry (1973: 19) considers what he terms 'change processes' as always a problem as they challenge the existing order and institutions in whatever context they are found and are often universally unpopular, something clearly apparent in maritime governance. Allen et al. (1985: 66) agree that change is constantly occurring with plans and designs need to be placed upon a 'shifting terrain' and not on an artificial 'static frozen reality'. However in contrast, Angelides and Caiden (1994: 228) are suspicious of the significance increasingly attached to change. They suggest that 'continuous change and historic breakpoints' might mask the presence of greater constancy and:

> serve as a useful reminder that even the most pronounced changes have antecedents, that the past cannot be ignored, and that there is always a danger of mistaking mere commotion for turbulence and thus exaggerating the depth and breadth of changes, Rosenau (1990: 69).

They see change as subjective with a different meaning depending upon the individual and context and a different reception also common, some welcoming it, others not. Policy-makers commonly ignore the prospect for change, happier assuming that nothing will happen to their familiar world or, that they will be gone before the changes they do expect to occur actually happen. Most policy-makers actually assume stability because it is convenient and comforting. Maritime policy-making is no exception.

Cheng and Van de Ven (1996: 593–5) spend considerable time looking at the relationship between innovation and change and find that process is a central component. Taking the example of organisations, they see that this 'innovation journey' always consists of a series of events that create and transform an idea into a reality. Uncertain, dynamic and often random (Quinn 1985; Kanter 1988; Jelinek and Schoonhoven 1990) it is always a journey into the unknown through which innovation and change occurs.

Two views of the innovation and change process were identified—one cyclical characterised by trial and error learning (March and Olsen 1975; Cohen and Sproull 1991) with change occurring through the feedback between 'the actions people take and the outcomes experienced from their actions' (Cheng and Van de Ven 1996: 594). The other sees change as 'an evolutionary metaphor… a stochastic process of exogenous random events (Cohen et al. 1972; Hannan and Freeman 1989; Tushman and Anderson 1986). The source of change is assumed to be largely a consequence of some event which is outside the decision-making process and is thus random, and often associated with 'key' exogenous events. In both approaches, the relationship between process and change is emphasised, and in fact, a combination of the two would seem to be more appropriate, varying in balance with context and issue.

Gemmill and Smith (1998: 752) also note the significance of systemic change as a process in organisations and institutions, whilst Mihata (1997: 30–33) citing Arthur E. Murphy's interpretation of G.H. Mead's *The Philosophy of the Present* (1932) reminds readers that process has a specific meaning in social thought. The world is inherently complex, dynamic and multidimensional, and the use of processes and emergence can help to conceptualise the complexity.

Using the concept of emergence which represents our conceptualisation of change and the processes that make it come about, Mihata goes on to suggest that it is a complex phenomenon that commonly results in outcomes that are unexpected and also very different from its original constituent parts. The process of change may also generate entirely new effects that evolve from existing conditions, and consequently, new policies may have to contemplate new problems and issues that could not have been (easily at least) predicted (Nagel 1979: 374).

Collier and Esteban (1999: 176) also looked at organisations and suggest that those which only functioned when the environment was static and under control were anachronistic in a world which is now turbulent, and characterised by flux and transformation (Morgan 1986). Citing Ghoshal and Bartlett (1998), they conclude that 'change has become overwhelming to the point where traditional linear organisations have become dysfunctional'. This is undoubtedly true in the shipping sector where 'increased communication possibilities, technological change and financial innovation have introduced flexibility into world economic structures' to an unparalleled extent.

Shields (1997: 4) considers the characteristics of change and how they have a close relationship to flows. He notes the 'channelling factors' which place curbs on desires, which in his terms have impact upon the changes which take place in the will to live, love and sociality:

> Change no longer refers to an underlying genealogy, a history or a transcendent principle, for it does not repeat an identity or a law. Instead, change changes in relation to other changes; it has an absolute, intensive speed, which occupies a smooth space in the manner of a vortex, always repeating that which differs. Deleuze and Guattari (1988: 381).

Urry (2000: 27) continues the emphasis on change by quoting Derrida (1987: 27):

> *Differance* is incompatible with the static, synchronic, taxonomic, ahistoric motifs in the concept of *structure*.

Le Poideven (2003: 26) in his consideration of space and time devotes considerable energy to the issue of change and concludes that its process is fundamental to the universe and once started can never cease. 'Change thus begets change'.

Bird (1981: 129) had already emphasised the relationship of science to process and change suggesting that:

> Modern philosophers of science would probably agree that any account of the scientific method should include an in-built mechanism for change; indeed scientific successes usually involve changes in our beliefs about the world. While some students of scientific method have asserted that change is periodic via successive dominating paradigms, others believe in 'revolution in permanence' (Popper 1974: 2 and 1147); 'we can break out of our framework at any time' (Popper 1970: 56); a 'constant-revision' view of our knowledge of nature (Rescher 1978: 51).

Urry (2005: 236) takes this further and comments with respect to science, process and change noting Rifkin's (2000) suggestion that contemporary science no longer saw phenomena as 'static, fixed and given'. The observer is considered as always changing what is observed, and apparently, stable entities are in fact unstable as there is never a structure which can be seen as separate from process.

Kennedy (2007: 272) interprets the significance of process and change through the concept of spatiality and the way that globalisation has changed its very nature and form and places much more emphasis upon the movement of businesses, migrants, cultures, money, ideas, goods and so on. This focus upon mobilities can be seen in the work of Appadurai (1990) and Scholte (2000) on despatialization and deterritorialisation; Harvey (1989) and time space compression; scapes (Appadurai 1990); the space of flows (Castells 1996); diverse mobilities, global fluids and complex human and inhuman hybrids (Urry 2003); the ubiquity of nomadism and place polygamy (Beck 2000); and leading bi- and multifocal lives in several simultaneous locations through transnational migration (Basch et al. 1994; Vertovec 1999). Inherently, all these interpretations are centred upon mobility itself conceptualised in change and process.

Neo and Chen (2007: 16) emphasise the need to move towards what they call 'dynamic governance' which can accommodate not only 'defined processes' but would also permit agility to be expressed and exercised through a continuous review and redesign.

At the same time, Taleb (2007) introduces the idea of a 'Black Swan', as a highly improbable event which has specific impact depending on context, history and much else. Using this metaphor (based on the discovery unexpectedly of black swans in a location where they were unanticipated), he goes on to emphasise the relativity of change. Taking another metaphor (this time the turkeys at Christmas we noted earlier) and at risk of becoming obsessed by fowls:

> From the standpoint of the turkey, the nonfeeding of the one thousand and first day is a Black Swan. For the butcher it is not, since its occurrence is not unexpected. So you can see here that the Black Swan is a sucker's problem. In other words, it occurs relative to your expectation.
>
> Note that these events do not have to be instantaneous surprises... Some Black Swans can come from the slow building up of incremental changes in the same direction, as with books that sell large amounts over years, never showing up on the best-seller list, or from technologies that creep up on us slowly, but surely. ... Matters should be seen on some relative, not absolute, timescale; earthquakes last minutes, 9/11 lasted hours, but historical

> changes and technological implementations are Black Swans that can take decades. In general positive Black Swans take time to show their effects whilst negative ones happen very quickly - it is much easier and much faster to destroy than to build. Taleb (2007: 44–45).

Change is thus relative both temporally and in terms of its impact upon different stakeholders. Effective maritime governance needs to recognise these relative issues and be designed to accommodate time and the continuous fluctuation in stakeholder impact and involvement if it is to be sensitive to the problems it addresses. Thus, for example, the impact of containerised shipping has been much slower than that of some aspects of tanker safety and the security of containers themselves, whilst the significance of all of these issues varies considerably depending upon whether an individual is employed in container shipping, freight forwarding, port operations, ship finance, agencies or wherever.

Process and Time

> An observer cannot even describe a cultural custom unless he has observed the behaviour at least twice; two observations at Time1 and Time2 are an absolute minimum – otherwise he is not certain whether or not he has observed an idiosyncratic response that may never occur again. Fortes (1949: 342) has also stated that social structure must be 'visualized' as a 'sum of processes in time'. (Vogt 1960: 22).

Davis (1899: 482) was a very early commentator on the relationship of process to time using the changes he had observed in landforms as his example of how processes acted. 'Process cannot however, complete the work instantly, and the amount of change from initial form is therefore a function of time'.

Meanwhile, Reichenbach (1958: 116, 117) was convinced: 'every lapse of time is connected with some process, for otherwise it could not be perceived at all'; and 'processes of nature thus determine the flow of time'. Janelle (1969: 348) took this further in linking process and time with space: 'spatial reorganization (is) a process by which places adapt both the locational structure and the characteristics of their social, economic and political activities to changes in timescale connectivity'.

The relationship between process and time was also identified by Blaut (1961: 1) in discussing the significance of space and its relationship to process. He noted the work of Leibnitz (cited in Riechenbach 1958, 20) who considered the relative concept of space as a relationship between events and as a result inevitably bound to time and process. Blaut went on (1961: 2) to suggest that 'relative space is inseparably fused to relative time, the two forming what is called the space-time manifold, or simply process. Nothing in the physical world is purely spatial or temporal; everything is process'. Time is always there, even if only implied, and the spatial dimension cannot exist except temporally—whether considering the location of ports (which changes over time); the location of commodities and cargoes (which also change location both physically and in terms of relative significance); and the location of financial accessibility (the City of London, e.g.—which has also moved over time both actually to

London Docklands and in abstract form taking on a virtual presence over time as communications have changed). Time and space together form a process.

Harvey (1969: 423) warns of taking time as some sort of independent variable—which it certainly is not—rather than a parameter to be estimated which needs an appropriate measure if the processes which are going on are to make any sense. Commonly, stages are invented [see, e.g. Rostow's (1960) economic-growth stages; or Taylor's (1937) zones and strata technique] to provide a sequential framework which sometimes then takes on a life of its own in implying that it derives from some sort of mechanism. The danger in all this is that the stages take on an artificial and unwarranted responsibility for generating change rather than just providing a structure against which change can be assessed.

Roseman (1971) provides a wide discussion of time and space in migration from a process perspective, whilst Thrift (1977: 65) is emphatic in insisting that time is an essential part of spatial studies and that without both changes could not be understood. Processes are the underlying causes of these changes, and therefore, the relationship between process and time is both essential and inevitable.

Thornes (1979) notes that processes occur at varying rates and it is important to recognise the impact discipline on the length of study. Processes may even cease and then restart; they may last for long or short periods; their impact may be instantaneous or slow—and so the relationship between process and time is a fundamental one. Maritime policy-making clearly should accommodate the ability of the sector to make quick changes; or none at all. To react over a considerable period of time or instantaneously. Thus, the impact of (say) changes in labour regulations, environmental standards or the scrapping of vessels will be different between the sectors, in some cases much quicker than others, and in others much more irregular. Policies (and their governance) need to accommodate this irregularity (and even unpredictability) if they are to be effective. The expectation of results from the introduction of a new policy can generate disappointment if the change does not occur immediately as anticipated and the consequences of this can be policy abandonment or neglect.

Shotter (1983: 20) emphasises the important role of time in the processes of growth and development and the contribution of Prigogine (1980) to this debate and in particular the contrast that exists between machines and organisms, between stasis and evolving totalities and the tendency for time to be marginalised.

Thietart and Forgues (1995: 21–22) consider the relationship of organisations to process and time noting the importance of their dynamic character. Organisations are in effect processes manifesting themselves as forms. Thompson (1967: 6) sums it up: 'the complex organisation (of which shipping is undoubtedly part) is a set of interdependent parts which together make up a whole in that each contributes something and receives something from the whole, which in turn is interdependent with some larger environment'. The complexity of multiple actors, both internal and external to the organisation, working both with and against each other and with temporal ramifications both in historical determination and future implication is immense. Processes are rarely linear, wholly predictable or rational.

Many are muddles, hesitant, irrational and intuitive but almost always have a serious temporal dimension.

Griffin (1998: 2) provides a philosophical interpretation of time and process seeing everything as a momentary event rather than Whitehead's (1925: 72, 1929: 41) view of 'temporal extensiveness'. Sabatier (1998: 102) takes the importance of stakeholders in policy-making and suggests that their substantial influence is always effective over extended periods (over a decade commonly) through the application of a series of integrated processes and through a complex of organisations.

Meanwhile, Rescher (1999: 38) is emphatic that processes are always temporal as for a process to 'exist (to actually be realized) it must exist in time'. Meanwhile, Wolch and De Verteuil (2003: 161) use Hagerstrand's conception of time geography to clarify the relationship between process and time. We return to time geography in a later chapter but for the time being the idea that individuals use time and space to fashion the processes that activate their lives is one that has gained common acceptance. Pred (1996) notes how life trajectories follow paths or processes over varying time periods of hours, days, weeks and so on illustrating the close relationship between each of the concepts.

Both Kent (2003: 112) and Matthews and Herbert (2008: 82) note the close relationship that exists between geographical process and time, whilst Rhoads (2005: 137) continues in the context of geomorphological change stressing the significance of process and form interaction which always occurs over time. This might lead to no changes at all if a steady state is produced by process and form coming together, but inevitably, even if this is the case, a time dimension will always be a feature.

Le Poidevin (2003: 14–17) discusses the relationship between time, change and process, issues which are taken up by Taylor (2003: 151) contrasting static and dynamic models in interpreting society, whilst Richards et al. (2004: 328) see 'space-time (as) simply a frame of reference within which mechanisms become processes that cause change in various phenomena or structures'. Dennis and Urry (2009: 52) in their consideration of a life without cars emphasise the significance of time in all the systems that form the foundations of society. Systems move through time and in their terms 'are in process'. Using thermodynamics as their example, they go on to illustrate the potential complexity of the systems process as it changes over time with an 'arrow or flow of time that results in futures that are unstable, relatively unpredictable and characterized by various possibilities'.

Process and Governance

Process, governance and policy-making are intrinsically linked, and to achieve any form of effective governance in the maritime sector, it will be necessary explicitly to accommodate the dynamic features of policy-making. There have been a number of studies that reaffirm this—Sabatier (1991) looks at the relationship of process to government policy-making; Sutton (1999) considers the relationship in some depth in relation to overseas development; and Brooks and Pallis (2003) do the same but in consideration of the emergence of port policies.

The fact that policies can be viewed as processes is evidenced by the extensive literature that links them. This includes Lowi (1964), Dye (1966), Sharkansky (1968), Niskanen (1971), Mayhew (1974), Jones (1977), Eyestone (1978), Anderson (1979), Wilson (1980), Gormley (1983), Kingdon (1984), Salisbury (1986), Calvert et al. (1989), McCubbins et al. (1989), Wood and Waterman (1991), Macey (1992), Baumgartner and Jones (1993), Rosenthal (1993), Schneider et al. (1995), Borgese (1998: 138, 186–189) who even places it in a maritime context, Braganza and Korac-Kakabadse (2000: 51), Cashore and Howlett (2007: 332), Lazarus (2009: 1) and Seldadyo et al. (2010: 626). Schlager and Blomquist (1996, 653–659) identify three main processes used in policy-making and analysis—Institutional Rational Choice (Ostrom 1990); the Politics of Structural Choice (Moe 1990); and Sabatier's (1988) Advocacy Coalitions. Without suggesting this is comprehensive, it does indicate the strong relationship between the processes of policy-making, the policies produced and their governance, something also noted by Borzel and Risse (2010: 114, 115).

Kasperson and Minghi (1969: 200) link communications, policies and processes through the vehicle of political geography stressing the 'distribution of political influence, both vertical and spatial, communications, linkages and policy decisions' and going on to point out the significance of time (and its process implications) in policy-making and governance. Meanwhile, Hagerstrand (1973: 85) hints at the relationship between politicians and policy-making and the processes that characterise social space and the lives of those who live in it. His work with time geography encompasses these relationships in defining the limits and possibilities within which individuals are active, suggesting that life processes, governance and policy-making are intrinsically connected.

Berry (1973: 79) provides anecdotal evidence of the importance of process to policy-making and how the two are inseparable. A special advisor to the US President had commented on the difficulties of generating meaningful policy in urban areas of the USA:

> The knowledge crisis facing the policy-maker is very real because he cannot get adequately past Question Number One, 'What do we know now?' The task of 'synthesis' is not being done. Instead he confronts fragmentation, static rather than dynamic images under conditions of accelerating social and technological change, unexplained processes depicted as isolated facts, and the fact his point of intervention and what would be the consequences of alternative actions are not given him by the knowledge community. Because Question Number One has not been answered, he is even more vulnerable on Question Number Two, 'What do we need to know?' He does not have a Mendeleevyan chart that by its gaps shows what must be sought. Instead he confronts a host of competitive research proposals that relate to no universe of integrated knowledge, and generally do not even advance a hypothesis to relate the proposal to the existing body of knowledge. And further the research proposer is generally ignorant of the operational knowledge requirements of the policy-maker – because there is no scientific base of knowledge to which decisive appeal can be made.

Kobayashi (1989: 164-165) likens process to that of a dance, and although considering this in the context of landscape, the analogy can be taken further to be akin to that of policy-making.

> In dance, there is a long journey from the *barre* to centre stage. Along the way, the discrete *pliez* and *relevez* of the *classe* give way to undifferentiated movement. The dance cannot be reduced to its basic elements but extends beyond itself as a tantalization that includes its history and its potential. It releases time, space, form and movement, to bring dancers and audience into a common sphere of expression. And yet we are fooled. For, the moment when the *pas de deux* reaches its exquisite climax, when the world seems to begin and end with a single subtle gesture, that moment, could never exist – or could never be the same – without each agonizing *pliez* that has gone before, without every dedicated encounter with the cold reality of the *barre*.

In a similar way, policies can be viewed as this dance. They are organised and choreographed; they can produce emotional responses and in some ways policies require expert interpretation to appreciate their finer points. Policies are 'irreducible to their constituent elements' or at least in a meaningful way. Like dance, policies are a form of language and to understand a policy it is necessary to know its history, structure and syntax.

Walt and Gilson (1994: 354) identify policy-making (albeit in the health sector in developing countries) as a profoundly political *process* and that its governance inappropriately tends to focus upon its content. Angelides and Caiden (1994: 228) regard policy-makers as tending to assume that the world as they know it will continue indefinitely. Whilst this is understandable and reflects political and personal ambitions to which stability may be beneficial, the need to be able to conceive and plan for sudden change in processes and the context for those changes is fundamental to good governance (Sztompka 1984).

Schneider and Ingram (1998: 62) examine the methods commonly used to design policies outside of any particular sector and find that it is more of a case of troubling with ideas rather than selection with remarkably little effect. Significant in the cases where innovative designs did emerge was to accommodate effectively the processes used to generate ideas—including searching 'through large stores of information—make comparisons, find analogies and combine elements cafeteria-style to create' proposals. One common approach was 'policy copying' (or its equivalent under another name) to which we return later (Walker 1969).

Sabatier (1998: 98) develops his own Advocacy Coalition Framework dating from the decade before and with the goal of providing a 'coherent understanding of the major factors and processes affecting the overall policy process including problem definition, policy formulation, implementation and revision in a specific policy domain'. The emphasis on policy process is clear.

Collier and Esteban (1999: 180) spend some time relating governance models in what they termed 'participative organisations', to change and instability and in so-doing closely linked process and 'cybernetic governance'. Meanwhile, Braganza and Lambert (2000: 180–181) identify a close relationship between process and governance and use a Process–Governance Framework to clarify the relationship's structure and suggest where decisions are made and how effective they might be (Fig. 4.2).

	Defines→	Have→	Addressed By→	Defines→	
Process Perspective	**Business Strategy**	**Stakeholders**	**Expectations**	**Processes**	**Activities**
Governance Perspective	Monitor. Evaluate. Agree. Prioritise. Quantify. Joint. Communicate.	Identify. Prioritise.	Understand. Assess conflict. Categorise.	Define. Align with expectations and objectives. Set performance goal. Classify. Balance responsibility for process and function.	Define the activities. Identify degree of discretion or prescription. Align roles, responsibilities, targets and rewards to function and process identity information requirements. Define ways of working. Assess resource requirements.
Who Is Responsible	Business leaders	Business leaders and sector managers	Senior and middle managers	Business leaders and senior managers	Senior managers and managers

Fig. 4.2 The process–governance framework. *Source* Braganza and Lambert (2000: 181)

Boons et al. (2009: 231–2) define governance processes as 'coordinated actions of public and private actors around collective issues'. Governance is considered an evolving process because of the 'dynamic interactions between self-organising participants in governance processes, management interventions and unmanageable internal and external dynamics'. They are difficult to control and do not align themselves clearly with the classical steering mechanisms of markets, networks and hierarchies. Boons et al. (2009: 241) also note that under the very traditional jurisdictional hierarchy that the maritime sector exhibits, the 'capriciousness' of the processes that are characteristic of the sector makes policy-making difficult (if not meaningless) and the results of such collective action 'difficult if not impossible to achieve'.

Klijn and Snellen (2009: 22–23) note Klijn's (1996: 116) comments:

> The recent interest in the concept of policy networks can be seen as an attempt to 'contextualise' the process approach. Not only does policy making take place in settings where there are many actors and there is ambiguity regarding preferences, information and strategies chosen, but it also occurs within certain inter-organizational networks of a more lasting nature. The policy network approach thus takes up where the process approach leaves off. Problems, actors and perceptions are not chance elements of policy processes but are connected with the inter-organizational network within which these processes occur.

Meanwhile, Borzel and Risse (2010: 114) define governance as being constituted of both structure and process. The latter was felt to be an appropriate term because it 'pinpoints the modes of social coordination by which actors engage in rule-making and implementation and in the provision of collective goods'.

Process Models

Blaut (1962: 2) was one of the earliest commentators to note the importance of process models suggesting that earlier debate about process philosophy (to which we return in the next chapter) had now been replaced by process modelling. These models he suggest are characterised by 'systems of interacting interpenetrating, part-processes'.

Some years later, Harvey (1967: 564–566) considers process models within a broader discussion of models in general and notes how ignorant at that time geographers were of process modelling and its relationship to the spatial patterns that emerge. This particularly applied where more than one process is going on at any time and attempts to model collections of simultaneous processes were rare. He identifies a number that had been made by then: process models could be found that considered town growth (Muth 1961; Winsborough 1962); industrial location change (Fuchs 1962); information diffusion (Dodd 1950); migration (Wendel 1953; Lovgren 1956; Hagerstrand 1957; Nelson 1959; Sjaastad 1960; Raimon 1962; Olsson 1965a, b; Morrill 1965); colonisation (Bylund 1956, 1960); transport (Taafe et al. 1963; Kansky 1963).

Berry (1973: 16) adopts a process model in his examination of geographical planning using the example of ecosystems. Figure 4.3 shows the generic principles that lie behind his design based on the identification of a series of common approaches to planning. These models are focussed upon processes that underlie planning decisions and, although rather crude, indicate the significance that process modelling can take on.

Van de Ven (1992: 171) provides a detailed summary of a variety of process models used in strategic management and taken from a number of sources (Fig. 4.4) (Scott 1971; Greiner 1972; Cohen et al. 1972; Mintzberg et al. 1976; Quinn 1980; Gluck et al. 1980; Lorange 1980). Each focuses on progression within organisational activity (nature, sequence and order) over time rather than the variables to which these are addressed.

One significant example is the garbage can model summarised neatly by Olsen (2001: 191) as the assumption that 'problems, solutions, decision-makers, and choice opportunities are independent, exogenous streams flowing through a system'. Takahashi (1992, 1997: 92) suggests that it can be used as a 'general frame within which to describe almost any decision-processes in many kinds of organizations ranging from schools to navies' citing Lynn (1982) and himself (1992) as examples. They both go on to stress that there are a large number of garbage can models which explore the different circumstances under which each model operates (March and Olsen 1986) commonly focussing upon the different ways in which the streams are coupled. Others look at the management of garbage can situations (Cohen and March 1974: 305–315), whilst Mucciaroni (1992: 461) emphasises the approach's advantages over the more commonly applied rigid models used to understand political processes.

	PLANNING FOR PRESENT CONCERNS	PLANNING FOR THE FUTURE	PLANNING FOR THE FUTURE	PLANNING FOR THE FUTURE
	Reacting to Past Problems	*Responding to Predicted Futures*	*Responding to Predicted Futures*	*Creating Desired Future*
	AMELIORATIVE PROBLEM SOLVING	ALLOCATIVE TREND MODIFYING	EXPLOITIVE OPPORTUNITY SEEKING	NORMATIVE GOAL ORIENTED
	Planning for the Present	*Planning Toward the Future*	*Planning with the Future*	*Planning from the Future*
Planning Mode	Analyse problems, design interventions, allocate resources accordingly	Determine and make the best of trends and allocate resources in accordance with desires to promote or alter them	Determine and make the most of trends and allocate resources so as to take advantage of what is to come	Decide on the future desired and allocate resources so that trends are changed or created accordingly. Desired future may be based on present, predicted or new values
Present or Short Range Results	Ameliorate present problems	A sense of hope. New allocations shift activities	A sense of triumphing over fate. New allocations shift activities	A sense of creating destiny. New allocations shift activities
Future or Long Range Events	Haphazardly modify the future by reducing the future burden and sequence of present problems	Gently balance and modify the future by avoiding predicted problems and achieving a balanced progress to avoid creating major bottlenecks and new problems	Unbalance and modify the future by taking advantage of predicted happenings, avoiding some problems and cashing in on others without major concern for emergence of new problems	Extensively modify the future by aiming for what could be. Change the predictions by changing values or goals, match outcomes to desires, avoid or change problems to ones easier to handle or tolerate.

Fig. 4.3 Differing models of planning. *Source* Berry (1973: 16)

There are a considerable number of others who have commentated on the approach to understanding process including Cohen et al. (1972), Padgett (1980), Alexander (1982), March and Olsen (1983), and Masuch and LaPotin (1989). Meanwhile, its use has been linked to temporal order and the importance of time in process (Huggett and Perkins 2004: 231; March and Olsen 1984: 746), metaphors and their use in process analysis (Bendor et al. 2001: 188), atomism (Fioretti and Lomi 2008: 193), institutions and garbage can modelling (Levitt and Nass 1989),

Model	Beginning	<----------	Activity Phases	---------->	End
Strategic Decision Models					
Mintzberg et al. (1976) Field study of 25 strategic unstructured decision processes.	*Identification phase* Decision recognition routine. Diagnosis routine.	*Developmental phase* Search routine. Design routine.	*Selection phase* Screen routine. Evaluation-choice routine. Authoriz-ation routine.		
Cohen, March and Olsen (1972) Garbage can model of decision-making.	Choices. Problems. Solutions. Energy of Participants.	>>>>>>>>>>>	>>>>>>>>	>>>>>>>>	>>>>>>>
Quinn (1980) Case studies of nine major corporations..	Stages include: Sense need.	Develop awareness and understanding.	Increase support.	Build consensus.	Formal commit-ment.
Strategic Planning Models					
Gluck, Kaufman and Walleck (1980) Study of formal planning systems in 120 companies.	Basic financial planning.	Forecast based planning – predict the future.	Externally orientated planning – think strategic-ally.	Strategic manage-ment – create the future.	
Lorange (1980) Normative model of corporate strategic planning.	Objective setting – identify relevant strategic alternatives.	Strategic programming – develop programs for achieving chosen objectives.	Budgeting – established detailed action program for near-term.	Monitoring – measure progress toward achieving strategies.	Rewards – establish incent-ives to motivate goal achieve-ment.
Organization Development Models					
Scott (1971) Stages of corporate development.	Single products, channel and entrepreneurial structure.	Single product, channel and functional structure.	Multiple products, channels and divisional-ised structure.		
Greiner (1972) Stages of organizational growth through evolution and revolution.	Growth through creativity – leadership crisis.	Growth through direction autonomy crisis.	Growth through delegation – control crisis.	Growth through coor-dination – red tape crisis.	Growth through collabora tion – crisis of?

Fig. 4.4 Development process models in strategic management. *Source* Van de Ven (1992: 171)

and policy-making (Lipson 2007: 83–84). Many of these issues have been raised in our discussion of the wider context for maritime governance.

Four broad types of process model are identified.

Life cycle process models assume that 'change is immanent;...the developing entity contains within it an underlying logic, program, or code that regulates the process of change and moves it from a given point of departure toward a subsequent end which is already prefigured in the present state'. Van de Ven and Poole (1988: 38) suggest that 'what lies latent rudimentary or homogenous in the embryo or primitive state, becomes progressively more mature, complex and differentiated. External environmental events and processes can influence how the immanent expresses itself, but they are always mediated by the imminent logic, rules or programs that govern development'.

This type of model is reflected to a certain extent in the international shipping industry with a series of rules, and regulations, norms and expectations that have influence upon the industry's development but fail to control it and have little influence upon underlying fundamental trends and urges such as those to maximise income, minimise regulation and retain a liberal image and expression. Examples of applications to many other sectors come from developmentalism (Nisbet 1970), biogenesis (Featherman 1986), ontogenesis (Baltes et al. 1986), child development (Piaget 1975), human development (Levinson 1978), moral development (Kohlberg 1969), organisational development (Greiner 1972, Kimberly and Miles 1980), group decision-making (Bales and Strodtbeck 1951; Poole and Roth 1989; Gersick 1988); and new venture development (Burgelman and Sayles 1986).

Teleology process models do not presume any sequence of events that has to occur nor any necessary and logical direction. However, it is still possible to assess progress towards an envisioned end state and therefore how it is becoming more (or less) integrated, or growing more complex. Teleological models look towards an anticipated conclusion (permanent or temporary), but there are commonly a variety of ways of achieving it. Once again international shipping can be seen to fit in with this whereby commonly, logically desired endpoints (profit, safety, environmental cleanliness, secure transport) are agreed by all, but the process towards achieving these can be varied and even contradictory—for example through regulation and liberation; nationalisation and privatisation; and subsidy and competition. There are many examples of its application including functionalism (Merton 1968); decision-making (March and Simon 1958); epigenesist (Etzioni 1963); enactment (Weick 1979); voluntarism (Parsons 1951); adaptive learning (March and Olsen 1976); and strategic planning (Chakravarthy and Lorange 1991).

Dialectic process models assume that 'the developing entity exists in a pluralist world of colliding events, forces or contradictory values which compete with each other for domination and control'. Stability and equilibrium in the processes which are acting contradictorily are achieved through the balance of opposing forces and change only occurs when the balance is upset. Thus, international shipping displays stability, whilst issues such as the environment and safety can be made to balance with profit and efficiency but when one or any of these factors becomes dominant or uncontrollable, then a change in the system has to occur. Dialectic models are thus characterised by oscillation and chaos. Equilibrium has been discussed by Blau (1964), French and Bell (1978), Pfeffer (1981) and Astley and Zajac (1991); oscillation by Schumpeter (1942), Lindblom (1965), Quinn (1980) and Masuch (1985); and chaos by Zeeman (1976), Prigogine and Stengers (1984) and Martin (1990).

Evolution process modelling refers to 'cumulative changes in structural forms of population entities across communities, industries or society at large' (Van de Ven 1992: 179). Change is reflected in continuous 'variation, selection and retention' and reflects the characteristics of biological evolution. In international shipping, this is shown in the continuous adaption of the sector to economic, social, legal and financial changes that take place and which have resulted, for example, in the spontaneous development of the container shipping sector and the adoption of flags of convenience as a response to changes in the sector. Only those players who have adequately adapted to such activities have survived. Generic examples come from Campbell (1969), Hannan and Freeman (1977) and Aldrich (1979), whilst more specific applications are from global population change (Carroll and Hannan 1989), organisational strategy (Singh 1990, Burgelman 1991) and socio-psychology (Weick 1979; Gersick 1991).

Conventional linear modelling of organisations is inadequate and in the few situations where they remain they are 'dysfunctional' (Ghoshal and Bartlett 1998). The combination of constant change (e.g. in cargo ownership, flags, sources of finance, and bunkers and seafarers) with the flexibility to locate almost anywhere has 'shifted the centre of organisational gravity to the boundaries of organizations' and as such their modelling has to centre around the processes going on.

Collier and Esteban (1999: 176) consider the characteristics of organisations and conclude that the attempts to model them are inadequate because they assume that the 'environment is static and control is complete'. Their view is that this is anachronistic because organisations now operate in a world that is dynamic. In truth that has always been the case but until fairly recently, the level of dynamism was relatively slow and an assumption of stasis was not unreasonable in many cases. This is true of shipping as much as any other industry, but with the intensification of globalisation, the characteristics of most industries and their organisations have changed. To model them then needs to take account of the processes that are going on and accommodate the turbulence, flux and transformation that Morgan (1986) considers is generated. Collier and Esteban (1999: 176) continue:

> In a changing environment mechanistic linear models of organization and associated understandings of governance based on Newtonian logic are no longer appropriate. Environmental turbulence has forced organizations to abandon the strategic 'linear' planning which relies on forward projection – in particular the re-engineering of bureaucratic structures – because it has been proved to be ineffective in a climate of continual change. Organizations now have to learn and plan as 'open systems' in order to survive and this requires a shift from 'tightly coupled' to 'loosely coupled' organisational structures – in other words to greater flexibility.

Conclusions

The issue of process is fundamental to ensuring effective governance in any sector, but perhaps as much as in any, the maritime sector exhibits characteristics that make the incorporation of a dynamic process into its governance and policy-making almost essential. The debate on the role of process has been long and full of contention but never is it far away even in the deliberations of those who consider that form is of major significance. Maritime governance should see process as a central plank of effectiveness in policy-making as it relates directly to the dynamic nature of the industry and the environment in which it works. For maritime policy-making to have any realistic impact, it needs to be dynamic and thus incorporate the processes that are going on (in finance, operations, trade, negotiations, politics and many more) effectively.

Process has not been ignored by any means in the debate within the broader transport sector as a whole and also within the wider social science remit. Schumm and Lichty (1965) provide an extensive debate on the importance of geomorphological processes, subsequently backed up by Unwin (1992: 117), whilst Sauer (1941: 2) emphasises how 'geography, in any of its branches, must be a genetic science, that is, account for origins and processes'. Berry (1973: 19) considers the role of processes in his debate on metageography and how this process view focuses upon becoming with its continuous view of events. He places this in a continuum of history with process repeating and reiterating spatial behaviour. Gregory (2000; 105) continues the geographical theme with consideration of both the value of a formal interpretation of science and that of a procedural one using primary examples of the former from Wooldridge (1958: 31)—'I regard it as quite fundamental that geomorphology is primarily concerned with the interpretation of forms, not the study of processes'—and the latter from Stoddart (1997: 384) who considers Wooldridge 'remarkably out of touch' and Strahler (1952: 924) who stresses the need to search for 'vitality'. Rhoads concurs (2005: 134–5) quoting Clark (1954: 71): 'whatever interests us in the contemporary scene is to be understood only in terms of the processes at work to produce it'.

Extensive examples also come from the transport literature. Bird (1981: 137) emphasises the contradictions of decision-makers in port planning between satisficing and optimising which is reflected in 'process planning' rather than 'structural planning' (Simon 1976; Jantsch 1980: 270–2). To get around this, there is a need to place process as the focus of planning so that the dangers of short-termism are avoided.

Baird (2000: 182–4) looks at port privatisation which he considers as a policy process and not something that can be introduced with a single event and which can then be assumed finished. Privatisation takes time, changes through time, needs to be recognisant of these changes and processes if it is to be effective. Policy (and its governance) must reflect this.

Continuing the port theme, Van Gils et al. (2009: 77) emphasise how governance is always complex and thus dynamic and take four cases studies from two ports—Hamburg and Rotterdam—to illustrate how emerging events generate dynamic processes:

> Harbour systems are excellent representatives of complex systems. The governance systems encompass local community planning as well as global logistics planning. As a result, considerable numbers of actors with different stakes, strategies and operating procedures attempt to influence port governance processes. The interrelationships between these sub-systems and their actions reinforce the dynamic and unpredictable course of development.

And Wolch and DeVerteuil (2003: 161) bring it all rather neatly together by referring to Hagerstrand's (1973) time geography, with its clear and close relationship to transport and the process dominated dynamics of society which we will discuss in some depth in a later chapter. Time geography is a:

> notation and heuristic device designed to contextualise collateral processes in time and space. Individuals use time and space as resources to fashion particular paths, a trajectory or movement that spans the daily to the yearly and even the lifetime scales (Pred 1996). In turn, these paths interact with larger time-space projects, that is the building of numerous individual time-paths towards a similar goal (Jackson and Smith 1984: 50).

There is clear application of approaches such as this to the dynamism of maritime governance with the specific aim to accommodate the life processes that dominate policy-making issues in shipping and ports. But now, it is time for us to move on and consider a wider interpretation of process and its relationship to maritime governance.

References

Abrams, P. (1983). *Historical sociology*. Ithaca NY: Cornell University Press.

Adgar, W. N. (2000). Social and ecological resilience: Are they related? *Progress in Human Geography, 24*(3), 347–364.

Agranoff, R., & McGuire, M. (2003). *Collaborative public management: New strategies for local governments*. Washington DC: Georgetown University Press.

Ahmed, S., Castaneda, C., Fortier, A., & Sheller, M. (Eds.). (2003). *Uprooting/regroundings: Questions of home and migration*. Oxford: Berg.

Aldrich, H. (1979). *Organizations and environments*. Englewood Cliffs NJ: Prentice-Hall.

Alexander, E. R. (1982). Design in the decision-making process. *Policy Sciences, 14*, 279–292.

Allen, P. M., Sengller, M., Engelen, G., & Boon, F. (1985). Towards a new synthesis in the modeling of evolving complex systems. *Environment and Planning B, 12*, 65–84.

Allmendinger, P. (2001). *Planning in postmodern times*. London: Routledge.

Amin, A., & Thrift, N. (2002). *Cities: Reimagining the Urban*. Cambridge: Polity Press.

Anderson, J. E. (1979). *Public policymaking*. New York: Rinehart and Winstone.

Angelides, C., & Caiden, G. (1994). Adjusting policy-thinking to global pragmatics and future problematics. *Public Administration and Development, 14*, 223–239.

Appadurai, A. (1990). Disjuncture and difference in the global cultural economy. In M. Featherstone (Ed.), *Global culture* (pp. 295–310). London: Nationalism, Globalization and Modernity, Sage.

Appadurai, A. (1996). *Modernity at large: Cultural dimensions of globalization*. Minneapolis MN: University of Minnesota Press.

Astley, W. G., & Zajac, F. (1991). Intraorganizational power and organizational design: Reconciling rational and coalitional models of organization. *Organization Science, 2*(4), 399–411.

Atkinson, A. A., Waterhouse, J. H., & Wells, R. B. (1997). A stakeholder approach to strategic performance measurement. *Sloan Management Review, 38*, 25–37.

Baird, A. J. (2000). Port privatisation: Objectives, extent, process and the UK experience. *International Journal of Maritime Economics, 2*(43), 177–194.

Bales, R. F., & Strodtbeck, F. L. (1951). Phases in group problem-solving. *Journal of Abnormal and Social Psychology, 46*, 485–495.

Baltes, P. B., Dittman-Kohli, F., & Dixcon, A. (1986). Multidisciplinary propositions on the development of intelligence during adulthood and old age. In A. B. Sorensen, P. E. Weinert, & L. R. Sherrod (Eds.), *Human development and the life course: Multidisciplinary perspectives* (pp. 467–507). Hillsdale NJ: Lawrence Erlbaum.

Basch, L., Schiller, N. G., & Blanc, C. S. (1994). *Nations unbound: Transnational projects, post-colonial predicaments and deterritorialized nation states*. New York: Gordon and Breach.

Baumgartner, F., & Jones, B. (1993). *Agendas and instability in American politics*. Chicago IL: Chicago University Press.

Beck, U. (2000). *What is globalization?*. Cambridge: Polity Press.

Bendor, J., Moe, T. M., & Shotts, K. (2001). Recycling the garbage can: An assessment of the research program. *American Political Science Review, 95*(1), 169–190.

Bennett, C. J. (1991). What is policy convergence and what causes it? *British Journal of Political Science, 21*, 215–233.

Benninghoff, A. (1938). *Lehrbuch der anatomie des menschen*. Munich: JF Lehmann Verlag.

Berger, P. L., & Luckmann, T. (1967). *The social construction of reality: A treatise in the sociology of knowledge*. Garden City NY: Anchor Books.

Berry, B. J. L. (1973). A paradigm for modern geography. In R. J. Chorley (Ed.), *Directions in geography* (pp. 3–21). London: Methuen.

Bhaskar, R. (1978). On the possibility of social scientific knowledge and the limits of neutralism. *Journal for the Theory of Social Behavior, 8*, 1–28.

Bhaskar, R. (1979). *The possibility of naturalism: a philosophical critique of the contemporary human sciences*. Brighton: Harvester Press.

Bigelow, J., Ellis, B., & Pargetter, R. (1988). Forces. *Philosophy of Science, 55*(4), 614–630.

Bird, J. (1981). The target of space and the arrow of time. *Transactions of the Institute of British Geographers NS, 6*, 129–151.

Blau, P. M. (1964). *Exchange and power in social life*. New York: Free Press.

Blaut, J. M. (1961). Space and process. *The Professional Geographer, XIII, 4*, 1–7.

Blaut, J. M. (1962). Object and relationship. *The Professional Geographer, XIV, 6*, 1–8.

Bohm, D. (1975). Human nature as a product of our mental models. In J. Bentall (Ed.), *The limits of human nature*. London: Allen Lane.

Bohm, D. (1980). *Wholeness and the implicate order*. London: Routledge.

Boons, F., Van Buuren, A., Gerrits, L., & Teisman, G. R. (2009). Towards an approach of evolutionary public management. In F. Boons, M. W. van Buuren, & L. Gerrits (Eds.), *Managing complex governance systems. Dynamics, self-organization and coevolution in public investments* (pp. 231–249). London: Routledge.

Borgese, E. M. (1998). *The oceanic circle: Governing the seas as a global resource*. Tokyo: United Nations University Press.

Borzel, T. A., & Risse, T. (2010). Governance without a state; can it work? *Regulation and Governance, 4*, 113–134.

Bourdieu, P. (1977a). *Outline of a theory of practice*. Cambridge: Cambridge University Press.

Bourdieu, P. (1977b). The economics of linguistic exchange. *Social Science Information, 6*, 645–668.

Bourdieu, P. (1984). *Distinction. A social critque of judgement*. Cambridge MA: Harvard University Press.

Braganza, A., & Korac-Kakabadse, N. (2000). Towards a function and process orientation: Challenges for business leaders in the new millennium. *Strategic Change, 9*, 45–53.
Braganza, A., & Lambert, R. (2000). Strategic integration: Developing a process-governance framework. *Knowledge and Process Management, 7*(3), 177–186.
Brenner, N. (1997). State territorial restructuring and the production of spatial scale. *Political Geography, 16*, 273–306.
Brenner, N. (1998). Between fixity and motion: Accumulation, territorial organization and the historical geography of spatial scales. *Environment and Planning D, 16*(4), 459–481.
Brenner, N. (1999). Beyond state-centrism? *Space, territoriality and geographical scale in globalization studies, Theory and Society, 28*(1), 39–78.
Brooks, M. R., & Pallis, A. A. (2003). Assessing port governance models: Process and performance components. *Maritime Policy and Management, 35*(4), 411–432.
Bulmer, S. (1994). The governance of the European Union: A new institutionalist approach. *Journal of Public Policy, 13*(4), 351–380.
Bunge, W. (1968). *Fred K. Schaefer and the science of geography, Harvard papers in theoretical geography*, special paper A. Cambridge MA: Harvard University Press.
Bunge, W. (1979). Schaefer and the science of geography. *Annals of the Association of American Geographers, 69*(1), 128–132.
Burgelman, R. A. (1991). Intraorganizational ecology of strategy-making and organizational adaptation: theory and field research. *Organization Science, 2*(3), 239–262.
Burgelman, R. A., & Sayles, L. R. (1986). *Inside corporate innovation strategy structure and managerial skills*. New York: Free Press.
Buttimer, A. (1976). Grasping the dynamism of lifeworld. *Annals of the Association of American Geographers, 66*(2), 277–292.
Bylund, E. (1956). *Kolonisering av Pite Lappmark*, t. O. M. dr 1867, (Uppsala).
Bylund, E. (1960). Theoretical considerations regarding the distribution of settlement in Inner North Sweden. *Geografiska Annaler, 42*, 225–231.
Calvert, R., McCubbins, M., & Weingast, B. (1989). A theory of political control and agency discretion. *American Journal of Political Science, 33*, 588–611.
Campbell, D. (1969). Variation and selective retention in socio-cultural evolution. *General Systems, 16*, 69–85.
Caporaso, J. (1996). The European Union and forms of state: Westphalian, regulatory or post-modern. *Journal of Common Market Studies, 34*(1), 29–52.
Carroll, G., & Hannan, T. (1989). Density delay in the evolution of organizational populations, A model and five empirical tests. *Administrative Science Quarterly, 34*(3), 411–430.
Cashore, B., & Howlett, M. (2007). Punctuating which equilibrium? Understanding thermostatic policy dynamics in Pacific Northwest forestry. *American Journal of Political Science, 51*(3), 532–551.
Castells, M. (1996). *The information age*: *Economy, society and culture* (Vol. 1). The Rise of the Network Society, Oxford: Blackwell.
Castells, M. (1997). *The information age*: *Economy, society and culture* (Vol. 2). The Power of Identity, Oxford: Blackwell.
Castells, M. (1998). *The information age*: *Economy, society and culture* (Vol. 3). End of Millennium, Oxford: Blackwell.
Castells, M., & Portes, A. (1989). World underneath: The origins, dynamics, and the effects of the informal economy. In A. Portes (Ed.), *The informal economy: studies in advanced and less developed countries* (pp. 11–37). Baltimore MD: John Hopkins University Press.
Chakravarthy, B. S., & Lorange, P. (1991). *Managing the strategy process*. Englewood Cliffs NJ: Prentice Hall.
Cheng, Y.-T., & Van de Ven, A. H. (1996). Learning the innovation journey: Order out of chaos? *Organizational Science, 7*(6), 593–614.
Chorley, R. J. (1973). Geography as human ecology. In R. J. Chorley (Ed.), *Directions in geography* (pp. 155–169). London: Methuen.

Clark, A. H. (1954). Historical geography. In P. E. James & C. F. Jones (Eds.), *American geography; inventory and prospect* (pp. 70–105). Syracuse NY: Syracuse University Press.
Clifford, J. (1992). Travelling cultures. In L. Grossberg, C. Nelson, & P. A. Treichler (Eds.), *Cultural studies* (pp. 96–116). New York: Routledge.
Clifford, J. (1997). *Routes*. Cambridge MA: Harvard University Press.
Cohen, M. D., & March, J. P. (1974). *Leadership and ambiguity*. New York: The American College President, McGrawHill.
Cohen, M. D., March, J. G., & Olsen, J. P. (1972). A garbage can model of organisational choice. *Administrative Science Quarterly, 17*, 1–25.
Cohen, M. D., & Sproull, L. S. (1991). Editors' introduction. *Organizational Science, 2*(1), 1–13.
Coleman, S., & Crang, M. (Eds.). (2002). *Tourism between place and performance*. Oxford: Berghahn Books.
Collier, J., & Esteban, R. (1999). Governance in the participative organisation: Freedom, creativity and ethics. *Journal of Business Ethics, 21*(2/3), 173–188.
Craig, J., & Yetton, P. (1993). Business process redesign: A critique of process innovation by thomas davenport as a case study in the literature. *Australian Journal of Management, 17*, 285–306.
Cram, L. (2011). The importance of the temporal dimension: New modes of governance as a tool of government. *Journal of European Public Policy, 18*(5), 636–653.
Cresswell, T. (Ed.). (2001). Mobilities. *New Formations*, 43.
Crouch, D., & Lubbren, N. (Eds.). (2003). *Visual culture and tourism*. Oxford: Berg.
Crowston, K. (1997). A coordination theory approach to organizational process design. *Organization Science, 8*, 157–175.
Dacey, M. F. (1964). Two dimensional random point patterns. *Papers in Regional Science, 13*(1), 41–55.
Davenport, T. H. (1993). *Process innovation: Reengineering work through information technology*. Boston MA: Harvard Business School Press.
Davies, T. R. V. (1991). Information technology and white collar productivity. *Academy of Management Executive, 5*, 55–67.
Davis, W. M. (1899). The geographical cycle. *The Geographical Journal, 14*(5), 481–504.
Davis, J. B. (1989). Keynes on atomism and organicism. *The Economic Journal, 99*, 1159–1172.
De Bruijn, H., Teisman, G. R., Edelenbos, J., & Veeneman, W. (Eds.). (2004). *Multiple land use and the management of complex projects*. Utrecht: Lemma. (in Dutch).
Degen, M., & Hetherington, K. (2001). Hauntings. *Space and Culture, 11*(12), 1–6.
Deleuze, G., & Guattari, F. (1988). *A thousand plateaus*. Capitalism and Schizophrenia, Minneapolis MN: Minnesota University Press.
Dennis, K., & Urry, J. (2009). *After the car*. Cambridge: Polity Press.
Der Derian, J. (1990). The (s)pace of international relations: Simulation, surveillance and speed. *International Studies Quarterly, 34*(3), 295–310.
Derrida, J. (1987). *Positions*. London: Athlone Press.
Dodd, S. C. (1950). The interactive hypothesis: A gravity model fitting physical masses and human groups. *American Sociological Review, 15*, 245–256.
Dodge, S. D. (1936). The chorology of the claremont-springfield region in the upper connecticut valley in New Hampshire and Vermont. *Papers of the Michigan Academy of Sciences, Arts and Letters, 22*, 335–353.
Dooley, K. J., & Van de Ven, A. H. (1999). Explaining complex organizational dynamics. *Organization Science, 10*(3), 358–372.
Doolittle, W. E. (1984). Agricultural change as an incremental process. *Annals of the Association of American Geographers, 74*(1), 124–137.
Dopfer, K. (1991). Toward a theory of economic institutions: Synergy and path dependency. *Journal of Economic Issues, XXV*, 2, 535–550.
Dreyfus, H. (1991). *Being-in-the-world: A commentary on Heidegger's being and time, division 1*. Cambridge MA: MIT Press.

Durkheim, E. (1895). *Les Regles de la Methode Sociologique*. Paris: F. Alcan.
Dyck, I. (1990). Space, time and renegotiating motherhood: An exploration of the domestic workplace. *Environment and Planning D, 8*, 459–483.
Dye, T. (1966). *Politics, economics and the public*. Rand McNally: Chicago IL.
Edelenbos, I., Klijn, E.-H., & Kort, M. (2009). Managing complex process systems. Surviving at the edge of chaos. In G. Teisman, A. Van Buuren, & L. Gerrits (Eds.), *Managing complex governance systems* (pp. 172–192). London: Routledge.
Edwards, C., & Peppard, J. (1994). Forging a link between business strategy and business reengineering. *European Management Journal, 12*, 407–416.
Eichenbaum, J., & Gale, S. (1971). Form, function and process: A methodological enquiry. *Economic Geography, 47*(4), 525–544.
Elias, N. (1984). *Uber die Zeit*. A zur Wissenssoziologie II, Berlin: Suhrkamp Verlag.
Emerson, R. W. (1841). *Essays: First series*, createspace independent publishing platform.
Engelen, G. (1988). The theory of self-organization and modelling complex urban systems. *European Journal of Operational Research, 37*, 42–57.
Ettlie, J. E., & Reza, E. M. (1992). Organizational integration and process innovation. *Academy of Management Journal, 35*, 795–827.
Etzioni, A. (1963). The epigenesis of political communities at the international level. *American Journal of Sociology, 68*, 407–421.
Euske, K. J., & Player, R. S. (1996). Leveraging management improvement techniques. *Sloan Management Review, 38*, 69–79.
Eyestone, R. (1978). *From social issue to public policy*. New York: Wiley.
Featherman, D. L. (1986). Biography, society and history: Individual development as a population process. In A. B. Sorensen, F. E. Weinert, & L. R. Sherrod (Eds.), *Human development and the life course: Multidisciplinary perspectives* (pp. 99–149). Hillsdale NJ: Lawrence Erlbaum.
Fioretti, G., & Lomi, A. (2008). The garbage can model of organizational choice: An agent-based reconstruction. *Simulation Modelling Practice and Theory, 16*, 192–217.
Fortes, M. (1949). *The web of kinship among the tallensi; the second part of an analysis of the social structure of the Trans-Volta tribe*. London: Oxford University Press.
Francke, M., & Ham, E. (2006). Space of flows. *Manuel Castells, Geographical Approaches, 26*, 01.
French, W. L., & Bell, C. H, Jr. (1978). *Organization development: Behavior science interventions for organization improvement*. Englewood Cliffs NJ: Prentice-Hall.
Friedmann, J. (2000). Reading Castells: Zeitdiagnose and social theory. *Environment and Planning D, 18*, 111–120.
Fuchs, V. R. (1962). *Changes in the location of manufacturing in the United States since 1929*. New Haven CT: Yale University Press.
Gage, R. W., & Mandell, M. P. (Eds.). (1990). *Strategies for managing intergovernmental policies and networks*. New York: Praeger.
Gemmill, G., & Smith, C. (1998). A dissipative structure model of organization transformation. *Human Relations, 38*(8), 751–766.
Gersick, C. J. (1988). Time and transition in work teams: Toward a new model of group development. *Academy of Management Journal, 31*(1), 9–41.
Gersick, C. J. (1991). Revolutionary change theories: A multilevel exploration of the punctuated equilibrium paradigm. *Academy of Management Review, 16*(1), 10–36.
Gertler, M. S. (1988). Some problems of time in economic geography. *Environment and Planning A, 20*(2), 151–164.
Ghoshal, S., & Bartlett, C. A. (1998). *The individualized corporation: A fundamentally new approach to management*. London: Heinemann.
Giddens, A. (1976). *New rules of sociological method*. London: Hutchinson.
Giddens, A. (1979). *Central problems in social theory: Action, structure and contradiction in social analysis*. Berkeley CA: University of California Press.

Giddens, A. (1981). *A contemporary critique of historical materialism. Power, property and the state* (Vol. 1). Berkeley CA: University of California Press.

Giddens, A. (1983). *Profiles and critiques in social theory*. Berkeley CA: University of California Press.

Ginsburg, F., Abu-Lughod, L., & Larkin, B. (2002). *Media worlds: Anthropology on New Terrain*. Berkeley CA: University of California Press.

Gluck, F. W., Kaufman, S. P., & Walleck, A. S. (1980). Strategic management for competitive advantage. *Harvard Business Review, 58*(4), 154–161.

Goodchild, M. F. (2004). GIScience, geography, form and process. *Annals of the Association of American Geographers, 94*(4), 709–714.

Goodwin, M., Duncan, S., & Halford, S. (1994). Regulation theory, the local state and the transition of urban politics. *Environment and Planning D, 11*, 67–88.

Goodwin, M., & Painter, J. (1996). Local governance: The crisis of Fordism and the changing geographies of regulation. *Transactions of the Institute of British Geographers NS, 21*, 635–648.

Gormley, W. (1983). *The politics of public utility regulation*. Pittsburgh PA: University of Pittsburgh Press.

Gregory, D. (1978). *Ideology, science and human geography*. London: Hutchinson.

Gregory, D. (1982a). *Regional transformation and industrial revolution: A geography of the yorkshire woollen industry*. London: Macmillan.

Gregory, D. (1982b). Solid geometry: Notes on the recovery of spatial structure. In P. Gould & G. Olsson (Eds.), *A search for common ground* (pp. 187–219). London: Pion.

Gregory, K. J. (2000). *The changing nature of physical geography*. London: Arnold.

Greiner, L. (1972). *Evolution and revolution as organizations grow* (pp. 165–174). Watertown, Massachusetts.: Harvard Business Review.

Gren, M. (2003). Time geography matters. In J. May & N. Thrift (Eds.), *Timespace* (pp. 208–225). London: Routledge.

Griffin, D. R. (1998). Process philosophy. In E. Craig (Ed.), *Routledge encyclopaedia of philosophy*. London: Routledge.

Grunbaum, A. (1971). The meaning of time. In E. Freeman & W. Sellars (Eds.), *Basic issues in the philosophy of time* (pp. 195–228). Chicago IL: Open Court.

Hagerstrand, T. (1957). Migration and area. In D. Hannerberg, T. Hagerstrand, & B. Odeving (Eds.), *Migration in Sweden: A symposium* (p. 13). Sweden: Lund Studies in Geography, Series B.

Hagerstrand, T. (1970). Regional forecasting and social engineering. In M. Chisholm, A. E. Frey, & P. Haggett (Eds.), *Regional forecasting* (pp. 1–7). London: Butterworths.

Hagerstrand, T. (1973). The domain of human geography. In R. J. Chorley (Ed.), *Directions in geography* (pp. 69–87). London: Methuen.

Hagerstrand, T. (1984). Presence and absence: A look at conceptual choices and bodily necessities. *Regional Studies, 18*, 373–380.

Hagerstrand, T. (2004). The two vistas. *Geografiska Annaler B, 86*(4), 315–323.

Hammer, M., & Champy, J. (1993). *Reengineering the corporation: A manifesto for business revolution*. London: Nicholas Brealey Publishing.

Hannan, M., & Freeman, J. (1977). The population ecology of organizations. *American Journal of Sociology, 82*, 929–964.

Hannan, M., & Freeman, J. (1989). *Organisational ecology*. Cambridge MA: Harvard University Press.

Hartshorne, R. (1939). The nature of geography; a critical survey in the light of the past. *Annals of the Association of American Geographers, 29*(3), 173–658.

Hartshorne, R. (1958). The concept of geography as a science of space, from Kant and Humboldt to Hettner. *Annals of the Association of American Geographers, 48*(2), 97–108.

Harvey, D. (1967). Models of the evolution of spatial patterns in human geography. In R. J. Chorley & P. Haggett (Eds.), *Integrated models in geography* (pp. 549–608). London: Methuen.

Harvey, D. (1969). *Explanation in geography*. London: Edward Arnold.
Harvey, D. (1982). *The limits to capital*. Chicago IL: University of Chicago Press.
Harvey, D. (1989). *The condition of postmodernity*. Oxford: Blackwell.
Harvey, D. (1993). From space to place and back again: Reflections on the conditions of post-modernity. In J. Bird, B. Curtis, T. Putnam, G. Robertson, & L. Tickner (Eds.), *Mapping the futures: Local cultures, global change* (pp. 3–29). London: Routledge.
Hawkins, R. (1986). A road not taken: Sociology and the neglect of the automobile. *California Sociologist, 9*, 61–79.
Heidegger, M. (1985). *History of the concept of time*. Bloomington IN: Indiana University Press.
Heisenberg, W. (1976). The nature of elementary particles. *Physics Today, 29*, 32–39.
Henry, J. (1955). Homeostasis, society and evolution: A critique. *The Scientific Monthly, 81*, 300–309.
Herskovits, M. J. (1955). *Cultural anthropology*. New York: Alfred A. Knopf.
Hesse, M. (1999). Globalocal. On the significance of space for time, speed and mobility. In J. Beckmann (Ed.), *SPEED. A workshop on space, time and mobility* (pp. 31–39). The Danish Transport Council: Copenhagen.
Hetherington, K. (2003). Moderns as ancients: Time, space and the discourse of improvement. In J. May & N. Thrift (Eds.), *Timespace* (pp. 49–72). London: Routledge.
Hettner, A. (1905). *Das Wesen und die Methoden der Geografie* (pp. 545–614). XI: Geographische Zeitschrift.
Hodgson, G. M. (1993). *Economics and evolution: Bringing life back into economics*. Ann Arbor MI: University of Michigan Press.
Huggett, R., & Perkins, C. (2004). Landscape as form, process and meaning. In J. A. Matthews & D. T. Herbert (Eds.), *Unifying geography: Common heritage, Shared future* (pp. 224–239). London: Routledge.
Jackson, P., & Smith, S. (1984). *Exploring social geography*. London: Arnold.
James, P. E., & Jones, C. F. (1954). *American geography. Inventory and prospect*. Syracuse NY: Syracuse University Press.
Janelle, D. G. (1969). Spatial reorganization: A model and concept. *Annals of the Association of American Geographers, 59*(2), 348–364.
Jantsch, E. (1980). *The self-organizing universe*. Oxford: Pergamon.
Jayanthi, S., & Sinha, K. K. (1998). Innovation implementation in high technology manufacturing. *Chaos-theoretic empirical analysis Management, 16*, 471–494.
Jelinek, M., & Schoonhoven, C. (1990). *The innovation marathon*. New York: Blackwell.
Jessop, B. (1997a). *Globalization and the national state; reflections on a theme from Poulantzas*. New York: Colloquium on Miliband and Poulantzas in Retrospect and Prospect.
Jessop, B. (1997b). The entrepreneurial city; re-imaging localities, re-designing economic governance, or re-structuring capital? In N. Jewson & S. MacGregor (Eds.), *Transforming cities: Contested governance and new spatial divisions* (pp. 28–41). London: Routledge.
Johnston, R. J., Gregory, D., Pratt, G., & Watts, M. (2000). *The dictionary of human geography*. London: Blackwell.
Jones, C. O. (1977). *An introduction to the study of public policy*. Boston: Duxbury.
James, C. F., & Jones, P. E. (1954). Introduction. In C. F. James & P. E. Jones (Eds.), *American geography inventory and prospect* (pp. 2–18). Annals of the Association of American Geographers, Syracuse NY: Syracuse University Press.
Jung, C. (1964). *Man and his symbols*. St Louis MO: Turtleback Books.
Kansky, K. J. (1963). *Structure of transportation networks*. Department of Geography, Research Paper: University of Chicago. 84.
Kanter, R. M. (1988). When a thousand flowers bloom; structural, collective and social conditions for innovation in organizations. In B. Staw & L. Cummings (Eds.), *Research in organizational behavior* (Vol. 10). Greenwich CT: JAI Press.
Kaplan, C. (1996). *Questions of travel*. Durham NC: Duke University Press.
Kasperson, R. E., & Minghi, J. V. (1969). Process: Introduction. In R. E. Kasperson & J. V. Minghi (Eds.), *The structure of political geography* (pp. 195–210). Chicago IL: Aldine Publishing.

Kaufmann, V. (2002). *Re-thinking mobility: Contemporary sociology*. Aldershot: Ashgate.
Kennedy, P. (2007). Global transformations but local, 'bubble' lives: Taking a reality check on some globalization concepts. *Globalization, 4*(2), 267–282.
Kent, M. (2003). Space: Making room for space in physical geography. In S. Holloway, S. P. Rice, & G. Valentine (Eds.), *Key concepts in geography* (pp. 109–130). London: Sage.
Kettinger, W. J., Teng, J. T. C., & Cuha, S. (1996). Information architectural design in business processing reengineering. *Journal of Information Technology, 11*, 27–37.
Kickert, W. J. M., Klijn, E. H., & Koppenjan, J. F. M. (1997). *Managing complex networks: Strategies for the public sector*. London: Sage.
Kimberly, J., & Miles, R. (1980). *The organizational life-cycle*. San Francisco CA: Jossey Bass.
Kingdon, J. (1984). *Agendas, alternatives and public policies*. Boston MA: Little Brown.
Klijn, E-H. (1996). Analyzing and managing policy processes in complex networks; a theoretical examination of the concept policy network and its problems. *Administration and Society, 28*(1), 90–119.
Klijn, E.-H., & Snellen, I. (2009). Complexity theory and public administration. In G. R. Teisman, A. Van Buuren, & L. Gerrits (Eds.), *Managing complex governance systems* (pp. 17–36). London: Routledge.
Kobayashi, A. (1989). A critique of dialectical landscape. In A. Kobayashi & S. Mackenzie (Eds.), *Remaking human geography*. Boston: Hyman Unwin.
Kohlberg, L. (1969). Stage and sequence: The developmental approach to socialization. In D. A. Goslin (Ed.), *Handbook of socialization theory and research* (pp. 347–480). Rand McNally: San Francisco CA.
Koput, K. (1992). *Dynamics of innovative idea generation in organizations: Randomness and chaos in the development of a new medical device*. Unpublished Ph.D. dissertation, School of Business, University of California, Berkeley CA.
Kosik, K. (1976). *Dialectics of the Concrete*, Boston Studies in the Philosophy of Science, 52.
Lash, S., & Urry, J. (1994). *Economies of signs and space*. London: Sage.
Layder, D. (1981). *Structure, interaction and social theory*. London: Routledge and Kegan Paul.
Lazarus, R. J. (2009). Super wicked problems and climate change: Restraining the present to liberate the future. *Environmental Law and Policy Annual Review, 2010*.
Le Poidevin, R. (2003). *Travels in four dimensions*. Oxford: Oxford University Press.
Leach, E. R. (1954). *Political systems of highland Burma*. Cambridge MA: Harvard University Press.
Lee, M. E. (1997). From enlightenment to chaos; toward nonmodern social theory. In R. A. Eve, S. M. Horsfall, & F. Lee (Eds.), *Chaos, complexity and sociology: Myths, models and theories* (pp. 15–29). Thousand Oaks CA: Sage.
Levinson, D. J. (1978). *The seasons of a man's life*. New York: Knopf.
Levitt, B., & Naas, C. (1989). The lid on the garbage can: Institutional constraints on decision-making in the technical core of college text publishers. *Administrative Science Quarterly, 34*, 190–207.
Lindblom, C. E. (1965). *The intelligence of democracy*. New York: Free Press.
Lipson, M. (2007). A garbage can model of UN peacekeeping. *Global Governance, 13*, 79–97.
Lista, G. (1986). *Futurism and photography*. London: Merrell Publishers.
Lorange, P. (1980). *Corporate planning. An executive viewpoint*. Englewood Cliffs NJ: Prentice Hall.
Lovgren, E. (1956). The geographical mobility of labour. *Geografiska Annaler B, 38*, 344–394.
Lowi, T. (1964). American business, public policy and political theory. *World Politics, 16*, 677–715.
Lynch, M. (1993). *Scientific practice and ordinary action*. Cambridge: Cambridge University Press.
Lynd, R., & Lynd, H. (1937). *Middletown in transition*. New York: Harvest.
Lynn, L. H. (1982). *How Japan innovates: A comparison with the US in the case of oxygen steel-making*. Boulder CO: Westview Press.

Macey, J. (1992). Organizational design and political control of administrative agencies. *Journal of Law Economics and Organization, 8*, 93–110.
Mackenzie, K. D. (1986). *Organizational design: The organizational audit and analysis technology*. Norwood NJ: Ablex.
MacLeod, G., & Goodwin, M. (1999). Space, scale and state strategy; rethinking urban and regional governance. *Progress in Human Geography, 23*(4), 503–527.
Mandell, M. P. (Ed.). (2001). *Getting results through collaboration: networks and network structures for public policy and management*. Westport CT: Quorum Books.
Mantel, S. J. (2005). *Core concepts on project management*. New York: Wiley.
March, J. G., & Olsen, J. P. (1975). The university of the past; organization and learning under ambiguity. *European Journal of Political Research, 3*, 141–171.
March, J. G., & Olsen, J. P. (1976). *Ambiguity and choice in organizations*. Bergen: Universititetsforlaget.
March, J. G., & Olsen, J. P. (1983). Organizing political life: What administrative reorganization tells us about government. *The American Political Science Review, 77*(2), 281–296.
March, J. G., & Olsen, J. P. (1984). The new institutionalism: Organizational factors in political life. *The American Political Science Review, 78*(3), 734–749.
March, J. G., & Olsen, J. P. (1986). Garbage can models of decision-making in organisations. In J. G. March & R. Weissinger-Baylon (Eds.), *Ambiguity and command, organizational perspectives on military decision-making* (pp. 11–35). Marshfield MA: Pitman.
March, J. G., & Simon, H. A. (1958). *Organization*. New York: Wiley.
Markusen, A. (2003). Fuzzy concepts, scanty evidence, policy distance: The case for rigour and policy relevance in critical regional studies. *Regional Studies* 37(6 and 7), 701–717.
Martin, J. (1990). Deconstructing organizational TABOOS: The suppression of gender conflict in organizations. *Organization Science, 1*(4), 339–359.
Massey, D. (1993). Politics and space/time. In M. Keith & S. Pile (Eds.), *Place and politics of identity* (pp. 141–161). London: Routledge.
Masuch, M. (1985). Vicious cycles in organizations. *Administrative Science Quarterly, 30*(1), 14–33.
Masuch, M., & LaPotin, P. (1989). Beyond garbage cans: An A1 model of organizational choice. *Administrative Science Quarterly, 34*, 38–67.
Matthews, J. A., & Herbert, D. T. (2004). Unity in geography: Prospects for the discipline. In J. A. Matthews & D. T. Herbert (Eds.), *Unifying geography* (pp. 369–393). London: Routledge.
Matthews, J. A., & Herbert, D. T. (2008). *Geography: A very short introduction*. Oxford: Oxford University Press.
May, J., & Thrift, N. (2003). Introduction. In J. May & N. Thrift (Eds.), *Timespace* (pp. 1–46). London: Routledge.
Mayhew, D. (1974). *Congress: The electoral connection*. New Haven MA: Yale University Press.
McCubbins, M., Noll, R., & Weingast, B. (1989). Structure and process, politics and policy: Administrative arrangements and the political control of agencies. *Virginia Law Review, 75*, 431–482.
McKelvey, B. (1999). Complexity theory in organization science; seizing the promise or becoming a fad? *Emergence, 1*(1), 5–32.
Mead, G. H. (1932). *The philosophy of the present*. New York: Prometheus Books.
Medd, W., & Marvin, S. (2005). From the politics of urgency to the governance of preparedness: A research agenda on urban vulnerability. *Journal of Contingencies and Crisis Management, 13*(2), 44–49.
Meentemeyer, V. (1989). Geographical perspectives of space, time and scale. *Landscape Ecology, 3*(3–4), 163–173.
Meredith, J. R., & Mantel, S. J. (2000). *Project management: A managerial approach*. New York: Wiley.
Merton, R. (1968). *Social theory and social structure*. New York: Free Press.

Mihata, K. (1997). The persistence of 'emergence'. In R. A. Eve, S. Horsfall, & M. E. Lee (Eds.), *Chaos, complexity and sociology: Myths, models and theories* (pp. 30–38). Thousand Oaks, CA: Sage.

Minkowski, E. (1933). *Lived time: Phenomenological and psychopathological studies.* Evanston IL: Northwestern University Press.

Mintzberg, H., Raisinghani, D., & Theoret, A. (1976). The structure of 'unstructured' decision processes. *Administrative Science Quarterly, 21*(2), 246–275.

Moe, T. (1990). The politics of structural choice: Toward a theory of public bureaucracy. In O. Williamson (Ed.), *Organization theory: From Chester Barnard to the present and beyond.* New York: Oxford University Press.

Mol, A., & Law, J. (1994). Regions, networks and fluids: Anaemia and social topology. *Social Studies of Science, 24*, 641–671.

Mol, A. P. J., & Spaargaren, G. (2005). From additions and withdrawals to environmental flows. *Organization and Environment, 18*(1), 91–107.

Morgan, G. (1986). *Images of organization.* Newbury Park CA: Sage.

Morrill, R. L. (1965). *Migration and the spread and growth of urban settlement.* Sweden: Lund Studies in Geography, Series B. 26.

Morris, M. (1988). At Henry Parkes motel. *Cultural Studies, 2*, 1–47.

Mucciaroni, G. (1992). The garbage can model and the study of policy making: A critique. *Polity, XXIV, 3*, 459–482.

Muth, R. F. (1961). Rural urban land conversions. *Econometrica, 29*, 1–23.

Nagel, T. (1979). *Mortal questions.* Cambridge: Cambridge University Press.

Nelson, P. (1959). Migration, real income and information. *Journal of Regional Science, 1*(2), 43–74.

Nelson, R. R., & Winter, S. G. (1982). *An evolutionary theory of economics.* Cambridge MA: Harvard University Press.

Neo, B. S., & Chen, G. (2007). *Dynamic governance.* Singapore: World Scientific.

Nisbet, R. A. (1970). Developmentalism: A critical analysis. In J. McKinney & E. Tiryakin (Eds.), *Theoretical sociology: Perspectives and developments* (pp. 167–206). New York: Meredith.

Niskanen, W. (1971). *Bureaucracy and representative government.* Chicago IL: Aldine.

Noe, T. H., & Rebello, M. J. (2007). *Dynamic governance.* Oxford: Oxford University Finance Workshop.

Norgaard, R. B. (1984). Coevolutionary development potential. *Land Economics, 60*(2), 160–173.

O'Riordan, T. (2004). Towards sustainability. In J. A. Matthews & D. T. Herbert (Eds.), *Unifying geography: Common heritage, shared future* (pp. 117–143). London: Routledge.

Odim, E. P. (1971). *Fundamentals of ecology.* Philadelphia PA: W.B. Saunders.

Olsen, J. P. (2001). Garbage cans, new institutionalism, and the study of politics. *American Political Science Review, 95*(1), 191–198.

Olsson, G. (1965a). *Distance and human interaction: A bibliography and review.* Philadelphia: Regional Science Research Institute, Bibliography Series B.

Olsson, G. (1965b). Distance and human interaction: A migration study. *Geografiska Annaler Series B, 47*(1), 3–43.

Ostrom, E. (1990). *Governing the commons.* New York: Cambridge University Press.

Padgett, J. F. (1980). Managing garbage can hierarchies. *Administrative Science Quarterly, 25*, 583–602.

Parsons, R. (1951). *The social system.* New York: Free Press.

Pascoe, D. (2001). *Airspaces.* London: Reaktion.

Peters, F. E. (1967). *Greek philosophical terms: A historical lexicon.* New York: NYU Press.

Peuquet, D. J. (2002). *Representations of space and time.* New York: Guilford Press.

Pfeffer, J. (1981). *Power in organizations.* Boston MA: Pitman.

Philo, C. (1984). Reflections on Gunnar Olsson's contribution to the discourse of contemporary human geography. *Society and Space, 2*, 217–240.

Piaget, J. (1975). *The child's conception of the world.* Totowa NJ: Littlefield Adams.
Pierson, P. (1993). When effect becomes cause; policy feedback and political change. *World Politics, 45*(4), 595–628.
Pierson, P. (2004). *Politics in time.* Princeton NJ: Princeton University Press.
Platt, J. (1970). Hierarchical growth. *Bulletin of the Atomic Scientists*, 26(9), 2–4, 46–48.
Poole, M. S., & Roth, J. (1989). Decision development in small groups V: Test of a contingency model. *Human Communication Research, 15*(4), 549–589.
Popper, K. (1970). Normal science and its dangers. In I. Lakatos & A. Musgrave (Eds.), *Criticism and the growth of knowledge* (Vol. 4, pp. 51–8). Proceedings of the International Colloquium in the Philosophy of Science: Cambridge, London.
Popper, K. (1974). Karl Popper. In P. A. Schilpp (Ed.), *The philosophy of Karl Popper.* La Salle IL: Open Court Publishing.
Pred, A. (1983). Structuration and place: On the becoming of sense of place and structure of feeling. *Journal for the Theory of Social Behavior, 13*, 45–68.
Pred, A. (1984a). Place as historically contingent process: Structuration and the time-geography of becoming places. *Annals of the Association of American Geographers, 74*(2), 279–297.
Pred, A. (1984b). Structuration, biography formation and knowledge: Observations on port growth during the late mercantile period. *Society and Space, 2*, 251–275.
Pred, A. (1985a). *Place, practice and structure: social and spatial transformation in Southern Sweden, 1750–1850.* Cambridge: Polity Press.
Pred, A. (1985b). The social becomes the spatial, the spatial becomes the social-enclosures, social change and the becoming of places in the Swedish province of Skane. In D. Gregory & J. Urry (Eds.), *Social relations and spatial structures* (pp. 337–365). London: Macmillan.
Pred, A. (1986). Power, practice and consciousness. *Svensk Geografisk Arsbok, 62*, 47–60.
Pred, A. (1996). Interfusions: Consumption, identity and the practices and power relations of everyday life. *Environment and Planning A, 28*(1), 11–24.
Prigogine, I. (1980). *From being to becoming: Time and complexity in the physical sciences.* San Francisco CA: Freeman.
Prigogine, I. (1997). *The end of certainty.* New York: Free Press.
Prigogine, I., & Stengers, S. (1984). *Order out of chaos.* New York: Heinemann.
Quinn, J. B. (1980). *Strategies for change: Logical incrementalism.* Homewood IL: Irwin.
Quinn, J. B. (1985). *Managing innovations: Controlled chaos* (pp. 73–84). Watertown, Massachusetts: Harvard Business Review.
Radaelli, C. M. (2000). Policy transfer in the European Union; institutional isomorphism as a source of legitimacy. *Governance, 13*(1), 25–43.
Raimon, R. L. (1962). Interstate migration and wage theory. *Review of Economics and Statistics, 54*, 428–438.
Reichenbach, H. (1958). *The philosophy of space and time.* London: Constable and Co Ltd.
Renwick, W. H. (1992). Equilibrium, disequilibrium, and nonequilibrium landforms in the landscape. *Geomorphology, 5*, 265–276.
Rescher, N. (1978). *Scientific progress.* Chichester: Wiley Blackwell.
Rescher, N. (1999). On situating process philosophy. *Process Studies, 28*(1/2), 37–42.
Rhoads, B. L. (2005). Process/form. In N. Castree, A. Rogers, & D. Sherman (Eds.), *Questioning geography* (pp. 131–150). Malden MA: Blackwell.
Richards, K., Bithell, M., & Bravo, M. (2004). Space, time and science. In J. A. Matthews & D. T. Herbert (Eds.), *Unifying geography* (pp. 327–352). London: Routledge.
Rifkin, J. (2000). *The age of access.* London: Penguin.
Riles, A. (2001). *The network inside out.* Ann Arbor MI: University of Michigan Press.
Rittel, H. W. J., & Webber, M. M. (1973). Dilemmas in a general theory of planning. *Policy Sciences, 4*, 155–169.
Robertson, R. (1992). *Globalization, social theory and global culture.* London: Sage.
Rojek, C., & Urry, J. (1997). Transformations of travel and theory. In C. Rojek & J. Urry (Eds.), *Touring cultures* (pp. 1–22). London: Routledge.

Roseman, C. C. (1971). Migration as a spatial and temporal process. *Annals of the Association of American Geographers, 61*(3), 589–598.
Rosenau, J. M. (1990). *Turbulence in world politics: A theory of change and continuity*. Princeton NJ: Princeton University Press.
Rosenthal, A. (1993). *The third house: Lobbyists and lobbying in the United States*. Washington DC: CQ Press.
Rostow, W. W. (1960). *The stages of economic growth*. Cambridge: Cambridge University Press.
Ruggles, R. (1998). The state of the notion: Knowledge management in practice. *California Management Review, 40*, 80–89.
Russell, B. (1918). The philosophy of logical atomism. *The Monist, 28*(4), 495–527.
Russell, B. (1926). *Our knowledge of the external world*. London: George Allen and Unwin.
Sabatier, P. A. (1988). An advocacy coalition framework of policy change and the role of policy-oriented learning therein. *Policy Sciences, 31*, 129–168.
Sabatier, P. A. (1991). Toward better theories of the policy process. *Political Science and Politics, 24*(2), 147–156.
Sabatier, P. A. (1998). The advocacy coalition framework; revisions and relevance for Europe. *Journal of European Public Policy, 5*, 98–130.
Sack, R. D. (1972). Geography, geometry and explanation. *Annals of the Association of American Geographers, 62*(1), 61–78.
Said, E. (1983). *The world, the text and the critic*. Cambridge MA: Harvard University Press.
Sailor, D. B. (1964). Moses and atomism. *Journal of the History of Ideas, 25*(1), 3–16.
Salisbury, R. (1986). Washington lobbyists: A collective portrait. In A. Cigler & B. Looms (Eds.), *Interest group politics* (pp. 146–161). Washington DC: CQ Press.
Sassen, S. (1994). *Cities in a world economy*. Thousand Oaks CA: Pine Forge.
Sauer, C. (1941). Foreword to historical geography. *Annals of the Association of American Geographers, 31*(1), 1–24.
Schaefer, F. K. (1953). Exceptionalism in geography :A methodological examination. *Annals of the Association of American Geographers, 43*(3), 226–249.
Schlager, E., & Blomquist, W. (1996). A comparison of three emerging theories of the policy process. *Political Research Quarterly, 49*(3), 651–672.
Schneider, A., & Ingram, H. (1998). Systematically pinching ideas: A comparative approach to policy design. *Journal of Public Policy, 8*(1), 61–80.
Schneider, M., Teske, P., & Minitrom, M. (1995). *Public entrepreneurs: Agents for change in American government*. Princeton NJ: Princeton University Press.
Scholte, J. A. (2000). *Globalization: A critical introduction*. Basingstoke: Macmillan.
Schumm, S. A., & Lichty, R. W. (1965). Time, space and causality in geomorphology. *American Journal of Science, 263*, 110–119.
Schumpeter, J. (1942). *Capitalism, socialism and democracy*. New York: Harper and Row.
Scott, B. R. (1971). *Stages of corporate development*. Boston MA: Harvard Business School.
Searle, J. (1983). *Intentionality. An essay in the philosophy of mind*. Cambridge: Cambridge University Press.
Seldadyo, H., Elhorst, J. P., & De Haan, J. (2010). Geography and governance: Does space matter? *Papers in Regional Science, 89*(3), 625–640.
Serres, M. (1995). *Angels: A modern myth*. Paris: Flammarion.
Sethi, J. D. (1985). Foreword. In R. Diwan & M. Lutz (Eds.), *Gandhian economics*. New Delhi: Gandhi Peace Foundation.
Sharkansky, I. (1968). *Spending in the American states*. Chicago IL: Rand McNally.
Sheller, M. (2003). *Consuming the Caribbean*. London: Routledge.
Sheller, M., & Urry, J. (2006). The new mobilities paradigm. *Environment and Planning A, 38*, 207–226.
Shields, R. (1997). Flow as a new paradigm. *Space and Culture, 1*, 1–4.
Shotter, J. (1983). 'Duality of structure' and 'intentionality' in an ecological psychology. *Journal for the Theory of Social Behaviour, 13*, 19–43.

Shrader-Frechette, K. (1977). Atomism in crisis: An analysis of the current high energy paradigm. *Philosophy of Science, 44*, 409–440.

Simon, H. A. (1976). From substantive to procedural rationality. In S. J. Latsis (Ed.), *Method and appraisal in economics* (pp. 129–148). Cambridge: Cambridge University Press.

Sims, P. (2003). Previous actors and current influences, trends and fashions in physical geography. In S. Trudgill & A. Roy (Eds.), *Contemporary meanings in physical geography, From What to why?* (pp. 3–24). London: Arnold.

Singh, J. V. (Ed.). (1990). *Organizational evolution: New directions*. Newbury Park CA: Sage.

Sjaastad, L. A. (1960). The relationship between migration and income in the United States. *Papers of the Regional Science Association, 6*(1), 37–64.

Slot, P. J., & Grabant, E. (1986). Extraterritoriality and jurisdiction. *Common Market Law Review, 23*, 545–565.

Snowden, D. (2005). From atomism to networks in social systems. *The Learning Organisation, 12*(6), 552–562.

Spaargaren, G., & Mol, A. P. J. (2008). Green global consumption: Redefining politics and authority. *Global Environmental Change, 18*, 350–359.

Stoddart, D. R. (1997). Richard J. Chorley and modern geography. In D. R. Stoddart (Ed.), *Process and form in geomorphology* (pp. 383–399). London: Routledge.

Strahler, A. N. (1952). Dynamic basis of geomorphology. *Bulletin of the Geological Society of America, 63*, 923–937.

Susskind, L., & Cruikshank, J. (1987). *Breaking Robert's rules: The new way to run your meeting, build consensus, and get results*. New York: Basic Books.

Sutton, R. (1999). *The policy process: An overview*, Overseas Development Institute, Working Paper, 118. London: ODI.

Swyngedouw, E. (1997). Neither global nor local: 'Glocalization' and the politics of scale. In K. Cox (Ed.), *Spaces of globalization, reasserting the power of the local* (pp. 138–166). Guilford: New York.

Sztompka, P. (1984). The global crisis and the effectiveness of the social system. In E. A. Tiryakian (Ed.), *International studies in sociology and social anthropology. the global crisis: sociological analyses and responses* (Vol. 34). Leiden: E.J. Brill.

Taafe, E., Morrill, R. L., & Gould, P. R. (1963). Transport development in underdeveloped countries. *Geographical Review, 53*, 503–529.

Takahashi, N. (1992). Decision-making by flight in Japanese firms. *Organizational Science, 26*(3), 21–32. (in Japanese).

Takahashi, N. (1997). A single garbage can model and the degree of anarchy in Japanese firms. *Human Relations, 50*(1), 91–108.

Taleb, N. N. (2007). *The black swan*. New York: Random House.

Taylor, G. (1937). *Environment, race and nation*. Toronto: University of Toronto Press.

Taylor, P. (2003). Time: From hegemonic change to everyday life. In S. Holloway, S. Rice, & G. Valentine (Eds.), *Key concepts in geography* (pp. 151–164). London: Sage.

Teisman, G., Gerrits, L., & Van Buuren, A. (2009a). An introduction to understanding and managing complex systems. In G. Teisman, A. Van Buuren, & L. Gerrits (Eds.), *Managing complex governance systems* (pp. 1–16). London: Routledge.

Teisman, G., Van Buuren, A., & Gerrits, L. (Eds.). (2009b). *Managing complex governance systems*. London: Routledge.

Thelen, K., & Steinmo, S. (1992). Historical institutionalism in comparative politics. In K. Thelen, S. Steinmo, & P. Longstreth (Eds.), *Structuring politics: Historical institutionalism in comparative analysis* (pp. 1–32). Cambridge: Cambridge University Press.

Thietart, R. A., & Forgues, B. (1995). Chaos theory and organization. *Organization Science, 6*(1), 19–31.

Thompson, J. D. (1967). *Organizations in action*. New York: McGraw-Hill.

Thornes, J. (1979). Introduction. In C. Embleton & J. Thornes (Eds.), *Process in geomorphology* (pp. 1–10). London: Edward Arnold.

Thrift, N. (1977). Time and theory in human geography. Part 1. *Progress in Human Geography, 1*(1), 65–101.
Thrift, N. (1983). On the determination of social action in space and time. *Society and Space, 1*, 23–57.
Thrift, N. (1996). *Spatial formations*. London: Sage.
Toulmin, S., & Goodfield, J. (1962). *The architecture of matter*. London: Hutchinson.
Touraine, A. (1977). *The self-production of society*. Chicago IL: University of Chicago Press.
Tushman, M. L., & Anderson, P. (1986). Technological discontinuities and organizational environments. *Administrative Science Quarterly, 31*, 439–465.
Unwin, T. (1992). *The place of geography*. Harlow: Longman.
Urry, J. (2000). *Sociology beyond societies*. London: Routledge.
Urry, J. (2003). *Global complexity*. Cambridge: Polity Press.
Urry, J. (2005). The complexities of the global. *Theory, Culture and Society, 22*(5), 235–254.
Urry, J. (2010). Mobile sociology. *The British Journal of Sociology, 61*, 347–366.
Van de Ven, A. H. (1992). Suggestions for studying strategy process: A research note. *Strategic Management Journal, 13*, 169–188.
Van de Ven, A. H., & Poole, M. S. (1988). Paradoxical requirements for a theory of organizational change. In R. Quinn & K. Cameron (Eds.), *Paradox and transformation: Toward a theory of change in organization and management* (pp. 19–63). New York: Harper Collins.
Van den Bergh, J. C. J. M., & Gowdy, J. M. (2000). Evolutionary theories in environmental and resource economics: Approaches and applications. *Environmental and Resource Economics, 17*, 37–57.
Van Gils, M., Gerrits, L., & Teisman, G. R. (2009). Non-linear dynamics in port systems. Change events at work. In G. R. Teisman, A. Van Buuren, & L. Gerrits (Eds.), *Managing complex governance systems* (pp. 76–96). London: Routledge.
Verstraete, G., & Cresswell, T. (Eds.). (2002). *Mobilizing place; placing mobility*. Amsterdam: Rodopi.
Vertovec, S. (1999). Conceiving and researching transnationalism. *Ethnic and Racial Studies, 22*, 447–462.
Virilio, P. (1975). *Bunker Archeologie*. Paris: Centre de Creation Industrielle.
Virilio, P. (1977). *L'Insecurite de Territoire*. Paris: Galilee.
Virilio, P. (1978). *Defense Populaire et Luttes Ecologique*. Paris: Galilee.
Virilio, P. (1980). *Esthetique de la Disparition*. Paris: Balland.
Virilio, P. (1983). *Pure War*. New York: Semiotext(e).
Virilio, P. (1984a). *Guerre et Cinema*. Logistique de la Perception, Paris: Editions de L'Etoile.
Virilio, P. (1984b). *L'espace critique*. Paris: Christian Bourgeois.
Virilio, P. (1984c). *L'Horizon Negatif.* Paris: Galilee.
Virilio, P. (1986). *Speed and politics*. New York: Semiotext(e).
Virilio, P. (1988). *La Machine de Vision*. Paris: Editions Galilee.
Virilio, P. (1997). *The open sky*. London: Verso.
Virilio, P. (2007). *The university of disaster*. Cambridge: Polity Press.
Vogt, E. Z. (1960). On the concepts of structure and process in cultural anthropology. *American Anthropologist NS, 62*(1), 18–33.
Von Bertalanffy, L. (1952). *Problems of life*. London: Watts and Co.
Walker, J. L. (1969). The diffusion of innovations among the American states. *American Political Science Review, 63*, 880–899.
Walt, G., & Gilson, L. (1994). Reforming the health sector in developing countries: The central role of policy analysis. *Health Policy and Planning, 9*(4), 353–370.
Webster, F. (2002). *Theories of the information society*. London: Routledge.
Weick, K. E. (1979). *The social psychology of organizing*. Reading MA: Addison-Wesley.
Wendel, B. (1953). *A migration schema; theories and observations*. Series B: Lund Studies in Geography. 9.
Whitehead, A. N. (1925). *Science and the modern world*. New York: Free Press.

Whitehead, A. N. (1929). *Process and reality: An Essay in cosmology*. New York: Free Press.

Whyte, L. L. (1944). *The next development in man*. London: The Cresset Press.

Whyte, L. L. (1954). *Accent on form*. Westport CO: Greenwood Press.

Williams, D. C. (1951). The myth of passage. *Journal of Philosophy, 48*(15), 457–472.

Williams, R. (1977). *Marxism and literature*. Oxford: Oxford University Press.

Wilson, J. Q. (1980). *The politics of regulation*. New York: Basic Books.

Winsborough, H. H. (1962). City growth and city structure. *Journal of Regional Science, 4*(2), 35–49.

Wolch, J. R., & De Verteuil, G. (2003). New landscapes of urban poverty management. In J. May & N. Thrift (Eds.), *Timespace* (pp. 149–168). London: Routledge.

Wood, B. D., & Waterman, R. (1991). The dynamics of political control of bureaucracy. *American Political Science Review, 85*, 801–828.

Wooldridge, S. W. W. (1958). The trend of geomorphology. *Transactions of the Institute of British Geographers, 25*, 29–35.

Zeeman, E. C. (1976). Catastrophe theory. *Scientific American, 234*, 65–83.

Zwart, P. J. (1973). The flow of time. In P. Suppes (Ed.), *Space, time and geometry* (pp. 131–156). Dordrecht: D. Reidel.

Chapter 5
Metaphor

Abstract The discussion of process and its significant relationship to meaningful and effective governance remains incomplete. Several issues that will help in the clarification of how governance processes can be improved can be identified and thereafter we can begin to move into others that have remained neglected by governance and policy-making in the maritime sector—flow and speed. Following that we can begin to draw together the threads of the discussion to see where maritime governance finds itself in the dynamic world of globalisation and pressures that exist that constrain and determine its effectiveness. In this chapter, we consider issues of process philosophy, process transfer, convergence and learning, the significance of metaphors in an understanding of process and particularly the role of nomadology, process complexity, dispersion and concentration, and the application of global fluids. Maritime governance may seem far from such issues but has a lot to take from the wider debate on dynamism and the concepts which characterise it, much as it might wish to hide away in the safety of commercialism and pragmatism.

> Whereas literal knowledge aspires to the inert status of information, metaphor works with indeterminacy to keep meaning safe from the final clarification that is its obituary. Meaning's play is not a game watched from the outside but one in which we live and throughout which we understand. We may fantasise about mastering literal knowledge, fixing it in our memories or reference books or filing cabinets, but metaphors in knowledges cannot be processed, always maintaining reserves of wisdom beyond our present understanding. When someone criticised the lack of likeness in Picasso's portrait of Gertrude Stein, Picasso advised the person to wait. In the same way, the meaning of rich metaphors keeps blooming; people think further by growing into them, awakening to their implications. Traditions of thought grow stale with the declining productivity of their key metaphors…
>
> Metaphoric activity is not the same as the culture's reality, but we are sceptical of the literal claim to re-present reality. Reality cannot really be seen, because we cannot see the world from the outside. Our knowledges are ours, mediated through us and projecting us into the world. We cannot fix or imitate the world as it really is. As Benjamin (1995) noted, 'Perhaps there is none of [man's] higher functions in which his mimetic faculty does not play a decisive role'… By letting us live (in) the world, metaphors enliven our understandings. Weber was too modest when claiming that the faculty for compassion or empathy lets us understand other people; it underlies all metaphoric truth.

M. Roe, *Maritime Governance*, DOI 10.1007/978-3-319-21747-5_5

> We do not come empty-handed to our performances and metaphors. When metaphor engages us, we respond through the emotions and memories that reverberate with the role.
>
> Game and Metcalfe (1996: 50–51) in Thrift (1999: 35–36).

The discussion of process and its significant relationship to meaningful and effective governance remains incomplete. Several issues that will help in the clarification of how governance processes can be improved can be identified and thereafter we can begin to move into others that have remained neglected by governance and policy-making in the maritime sector—flow and speed. Following that we can draw together the threads of the discussion to see where maritime governance finds itself in the dynamic world of globalisation and pressures that exist that constrain and determine its effectiveness.

In this chapter, we consider issues of process philosophy, process transfer, convergence and learning, the significance of metaphors in an understanding of process and particularly the role of nomadology, process complexity, dispersion and concentration, and the application of global fluids. Maritime governance may seem far from such issues but has a lot to take from the wider debate on dynamism and the concepts which characterise it, much as it might wish to hide away in the safety of commercialism and pragmatism.

Process Philosophy

> The point of philosophy is to start with something so simple as not to seem worth stating, and to end with something so paradoxical that no one will believe it. Bertrand Russell (1912: 514).

Process philosophy helps to underpin much of what else we have been considering in process and its relationship to governance. Hartshorne (1965: xiii) sees process philosophy as considering change as the 'successive becoming of events related to, but also differing from one another in some more or less abstract respects which interest us'. He sees change as the 'becoming of novelty' and process philosophy is central to this.

Meanwhile in the words of Browning (1965a: xxiii):

> Process philosophy is fundamentally a metaphysical position. Its basic doctrine is that the universe is essentially to be understood as creative, organic, and temporal. In this metaphysical soil a pragmatic epistemology and a teleological ethic seems to grow naturally, while a distinctive theology and philosophy of language may be easily nurtured.

The relationship of change to metaphor and its consequent role in process philosophy is considered by Huggett and Perkins (2004: 230). Process philosophy attempts to understand what change really means and considers such issues as what changes? Whether the items which are considered to have changed are part of a whole that has changed or whether they are separate and new items which have 'become' something else? Thus, as Hartshorne suggests, when the weather has changed from wet to dry, is there actually anything we can actually term 'weather' or is weather constituted by its wet and dry states? Process philosophy considers the change that has occurred rather than the items that change and the 'becoming' that this represents. It therefore helps to understand what change means in governance and what dynamic governance would attempt to accommodate—not some overall governance, nor individual items such as the environment or safety, but the becoming that is represented by the changes exhibited by each.

Chorley and Kennedy (1971: 251–252) note that there is far from universal agreement that 'becoming' is fundamentally different from 'being' finding that the distinction between what they term timeless and timebound changes impossible to sustain. They suggest that in this view, systems have a 'constant architecture in time' which is represented by a 'spiral of cause and effect' passing from 'becoming' (i.e. developing or evolving) at one integrative level of organisation, to 'being' (i.e. adopting a characteristic structure or morphology) at a higher level to 'behaving' (i.e. operating in a self-justifying, equilibrium manner) at a still higher level. Progressive integration inevitably leads to irreversible evolution.

Griffin (1998) considers that process philosophy refers to views where process or 'becoming' is always more fundamental than unchanged 'being'. He notes in particular the work of Samuel Alexander, Henri Bergson, John Dewey, Charles Hartshorne, William James, Lloyd Morgan, Charles Peirce and Alfred North Whitehead and roots in Hegel and Heraclitus. Whitehead (1967: 72, 1978: 41) in particular is seen as the most significant proponent suggesting that 'reality is the process' and 'an actual entity is process'. He sees the units that make up the universe as 'momentary' and 'occasions of experience'. Thus Whitehead (1978: 77) views:

> the most fundamental units of the world, the most fully actual entities, not as enduring individuals but as momentary events. Enduring individuals, such as electrons, molecules, and minds, are 'temporarily ordered societies' of these momentary events.

They are fundamentally temporal and as such cannot exist and exhibit the characteristics of a slice of time as time cannot be sliced, for this would demand momentary cessation of change which never occurs. However, for human convenience and better understanding, we allocate them a brief duration suggested by Bergson et al. (2002) to be less than a billionth to possibly a tenth of a second. In truth, these instants do not exist at all as everything is permanently in a process of becoming. In creating these artificial conditions of being the 'human intellect spatializes the universe… tends to ignore the fluency, and… analyze(s) the world in terms of static categories' (Whitehead 1967: 307).

Rescher (2000: 3) attempts to move the focus from Whitehead as process philosophy's prime promoter to the theory that lies behind it. Despite this, he begins by outlining Whitehead's fixation on process and that time and change are 'definitively central and salient metaphysical issues'. Temporality, historicity, change and passage are seen as 'fundamental facts to be reckoned with in our understanding of the world'. Whitehead (1929: 47) bases this on Leibnizian *appetition*—'the striving through which all things endeavour to bring new features to realization'. He sees the whole universe as characterised by an 'ongoingness', with generations of 'entities and occurrences succeeding one another without end'. Consequently 'nothing floats into the world from nowhere'. Process is movement and temporal change. Creativity provides an underlying structure which makes the movement intelligible (Eichenbaum and Gale 1971: 529).

Underlying this was Heraclitus and his notion that 'all things flow', where all was seen as strife and change (Platt 1970: 2). Harmony was the consequence of developing relationships between opposites and far from static, characterised by flux (Toulmin and Goodfield 1962: 47; Browning 1965b: 305, 306) and a rejection of a 'Parmenidean/Atomistic view that nature consists of the changeable interrelations amongst stable, unchanging units of existence'. Whilst change is everywhere there remains order within it (of which man is a central part) and nature reflects the constant process of interplay between the opposites that exist. From this emerges the concept of dialecticism which formed such a fundamental part of the early Soviet Communist interpretation of society:

> It is wise for those who hear, not me, but the universal reason, to confess that all things are one. This world, the same for all, neither any of the gods, nor any man has made, but it is always was, and is, and shall be a living fire, kindled in due measure and in due measure extinguished. Into the same river you could not step twice, for other waters are flowing. In change is rest. Craving and satiety. God is day and night, winter and summer, war and peace, plenty and want. For men to have whatever they wish, would not be well. Sickness makes health pleasant and good; hunger, satiety; weariness, rest. War is the father and king of all. It is hard to contend with passion, for whatever it craves it buys with life. Heraclitus, from Whyte (1944: 171–172).

> For there could be no harmony without sharps and flats, nor living beings without male or female, which are contraries. The harmony of the world is a harmony of oppositions, as in the case of the bow and the lyre. The unlike is joined together, and from differences results the most beautiful harmony, and all things take place by strife. Good and evil are the same. Unite whole and part, agreement and disagreement, accordant and discordant; from all comes one, and from one all. Heraclitus, from Whyte (1944: 172).

Despite living only two or three generations after Heraclitus, Plato presents an opposite view expressing the human demand for permanence in a sea of process (and change), something from which he never diverted in life as a 'pupil, lover, citizen and idealist' (Whyte 1944: 173–174). Taking on the Socratic ideal of moral and intellectual certainty, Plato does not reject the views of Heraclitus and process but restricts their application to the world of senses, contrasting them with a world of permanent ideas. This dualism was 'essential and permanent' but since only what was permanent could be granted 'real existence' the harmony of the 'ideal world was reality and the confusion of world of process was illusion' (Browning 1965b: 306). This approach was accepted for centuries with reality and its apparent permanence only ever questioned at times of severe emotional stress such as illness, love or impending death at which point the idea of impermanence (and hence an end to the stress) was bound to be appealing.

Meanwhile, Aristotle had the advantage of living after both Heraclitus and Plato. Whilst he accepted that change did take place and that motion was the key to understanding nature (Hagerstrand 2004: 315), his position was that anything that was immutable and hence immune to influence from other things, was superior to anything that changed or was dependent upon other things (Eichenbaum and Gale 1971: 529). Thus, process is relegated to second best and the classical view of the world begins to take shape characterised by stability and predictability, an ideal which was to last for some 2000 years.

Eichenbaum and Gale (1971: 529) also note the close relationship between the Buddhist approach of *dependent origination* with its clear association with relations and relativity and consequently process and becoming. 'What has an origin is relative to that origin; only what has always been as it is can be *absolute*, wholly independent of other things'.

Rescher (2000: 4) goes on to outline how processes were viewed from earliest times as important as well as 'persistent physical things'. He uses the idea of verbs rather than nouns—and the contrasting dynamic storms and heatwaves contrasting with dogs and oranges to make the point. Process philosophers will see the becoming as more important than the being with an understanding of reality only possible if the process of (say) maritime environmental degradation, seafarer safety deterioration or shipping economic inefficiency is appreciated as much as the existence of an environmental pollutant, death and injury or a liner cartel.

> The 'freshening' of the wind, the forming of waves in the water, the pounding of the surf, the erosion of the shoreline are all processes that are not really the machinations of identifiable 'things'. Consider such processes as 'a fluctuation in the Earth's magnetic field' and 'a weakening of the Sun's gravitational field'. Clearly such processes will make an impact on things (magnetic needles for example). But by no stretch of the imagination are these processes themselves the doings/activities of things/substances. There is not a thing 'magnetic field' or 'a gravitational field' that does something or performs certain actions - nor does the world or sun project such a field... As process philosophers see it, processes are basic and derivative because it takes a mental process (of separation) to extract 'things' from the blooming buzzing confusion of the world's physical processes. Traditional metaphysics sees processes... as the manifestation of dispositions... which must themselves be rooted in the stable properties of things. Process metaphysics involves an inversion of this perspective. It takes the line that the categorical properties of things are simply stable clusters of process-engendering dispositions. Rescher (2000: 7).

Perhaps Leonardo da Vinci (Undated but quoted in Richter 2008) can help us sum it all up:

> Nothing originates in a spot where there is no sentient, vegetable, and rational life; feathers grow upon birds and are changed every year; hairs grow upon animals and are changed every year, excepting some parts, like the hairs of beards of lions, cats, and their like. The grass grows in the fields and the leaves on the trees, and every year they are in great part renewed. So we might say that the earth has a spirit of growth; that its flesh is the soil, its bones the arrangement and connection of the rocks of which the mountains are composed, its cartilage the tufa, and its blood the springs of water. The pool of blood which lies around the heart is the ocean, and its breathing, and the increase or decrease of the blood in the pulses, is represented in the earth by the flow and ebb of the sea; and the heat of the spirit of the world is the fire which pervades the earth, and the seat of the vegetative soul is in the fires, which in many parts of the earth find vent in baths and mines of sulphur, and in volcanoes, and at Mount Etna in Sicily, and in many other places.

So where does that leave us in this consideration of process philosophy and its relationship with governance and specifically the maritime sector? Perhaps firstly it is important to recognise that the maritime sector is just one small component of society and as such it remains constrained, defined and stimulated by what surrounds it and the activities, ambitions and beliefs of the society it serves.

Emerson (1981) implies that change is a necessity of life and an appreciation of the dynamics needed for governance is no exception. Berry (1973: 8) continues by looking at the inadequacies of form as a driving force and the need to accommodate time within any systematic consideration of events and plans.

> To seek any fixed thing is to deal in false imagination, therefore, for all phenomenal existence is immediately also seen to be transitory when the dimension of time is added. No particular thing is 'real' in any absolute sense; it is passing into something else at every moment. Every individual, for example, is a progressively ageing, temporarily-organized 'bundle' of energy flows faced with ultimate disintegration.

Berry goes on to indicate that a continuous intellectual process is needed, recognising that every system and interpretation will always need reassessment in the light of what has happened; thus governance needs continuous reaffirmation if it is to be relevant. Static governance (maritime or otherwise) will fail. Man, as the central feature in governance design and application, must be viewed as:

> an information-processing, decision-making, cybernetic machine whose value systems are built up by feedback processes from his environment. These feedback processes are built into the most primitive forms of life, and they form a continuous spectrum all the way back through prehistory and to times when no life existed. Throughout this whole development of man's history, coming up through biological evolution and extending into cultural evolution, the essential message is one in which disorder or randomness is used to generate novelty, and natural selection then generates order. Van Potter (1971: 36).

Essentially, we are talking here of process metaphysics, derived from the work of the Ancient Greeks and as we noted earlier, the work of Heraclitus and his consideration of flows. The role of governance is to provide a basis for policy-making that is impacted by the environment within which it operates and the constant changes in the 'tidal waves of energy' that at times resemble stability but which in fact never cease changing (Toulmin and Goodfield 1962: 301). The world of maritime governance and policy-making should be looked at as a 'complex living system in which individuals, social groups and institutions are dynamically interrelated actors involved in continuous processes of decision-making' (Berry 1973: 9). The force fields that surround each of these actors determine their nature, purpose and meaning and equilibrium will only appear to exist when the forces are balanced; not because they are not there. Berry once more: 'in each case, of course, decisions are made in the relational context of perceived organization and structure, and processes set in motion by actions therefore reaffirm or reform the intrinsic self-organization of the system amidst the apparent disorder of myriad decisions and actions'.

Policy Transfer

> Each country has its problems, and each thinks that its problems are unique... However, problems that are unique to one country... are abnormal... confronted with a common problem, policy-makers in cities, regional governments and nations can learn from how their counterparts elsewhere responded. Rose (1991: 3).

> The most casual acquaintance with any important substantial area of regulation soon reveals that institutions and rules are widely imitated... Since regulation is typically begun under pressure of time, or in conditions of crisis, the incentive to imitate is great. The result is that 'early' regulators often provide a model for countries following later along the regulatory road.... (I)t is apparent that models emanating from countries exercising great economic and political power are most likely to be the objects of emulation. Hancher and Moran (1989: 285).

Policy transfer comes in many forms and the terminology used for the variety of processes that it represents can be confusing—policy convergence, policy diffusion, policy copying, policy learning.... All much the same thing with some local variation. For example, Radaelli (2000: 26) introduces us to the idea of policy transfer as outlined by Dolowitz and Marsh (1996) who defines it as:

> a process in which knowledge about policies, administrative arrangements, institutions and ideas in one political setting (past or present) is used in the development of policies, administrative arrangements, institutions and ideas in another policy setting.

It was originally developed to help understand the diffusion and adoption of policy in the USA (Walker 1969; Gray 1973) but has been adapted to look at learning (Rose 1993); convergence (Bennett 1991); democracy (O'Loughlin et al. 1998); a measure of transfer success (Dolowitz and Marsh 1996); political economy (Simmons and Elkins 2004); and institutional issues (Locke and Jacoby 1997). Rose (1991: 7) defines lesson drawing in the context of planning policy as 'an action-oriented conclusion about a programme or programmes in operation elsewhere'. This other programme was not spatially constrained in any way and could be in another region, city, nation or even global. The implication is always that the programme is not just reviewed but is always considered for application in some shape or form. He goes on to question why anyone should make comparisons in policy analysis:

> The major problems that face one government are often the same as those that face its neighbours... Although the existence of common or similar problems need not imply that every nation should or will respond in the same way, it does mean that each may draw lessons from the relevant experience of others. Rose (1988: 219).

Lesson drawing is distinctly different from the diffusion of public policy from innovation elsewhere to destination country, state or region (Walker 1969; Collier and Messick 1975; Savage 1985; Berry and Berry 1990). In diffusion studies, the focus is more on the adaptation by those receiving the new policy and the pattern of diffusion. Factors such as spatial propinquity, socio-economic resources and policy-maker characteristics are clearly important (Rose 1991: 9). Emphasis is upon diffusion sequence rather than content and form focusing on the nature of generators and absorbers rather than their characteristics.

Bennett (1991: 215) examines policy convergence which he sees as central to much of political science (and thus intrinsic to policy studies). He centres his definition of policy convergence on 'the tendency of societies to grow more alike, to develop similarities in structures, processes and performances' (Kerr 1983: 3). He identifies five key characteristics:

- A convergence of policy goals; a coming together of intent.
- It refers to policy content; convergence on statutes, rules, regulations, etc.

- Convergence on policy instruments—regulatory, administrative or judicial.
- Policy outcomes—including direct impacts or indirect consequences.
- Convergence on policy style; conflictual, consensual, incremental, reactive, etc. Bennett (1991: 218).

These may of course overlap, conflict and may or may not exist in each case.

Bennett and Howlett (1991: 275–277) continue this theme, also emphasising the relationship between policy learning and change and the wealth of research undertaken in particular between about 1972 and 1990 (Heclo 1974: 305; Walker 1974: 3; Hernes 1976; Weiss 1977a, b; Argyris and Schon 1978; Lynn 1978; Sabatier 1978; Lindblom and Cohen 1979; Etheredge 1981: 77–78; Nordlinger 1981; Hogwood and Peters 1983; Shrivastava 1983; Polsby 1984; Haas 1990; Bennett and Howlett 1991; Bennett 1991).

Evans and Davies (1999: 361) emphasise policy transfer's multidisciplinary characteristics whilst re-emphasising Dolowitz and Marsh's (1996: 344) definition of it as a voluntary process occurring as a result of free choices of political actors. Stone (1999: 52) agrees and suggests that policy transfer may include innovation, termination or convergence. Transfer objects might include policies, institutions, ideologies, attitudes, ideas or negative lessons (Dolowitz 1997). Those involved commonly include elected officials, political parties, civil servants, pressure groups, policy entrepreneurs/experts and supranational institutions (e.g. in particular the EU) (Dolowitz and Marsh 1996: 345).

Stone (2001: 2) considers policy transfer and lesson drawing as dynamic processes. They are not independent but linked closely with policy networks (Borzel 1998; Knoepfel and Kissling-Naf 1998) and shaped by these networks in the way they perform (Wolman 1992: 44).

Policy transfer is widely viewed as a process. Savage (1985: 6) quotes Farina and Kelly (1983: 25): 'clearly, the process of diffusion is a complex one'. There would be few who would argue with that from the maritime sector or elsewhere. Quite how policies emerge from experience elsewhere is thoroughly unclear even if it is equally as clear that they do. However, complex or not, diffusion is a significant feature. Bennett (1991: 217) in reviewing public policy convergence suggests that it occurs through a number of processes. Evans and Davies (1999: 367) for example see it as a process of policy change and understanding transfers requires the policy-maker to understand the underlying processes that are going on. Figure 5.1 outlines their interpretation of this process of transfer through a network of communications.

The concept of policy transfer, convergence, learning or any other of its variants is relatively recent yet it has emerged as a mainstream element of policy processes developing alongside greater interest in regional integration and in particular the emergence of the EU (Bennett 1991: 215). However, Stone (1999: 58) describes it as a 'coherent framework for thinking about what is an old practice' and 'one that was originally developed in the US as a means to explain policy adoption' in the federal system (Stone 2001: 3). It can be considered to be a consequence of industrialism which is commonly characterised by societal convergence

1 RECOGNITION Discourse Dissatisfaction Cyclical events Conflict Legitimation	2 SEARCH Regime International Transnational National Regional Local	3 CONTACT	4 EMERGENCE of information feeder network	5 COGNITION AND RECEPTION	6 EMERGENCE OF TRANSFER NETWORK
>>>>>>>>>>>>>>	>>>>>>>>>	>>>>>>>>>>>	>>>>>>>>>>>	>>>>>>>>>>>	>>>>>>>>>>>
7 ELITE and cognitive mobilization	8 INTER- ACTION	9 EVALUATION	10 DECISION enters policy stream	11 PROCESS	12 OUTCOME
>>>>>>>>>>>>>>	>>>>>>>>>	>>>>>>>>>>>	>>>>>>>>>>>	>>>>>>>>>>>	>>>>>>>>>>>

Fig. 5.1 The emergence and development of a voluntary transfer network. *Source* Evans and Davies (1999: 377)

Copying	Adoption more or less intact of a programme already in effect in another jurisdiction
Emulation	Adoption, with adjustment for different circumstances, of a programme already in effect in another jurisdiction
Hybridization	Combine elements of programmes from two different places
Synthesis	Combine familiar elements from programmes in effect in three or more different places
Inspiration	Programmes elsewhere used as intellectual stimulus for developing a novel programme without an analogue elsewhere

Fig. 5.2 Alternative ways of lesson drawing. *Source* Rose (1991: 22)

searching for an optimal arrangement, the rise of state planning, the increase in pragmatism at the expense of idealism and the 'levelling impact of technology'.

Rose (1991: 22) suggests that there are five ways in which drawing lessons from policies elsewhere can be applied (Fig. 5.2). Meanwhile, Dolowitz and Marsh (1996: 344) suggest it emerged with comparative political analysis in the 1960s starting with Walker's (1969) policy diffusion focussing in particular upon the causes of transfer and an explanation of its timing and likelihood. This slowly developed into transfer studies whereby politicians and planners began to examine the potential in policies used elsewhere and their adaptability to comparative situations.

Stone (2001: 4) emphasises its rapid development by noting the wide range of studies conducted since the 1980s and the variety of labels attached essentially to the same process. They include 'pinching ideas' (Schneider and Ingram 1988), 'external inducement' and 'policy band-wagoning' (Ikenberry 1990), 'lesson drawing' (Rose 1993), 'direct coercive transfer' (Dolowitz and Marsh 1996), 'policy borrowing' (Cox 1999), 'policy pushing' (Nedley 1999) and 'policy shopping'

(Freeman 1999). She settles on 'policy convergence' as it appears to stress the existence of 'structural forces' but the word 'semantics' also seems relevant here. However, there are undoubtedly some differences between each of the terms if only to a limited extent and the presence of so much activity focussed on one process suggests not just its existence but also its significance.

Maritime policy-making is no different in that it is centred around processes of policy formulation and application which have been adapted and borrowed from elsewhere in many circumstances. These include environmental policies for the control of emissions which have emerged from long debates more widely on climate change, maritime safety policies which are developments of industrial and technological innovation and research, security policies which have used experience gained recently from other transport sectors and more generic security debates, and economic efficiency policies which have a long history of policy learning, transfer and convergence from national and international economic policy-making. The degree of convergence and transfer is variable depending upon the specific nature of the issue (the policies on ballast water transfer for example, might have taken more or less from other policy sectors than policies on competition in the liner sector) and also the political and social context. Globalisation on the one hand has encouraged policy transfer and convergence by facilitating the cross-fertilisation of ideas through the development of governance regimes based on similar processes and structures but on the other has experienced difficulties as nation-states and shipowners see a true exchange of ideas and policies as threatening to their status (Bennett 1997: 214). Globalisation is not a requirement for transfer or convergence; but it helps (Stone 1999: 55).

Whatever terminology is used, these processes of policy transfer or convergence have advantages identified by Stone (1999: 53) and Evans and Davies (1999: 362). They include the fact that any search of policies and practices elsewhere and in particular in other countries and contexts must contribute to innovation and provide a stronger base than a local adjustment of policy taken from existing and familiar circumstances. As Schneider and Ingram (1988: 67) suggest:

> Cross national policy comparisons contribute to innovation. National governments are introverted and career officials identify with particular ministries. Unless the examples of other countries are brought to light through analysis, changes will be incremental.

Of course comparison does not necessarily mean adoption will occur, nor improvement of policies but the fact that this process has the potential to bring a range of changes from which improvements might be found must suggest it is the thing to do. In shipping, this is clearly the case. The international nature of the sector makes international policy convergence essential as national policies by definition would be narrow, introverted, lacking in experience and knowledge of circumstances elsewhere and ultimately conflictual permitting stakeholders to play national policies off against one another. The fact that this occurs already (cf. tonnage tax and flag-hopping) only goes to show that more convergence and transfer is needed. Thus, policy convergence and transfer gives nation-states the possibility of avoiding Scholte's (1996) 'methodological nationalism' which has increased in significance with the continued increase in the power of globalisation.

It is not all good news however; policy-makers need to ensure that transfer and convergence is not over-hasty and it is important to make comparisons across a variety of sources and contexts to ensure that the full range of opportunities and issues is understood. Some policy ideas are not transferable across context, jurisdiction or issue, and in this case, it may be detrimental even to try (Stone 1999: 54).

Policy convergence is seen by Bennett (1991: 219) as dynamic and relational and far from the static institutional assumptions of policy-making in the past—and which still characterises the maritime sector to an extent. Convergence he sees as a 'becoming' event rather than a condition of 'being' He quotes Inkeles (1999: 13–14): 'Convergence means moving from different positions towards some common point. To know that countries are alike tells us nothing about convergence. There must be movement over time toward some identified common point'. Convergence is a dynamic event which should be seen in terms of time rather than space (otherwise it is simply similarity).

Examples abound of policy transfer, convergence and the like. Crain (1966) who sees policy diffusion having its roots in primitive culture and so has always been with us provides an example applied to fluoridation in the USA. Rose (1991: 4) notes how the process goes on in employment policy in the EU (Carlson et al. 1986; Dommergues et al. 1989). Gray (1973: 1175) points out examples from rural sociology (the diffusion of hybrid seed corn amongst farmers) (Ryan and Gross 1943); medical sociology (drug adoptions by physicians) (Coleman et al. 1966); political science (the diffusion of city manager governments in the 48 US states) (McVoy 1940); and medicine (the contagion of a disease) (Bailey 1957; Rogers and Shoemaker 1971). Knoepfel and Kissling-Naf (1998: 344) note the rise in a 'continuous policy dialogue whereby the different actors introduce their views and reach a solution to the problem posed through an exchange of subjective ideas and transforming debate' and provide examples from Majone (1993), Sabatier (1993), Schon and Rein (1994), Fischer (1993), Fischer and Forester (1993), Singer (1993) and Nullmeier (1993). Coleman (1994) provides a discussion of banking and policy convergence whilst other commentators providing examples include Wolman (1992), Collier and Messick (1975), Eyestone (1977), Heclo (1974) and Savage (1985: 2–3) who cites Rogers (2003) in his identification of nine traditional areas for policy diffusion—rural sociology, communication, education, marketing, general sociology, public health and medical sociology, anthropology, geography and early sociology.

More specific in terms of governance is Collier and Messick's (1975: 1306) consideration of hierarchical diffusion where innovation occurs in the higher authorities and then filters down to the smaller and less advanced units—much as planned in the maritime sector. Hierarchical diffusion can be seen in examples from many sectors—radio stations, trolley cars, electronic technology, policy innovation in federal states, fire brigade operation and Rotary Club organisation in Chile are just some that have been cited. Hierarchical diffusion sounds logical—larger and more developed units have more resources; they may have access to more information; and smaller units tend to exhibit 'positioning behaviour' whereby they adapt to the desires of the higher units. All this suggests that the maritime governance model should work, but it may just be a reflection of a model of governance that in theory is unquestionable but in practice will always exhibit significant problems if the jurisdictional underpinning is inadequate. Meanwhile, Knoepfel and Kissling-Naf (1998: 347) emphasise the rise of network governance as a replacement for market and hierarchy which in turn stresses the importance of the exchange of ideas (O'Toole 1993: 53; Scharpf 1993: 125) something that fits well with the concept of policy transfer and convergence.

Evans and Davies (1999: 362) concur identifying the jurisdictional hierarchy as an 'essential element in policy transfer, that transfer can occur across all and any of the jurisdictional boundaries, and it is something that must be considered if it is to be understood adequately'. They continue by pulling together this discussion of policy transfer, convergence, learning and an abundance of the other categories by suggesting that as yet it fails to provide a full explanatory or theoretical basis for understanding the processes that underlie changes in policy but even so it is clearly fundamental to producing both adequate governance and polices. Gregor (1971: 193–194) sees policy transfer as a sustained metaphor or 'promissory note' for theories which have yet to emerge. As such, they are significant but remain undeveloped. Maritime governance might do well to accommodate this policy learning process as much as it can if it is to be effective and relevant.

Metaphor

> Invention flags, his Brain goes muddy,
> And Black Despair succeeds brown study.
>
> Congreve (1923: volume IV, 60).

> Vimes shrugged. 'That's it then,' he said, and turned away. 'Throw the book at him, Carrot.'
> 'Right, sir.'
> Vimes remembered too late.
> Dwarfs have trouble with metaphors.
> They also have a very good aim.
> *The Law and Ordinances of Ankh and Morpark* caught the secretary on the forehead. He blinked, staggered, and stepped backwards.
> It was the longest step he ever took. For one thing, it lasted the rest of his life. After several seconds they heard him hit, five storeys below.
>
> Pratchett (1989: 388).

> Thrown into the vast open sea with no navigation charts and the marker buoys sunk and barely visible, we have only two choices left: we may rejoice in the breath-taking vistas of new discoveries - or we tremble out of fear of drowning. One option not really realistic

> is to claim sanctuary in a safe harbour; one could bet what seems to be a tranquil haven today may soon be modernized, and a theme park, amusement promenade or crowded marina will replace the sedate boat sheds. Bauman (1998: 85).

> Virtually all statements – from quantitative descriptions of empirical data to the most vigorous mathematical formulations – are 'metaphors' which makes semiotics the queen of sciences. Artigiani (1987: 250).

Speaking of metaphors we now reach a core theme of governance and the need for change in the maritime sector and its approach to policy-making. Metaphors have been a central focus of discussion for as long as discussion of any issue has taken place (see for example McCloskey 1964; Weimar 1966; Sachs 1978; Ortony 1979; Livingstone and Harrison 1981: 96; Alvesson 1993; Cameron and Low 1999; Edwards et al. 2004). Traditionally, they have been drawn from fields of movement, and shipping has played a major role in this along with a variety of other transport and distribution-related examples including those cited by Edwards et al. focussing upon traffic congestion and movement (2004: 24, 31–32) and Law (1999: 573) notes the use of metaphors of mobility in social theory including those by Matless (1995), Wolff (1995), Hanson and Pratt (1995), McDowell (1996) and Cresswell (1997). In each case, the significance of change is obvious and consequently we can trace a close relationship between metaphor, movement and process which can help to indicate the importance of accommodating dynamism into the policy-making and governance areas. However, care needs to be taken:

> Metaphors are invaluable at the start of an inquiry. Ideas often come to us first as metaphors, a vague notion that x is like y. These can be creative insights; they may revolutionize a field. But they should not remain as they are born, for as scientific formulations, metaphors are flawed (Landau 1972). Bendor et al. (2001: 188).

Before we can begin to look closely at the contribution metaphor can make in the development of dynamic governance we must first be sure of what we mean by the term although we need to be careful as 'there is always a real possibility that one might drown in attempting to achieve hermetic sealing' (Edwards et al. 2004: 27). There are many definitions to choose from—take for example those provided by Livingstone and Harrison (1981: 95)—'a word which is usually applied to one sort of thing [is applied] to another sort of thing'—and Mills (1982: 237):

> It would be an error… to regard every metaphor as an explicitly formulated analogy, in which the words of comparison 'like', 'as' and so on, are omitted. This presupposes that the literal truth precedes the metaphor, which is thus always a conscious transference of the properties of one thing to another. But history shows that metaphors are generally older than expressed analogies… Metaphors may thus be viewed as expressing the vague and confused but primal perception of identity, which subsequent processes of discrimination transform into a conscious and expressive analogy between different things, and to which further reflection transforms into the clear assertion of an identity of common element (or relation) which the two different things possess. Vico (1968).

However, Morgan (1998: 4) provides us with all we need at this stage. A metaphor:

> … is a primal force through which humans create meaning by using one element of experience to understand another.

Morgan sees it as the primary tool for 'creating an understanding about what we now recognise as organisation and management' and as such has a central role in the development and understanding of governance in all fields, including shipping. Organisation, management and governance are themselves metaphors in that they are not 'shipping' itself but a way of viewing shipping so that understanding the process can be that much simpler and easier. It is a 'formative influence on language, on science, on how we think, how we see and how we express ourselves' (Morgan 1998: 5). As such, it can be incredibly valuable. Sfard (1998: 4) agrees: 'metaphors are the most primitive, most elusive and yet amazingly informative objects of analysis. Their special power stems from the fact that they often cross the borders between the spontaneous and the scientific, between the intuitive and the formal'.

Edwards et al. (2004) suggest that 'the essence of metaphor is understanding and experiencing one kind of thing in terms of another' (Lakoff and Johnson 1980: 5), and there are four ways of achieving this—metaphor, metonymy, simile and synecdoche. The first two involve replacement where 'one concept or idea is conceived in terms of another' (Edwards et al. 2004: 25). By so-doing, thinking is freed up and new perceptions of problems and issues can be derived. Movement metaphors and similes are very common and their application to an understanding of policies and their inadequacies is often derived through maritime scenarios and also can be applied readily to the maritime sector. Ships, the sea, safe harbours, storms, along with tourism, hotels, motels, vagabonds and nomads are common examples.

Meanwhile, metonymy and synecdoche require substitution as 'one idea or concept stands in for another with which it is related' (Edwards et al. 2004: 25). Nash (1989: 122) notes that metonymy substitutes 'a particular instance, property, characteristic or association for the general principle or function'. Thus, 'walking the plank' provides a representation of threat used to indicate the need for discipline.

Synecdoche is similar but substitutes a whole with a part or as Fraser (1979: 175) rather grandly puts it 'the substitution of one term for another within a predetermined hierarchy'. Thus, a captain of a ship may in an emergency do a 'headcount' of crew where heads refers to seafarers.

Metaphors are also frequently spatially characterised with clear relationship to the issues of maritime governance where space (and jurisdiction) remains a paramount issue. Paechter (2004) provides a typology:

- Area space—concerned with the drawing of boundaries; for example the field of study for maritime business, economics or finance.
- Movement through space—concerned with how for example, a seafarer moves through his or her career.
- Structural space—concerned with the foundations of the maritime sector and its policy-making institutions.
- Hierarchical space—concerned with assessment and attainment, for example top of his/her maritime career or profession.

- Distance space—concerned with institutional and individual relationships—for example the port/office—ship interface.

Metaphors clearly will cross categories such as these generating a 'metaphorical complex'.

Punter (2007: 2–3) provides numerous examples of the significance of metaphors in our daily lives. Take for example:

> a firm of London builders from Indian backgrounds, who some years ago painted on the side of their van the slogan: 'You've tried the cowboys: now try the Indians'. Punter (2007: 8)

He sees a metaphor operating 'by means of which one thing is made to stand in for another'. Metaphors are viewed as processes at work everywhere in language and by association, in organisation and activity as well including actions, thoughts, policies and ultimately in our case governance. It is inevitable in engagement and thus an essential part of the governance process.

As we have seen already, governance is a complex issue whose inadequacies are not resolved simply by laws and regulations. A much more complex design accommodating inference and opinion, social communication and political sensitivity is needed alongside. Urry (2004a) points out this complexity and its relationship to metaphor suggesting that it provides 'metaphors, concepts and theories essential for analysing' the disorderliness described by Gray (2001) as intractable. He also comments that existing processes for examining these complex systems are inadequate and lack the necessary characteristics to understand their complexity, richness and nonlinearity 'involving multiple negative and positive feedback loops'. Although referring directly to global systems, his comments could just as easily be applied to maritime governance—characterised by unpredictability and irreversibility, lacking equilibrium and order—or in the words of Prigogine (1997), 'pools of order that heighten overall disorder'. We return to complexity at a later stage.

Not everyone is convinced however. Bicchieri (1988: 102) for example suggests that:

> literal language is the only vehicle for expressing meaning and making truth claims, and metaphor is a deviant use of words other than in their proper places... figurative discourse is used only for rhetorical purposes or stylish embellishment; metaphor is denied any autonomous cognitive content.

Thus, metaphorical language is seen merely as ornamental rather than a way of better understanding the real world, including its role in policy-making. Taylor (1984) agrees suggesting that metaphors are seductively reductionist and the antithesis of the move towards accountability, quality control and common cores, themes central to maritime policy-making and governance from the 1980s onwards. Edwards et al. (2004: 27) point out criticisms of the use of metaphors particularly in an era of 'accountability' and 'quality control' as cited by Taylor (1984: 11). Cresswell (1997: 331) also notes David Harvey's comments quoted in Barnes and Duncan (1992: 10) that metaphors can 'hinder objective judgement', something that reflects the view of Locke quoted in Lakoff and Johnson (1980: 191) that:

> all the artificial and figurative application of words eloquence hath invented, are for nothing else but to insinuate wrong ideas, move the passions, and thereby mislead the judgement.

Urry (2000a: 26) points out the relationship between metaphor and globalisation, and the significance of which in our discussion of governance is clear. He sees that they have become socially powerful, and employed as 'the figurative use of exemplars, or icons or characteristics of mobility'. He suggests that there is no single globe but rather:

> different metaphors of the globe and globality. Central to notions of globalisation are various metaphors of the global which embody alternative presumptions of homogenisation/h etereogenisation, of simplicity/complexity, of movement/stasis, of inclusion/exclusion and so on.

Metaphors therefore have a significant role to play in understanding the processes of globalisation and the forces that are buffeting governance and their importance more generally lends substance to the concept—see for example the growth in interest in geography (Mills 1982) citing Weimar (1966), Kolodny (1973), Tuan (1978), Livingstone and Harrison (1980, 1981), Sitwell (1981), and Edwards et al. (2004: 152, 155) and particularly their metaphorical interpretation of globalisation as a 'moving practice' requiring an 'interpretative process'. Maritime governance is essentially global and consequently is particularly appropriate for a metaphorical interpretation. As we shall see in the coming sections, this has been a common approach with extensive use of movement metaphors (through nomads, hotels, motels, tourism and the like) across a range of issues and which might be particularly suitable to accommodate the process element of governance so sorely missing.

Urry (2000a: 27) provides some broad examples of application. These include the use of the sea, river, flux, waves and liquidity (Bachelard 1942), and the vagabond, nomad, pilgrim and motel (Deleuze and Guattari 1986; Braidotti 1994). The emphasis is on 'fluidity' or as Derrida (1987: 27) put it: '*Différance* is incompatible with the static, synchronic, taxonomic, ahistoric motifs in the concept of structure', something that rings particularly true in the context of our consideration of governance, process and form although this is dependent upon whether one associates the sea with freedom and space or with danger, illness and insecurity.

Clifford (1997: 3) emphasises the role of travel metaphors in the interpretation and understanding of culture which can also include governance and policy in its remit. This follows on from Buttimer's (1982: 91) understanding of root metaphors—forming, mechanism, organicism and contextualism—which could be used as the basis for analysing policy and governance. The latter is essentially culturally driven with the outcomes (policies, laws, regulations) merely the final expression of a long process of discussion dictated by cultural norms, which despite globalisation remain largely locally determined. Clifford suggests that although social existence is centred in circumscribed places—roots, gardens, dwellings—travel and all its related mobile facets have grown to be a:

> complex and pervasive spectrum of human experiences. Practices of displacement might emerge as constitutive of cultural meanings rather than as there simple transfer or extension… Cultural centers, discrete regions and territories, do not exist prior to contacts, but are sustained through them, appropriating and disciplining the restless movement of people and things.

Thus, maritime policies and governance, associated as they are with nation-states and their territories, are also influenced by this restless movement that defines the territories themselves—a succession of processes which can be interpreted through a variety of metaphors. This significance of movement in understanding society and with it the governance it needs has not been lost elsewhere. Cresswell (1997: 361) for example quotes Said (1994: 403):

> … surely it is one of the unhappiest characteristics of the age to have produced more refugees, migrants, displaced persons and exiles than ever before in history, most of them as an accompaniment to and, ironically enough, as afterthoughts of the post-colonial and imperial conflicts. As the struggle for independence produced new states and new boundaries, it also produced homeless wanderers, nomads, vagrants, unassimilated to the emerging structures of institutional power, rejected by the established order for their intransigence and obdurate rebelliousness.

He goes on:

> … it is no exaggeration to say that liberation as an intellectual mission, born in the resistance and opposition to the confinements and ravages of imperialism, has now shifted from the settled, established, and domesticated dynamics of culture to the unhoused, decentred and exilic energies, energies whose incarnation today is the migrant, and whose consciousness is that of the intellectual and artist in exile… And while it would be the rankest Panglossian dishonesty to say that the bravura performances of the intellectual exile and the miseries of the displaced person or refugee are the same, it is possible, I think, to regard the intellectual as first distilling them articulating the predicaments that disfigure modernity – mass deportation, imprisonment, population transfer, collective dispossession, and forced immigrations.

The relationship between movement, change, process and governance is clear and made clearer through the application of metaphor. Peters (1999: 31) quotes Emerson (1981: 335): 'Everything good is on the highway', although not all metaphors relating to movement are happy ones—'Every ship is a romantic object except that we sail in' (327) reflecting his view that nothing is so strange as the normal; nothing so terrifying as that which we inhabit since home is the place of 'language, sleep, madness, dreams, beasts, sex' (4). Homelessness is thus 'cheerful', travel embraces 'rapid domestication'. In Henry Thoreau's (2007) terms, the art is to travel without leaving home. We shall return to such notions of nomadism soon.

There are many examples of the use of movement metaphors that show their significance. Most appropriate for our purposes of exploring maritime governance are those relating to ships and shipping. Urry (2000a: 29, 2001: 239) for example, cites Gilroy's (1993: 4) use of the ship as a 'living, micro-cultural, micro-political system in motion'. Using the black slave trade as an example:

> ships were the living means by which the points within that Atlantic world were joined. They were mobile elements that stood for the shifting spaces in between the fixed spaces that they connected. Accordingly they need to be thought of as cultural and political units… they were… a means to conduct political dissent and possibly a distinct mode of cultural production. Gilroy (1993: 16–17).

An alternative has been the use of the motel as a metaphor and its characteristics of constant movement, change and flux. Urry (2000a, 2001: 240–241) once again provides examples. Clifford (1997) is cited by Urry (2000a: 30) although initially dismissing the traditional hotel which had been selected on the basis of being away from home, movements outside and a shelter from the unexpected. Home would be the opposite typified by fixture and stasis. He sees hotels as nostalgic and gentlemanly and as such rather too staid. Motels meanwhile are a different concept:

> The motel has no real lobby, it is tied into the network of highways, it functions to relay people rather than to provide settings for coherent human subjects, it is consecrated to circulation and movement, and it demolishes the sense of place and locale. Urry (2000a: 30).

Or in the words of Morris (1988: 3–5): 'motels memorialize only movement, speed and perpetual circulation'. They 'can never be a true place' and they are distinguishable from each other only in 'a high-speed, empiricist flash'. Like an airport transit lounge, they represent neither arrival nor departure but 'pause' (Urry 2000a: 30).

Alternative metaphors have used vagabonds, particularly suitable in understanding process, movement and change in that a vagabond has little idea of how long he/she will stay wherever he/she is and in addition often has no choice in the matter (Urry 2001: 240). Destinations are frequently unknown and certainly uncertain and the only thing certain is his/her temporality. Bauman (1999: 240) suggests that the vagabond is 'pulled forward by hope untested, pushed from behind by hope frustrated… (he/she) is a pilgrim without a destination'. Space is unstructured with trails marked only by their own footprints. He/she 'structures the site he (*sic*) happens to occupy at the moment only to dismantle the structure again as he leaves. Each successive spacing is local and temporary—episodic'.

Bauman also talks of tourists as metaphors which he feels are a product of a postmodern interpretation. Tourists are like vagabonds in that they know that their stay is relatively short and similarly 'he has only his own biographical time to string together the places he visits; otherwise nothing orders them in this rather than another temporal fashion'. The tourist is thus in charge almost entirely of his life world, dictating the changes and processes that affect and are experienced by him. Bauman (1999: 241) says it all rather well:

> It is the tourist's aesthetic capacity – his or her curiosity, need of amusement, will and ability to live through novel, pleasurable, and pleasurably novel experiences – which appears to possess a nearly total freedom of spacing the tourist's life-world; the kind of freedom which the vagabond, who depends on the rough reality of the visited places for

> his livelihood and who may only act to avoid displeasure by escaping, can only dream of. The tourists pay for their freedom; the right to disregard native concerns and feelings, the right to spin their own web of meanings, they obtain in a commercial transaction... Like the vagabond the tourist is extra-territorial; but unlike the vagabond, he lives his extraterritoriality as a privilege, as independence, as the right to be free, free to choose; as a licence to restructure the world.

It is this freedom to restructure and the desire and need to do so that rings so true in contrast to the inadequacies of maritime governance where the whole idea of encouraging restructuring and the changes this implies is an anathema (Urry 2001: 240). The tourist (and vagabond) provides a contrasting metaphorical interpretation of the problems of governance and the need for a dynamic re-interpretation of the frameworks for policy-making that exist in the maritime world. To quote Bauman (1999: 241) further:

> One more feature unites the lives of the vagabond and the tourist. They both move through the spaces other people live in... like a theatre performance, the most dramatic and impressive of contacts are securely encased between the wings of the stage and between the rise and the fall of the curtain – inside the time and place designated for the 'suspension of disbelief' – and guaranteed not to leak through them and spill over.

In this sense, the tourist represents the inadequacies of current maritime governance, formulated in a confined space and time which takes no account of (and commonly wishes not to) the changes occurring all around: a static image within a dynamic context.

And that takes us to smells. Well it does not really but the application of perhaps the least fashionable of senses as a metaphor in movement does provide some interesting interpretations. Rojek and Urry (1997a, b: 8–9) introduce us to the idea through the work of Porteus (1985) and Stallybrass and White (1986: 139).

Smell was always the primary target of social reformers as it has a 'pervasive and invisible presence difficult to regulate' (Rojek and Urry 1997a, b: 8). Key issues in nineteenth-century urban redevelopment centred upon smells, sewers and rats, slum-dwellers, prostitutes and kneeling maids, etc. In turn these concepts drove wealthier city dwellers out to newly forming suburbs.

However, these (semi) tangible smelly issues have been incorporated within metaphor and combine with movement to form a distinctive mechanism for interpreting and understanding the need for change and policy dynamics. Bauman (1999: 24) expressed it well in considering the static modernity that was envisaged and its relationship to unwanted smells:

> Modernity declared war on smells. Scents had no room in the shiny temple of perfect order modernity set out to correct.

Bauman's argument is that smell is 'subversive' in that it is almost impossible to remove completely and control and thus any attempt to create a 'pure rational order of things' is doomed to failure (Rojek and Urry 1997a, b: 8). Bad smells can be seen as an incentive (if not a requirement) to keep on the move, to absorb and accept change and changing circumstances and they are unavoidable. Hence, the governance of anything needs to also accept change, to accommodate process, and to sustain the dynamic.

Finally in our consideration of the variety of movement metaphors that have been successfully applied to issues of process, change and dynamism, we can look at the use of what Cresswell (1997: 339) terms bodily secretions and in particular fluids. Kristeva (1982: 69) is clear in her consideration of normality and how it is relatively defined according to context with relevance to space and location policy:

> Why does corporeal waste, menstrual blood and excrement, or everything else that is assimilated to them, from nail-pairings to decay, represent – like a metaphor that would have become incarnate – the objective frailty of symbolic order?

Mol and Law (1994) take the characteristics of human blood and the system it inhabits to 'interrogate the diverse spatial forms of social life' (Urry 2000a: 30). Blood they see as not conforming to the 'structures and regions of conventional anatomy'. Blood remains a fluid that moves through almost all parts of the human body and consequently is both everywhere and nowhere as it never rests. There is no definite structure only a structure for the vessels through which it passes.

Using the specifics of blood, Mol and Law apply it as a metaphor for the 'diverse forms of social life' and in turn it can be applied to our case of maritime governance which in itself is nothing more than a structured form of social activity. They identify three metaphors of social and spatial topologies: regions, which are territorialities with clustered patterns and boundaries to confine them; networks, where the relative distance between the features making up the network define its characteristics; and regional boundaries commonly may be crossed. Meanwhile, fluids (e.g. blood) differ substantially from regions and networks in that:

> neither boundaries nor relations mark the difference between one place and another. Instead, sometimes boundaries come and go, allow leakage or disappear altogether, while relations transform themselves without fracture. Sometimes, then, social space behaves like a fluid. Mol and Law (1994: 643).

In maritime governance, the existing framework is characterised by regions and networks, territories and boundaries, manifesting themselves in the nation-state above all others. The nation-state is an anachronistic feature of governance and as such the use of regional and network structures themselves are anachronistic as well. Meanwhile, fluids in their continuous movement, cross-boundary features and inherent dynamism are more suited to the need of globalised governance that characterises the maritime sector. Hence, the blood metaphor is wholly appropriate. Taking the case of anaemia in Africa, where it is extremely common, they argue that:

> We're looking at *variation without boundaries* and *transformation without discontinuity*. We're looking at flows. The space with which we are dealing is *fluid*. Mol and Law (1994: 658, original emphasis).

How wholly appropriate for dynamic maritime governance, where boundaries have become progressively less relevant, discontinuities less significant, and the policy space which is considered, acts like a fluid. Anaemia is seen by Mol and Law as like blood:

> flowing in and out of different regions, across different borders, using diverse networks. It changes as it goes, although this is often in ways which are more or less imperceptible at the time. Anaemia as an illness is like a fluid, like blood, and is subject to transformation.

> Fluids are subject to mixtures and gradients with no necessarily clear boundaries. The objects generated may not be clearly defined. Normality is a gradient and not a clear absolute. In a fluid space it is not possible to determine identities nice and neatly, once and for all; nor to distinguish inside from outside. Various other fluids may not be able to combine together with each other. Urry (2000a: 31).

Mol and Law (1994: 660 and 664) sum it up:

> (A) fluid world is a world of mixtures.
>
> The study of fluids, then, will be a study of relations, repulsions and attractions which form a flow… as it moves, it changes its shape and character.

We shall return to the issue of fluids as metaphors for progress and change in maritime policies and governance under globalisation in the coming sections. Globalisation (and hence maritime policy and governance) is susceptible to 'interpretative processes' (Edwards et al. 2004: 155) and as such the value of metaphor is clear. Meanwhile, we can finish this section by quoting from Lury (1987: 90–91) who brings together a whole range of related issues. She considers the relationship of globalisation and culture and sees it as:

> a space of flows in which time-space compression occurs, in which objects and people are dissected by the cut'n'mix of boundary crossing and return, in which culture as technology refers back into and outside itself, creating environments by design, and objects come to take on new capacities. It is a space that which is not homogenous, but its heterogeneity is not unplanned; rather it is a space in which subjects and objects do not come face-to-face, but inter-face. The possibilities for tourism of this new space of flows are only just beginning to be explored, but they have the potential both to expand the kind of journeys possible – through the incorporation of time-space compression and the multiplication of perspectives – and to provide the basis for new kinds of hierarchy among both travelling people and objects.

Lury focuses upon a range of concepts that are relevant to our discussion of metaphor, process, change and governance—in particular spaces of flows (which we return to in a later chapter) and time-space compression which underlies much of the discussion on globalisation, jurisdiction and the problems of the maritime sector. Meanwhile, McCloskey (1964: 215) helps to place the whole issue of metaphor into context:

> We look for, example, from the top of a building at the busy street below: and say 'Ants!'. This is to call what other people and we ourselves at other times would call 'people', by another word 'ants'. The new application of the word is given by the behavioural or literary context. We either go up a building and look down together at the street; or I write and you read, 'Beneath me I saw the people coming and going, busy steams of ants hurrying backwards and forwards across the pavements.

Nomads

> When I first came to that quiet corner of the Nile Delta I had expected to find on that most ancient and most settled of soils a settled and restful people. I couldn't have been more wrong. The men of the village had all the busy restlessness of airline passengers in

> a transit lounge. Many of them had worked and travelled in the sheikdoms of the Persian Gulf, others had been in Libya and Jordan and Syria, some had been to the Yemen as soldiers, others to Saudi Arabia as pilgrims, a few had visited Europe: some of them had passports so thick they opened out like ink-blackened concertinas…
>
> And none of this was new; their grandparents and ancestors and relatives had travelled and migrated too, in much the same way as mine had, in the Indian sub-continent – because of wars, or for money and jobs, or perhaps simply because they got tired of living always in one place. You could read the history of this restlessness in the villagers' surnames; they had names that derived from cities in the Levant, from Turkey, from faraway towns in Nubia; it was as though people had drifted here from every corner of the Middle East. The wanderlust of its founders had been ploughed into the soil of the village; it seemed to me sometimes every man was a traveller. Ghosh (1986: 135).

> It's great to have roots as long as you can take them with you. Gertrude Stein.
>
> …nomadic consciousness is an epistemological and political imperative for critical thought at the end of this millennium. Braidotti (1994: 1).

Nomads provide perhaps the most evocative of all metaphors that derive from our perception of change and movement and which can help to understand the relationship between governance, process and movement, issues central to the failure of the maritime policy-making process in a time of extended globalisation. The concept of nomadology has been widely discussed (see for example Morris 1988; Sheller and Urry 2006: 210; Kennedy 2007: 272) and remains both fashionable and central to Western society (Peters 1999: 17–18).

> Otherness wanders through its center. Exile is, perhaps, the central story told in European civilisation: the human estate as exile from God, the Garden of Eden, the homeland, the womb, or even oneself. Thus Eisen (1986: xi) recasts the opening line of the Book of Genesis. 'In the beginning, there was exile'.

Or even more evocative Gay et al. (1996: 23–24) use the nomadic metaphor to bring us right up to date (or almost—just substitute 'iPhone' for 'Walkman' and you have it) suggesting the Sony Walkman is:

> virtually an extension of the skin. It is fitted, moulded, like so much else in modern consumer culture, to the body itself… It is designed for movement – for mobility, for people who are always out and about, for travelling light. It is part of the required equipment of the modern 'nomad'… it is testimony to the high value which the culture of late modernity places on mobility.

Hannam (2009: 101) takes Calhoun's (1995: 4–5) definition of nomadology as 'a way of socially constructing reality and asking new questions'. D'Andrea (2006: 107) adds:

> Nomadology refers to a style of critical thinking that seeks to expose and overcome the sedentary logic of the state, science and civilization… It denounces a categorical binary of civilization whereby the dweller is positively assessed over the wanderer, seen as menace, distortion and problem… The privilege of fixity over mobility – of roots over routes – hinges on the issue of conventional modes of subjectivity; a dialectic of identification/alterity sustains a model of identity that constrains the self within rigid and exclusionary boundaries.

Cresswell (2006: 43) also contributes to the identification of nomadology which focuses upon:

> movement, mobility, and contingent ordering, rather than upon stasis, structure and social order… (the) corporeal, imagined and virtual motilities of people.

with increasing focus upon the transnational, global, forms of governance.

Salo (1986: 7) discusses some problems in identifying nomads which in turn leads to questions about their value as a metaphor but nevertheless nomads themselves (rather than the theory of nomadology) retain a positive metaphorical image despite their rather dubious connotations at times (vagrants, tinkers, homeless, etc.). Cresswell (1997: 367) says it all:

> …the nomad becomes an exhilarating character. Everywhere the nomad goes new freedoms and opportunities follow. Modernist thought, the doomed search for Truth established through the mythical 'view from nowhere', is characterized by stabilities and certainties. The familiar dualism of man/woman; white/black; true/false are all tethered to the geography of here and there. The movements of the nomad, on the ground and in the head, cannot help but transgress such simplicities.

Rojek and Urry (1997a, b: 10) cite a number of examples of the use of nomads as metaphors by other authors: Braidotti (1994) and nomadic consciousness; Clifford (1992: 101) and hotel lobbies setting a time and place for nomadic intercourse; Morris's similar approach using motels; and Jokinen and Veijola (1994) who place the nomadic metaphor into a sexual context suggesting its dominant maleness. Kaplan (1996: 63) cites Eisenstein (1969) who described the early days of Soviet Russia as when cinema was 'discovered' as a 'place with unimaginably great possibilities'. 'We came like Bedouins' and pitched our tents' Kaplan sees Eisenstein's words as distinguishing the new medium from the traditional theatre using the metaphors of desert and Bedouin to emphasise 'values of freedom, hybridity and modernity'. Such nomadic clichés are commonly identified with progress, change and dynamism, values we see repeated in our understanding of the problems that besiege maritime governance.

Malkki (1992: 37) adds that nomadology rethinks identity as:

> Always mobile and processual, partly a condition, a status, a label, a weapon, a shield, a fund of memories, etc. It is a creolized aggregate composed through bricolage.

Despite this not everyone is in favour. For example:

> The feudal ownership of land did bring dignity, whereas the modern ownership of movables is reducing us again to a nomadic horde. Forster (1931: 141).

However, first to nomads themselves and the variety of interpretations that have been made.

De Lange (2009: 3–4) in an appreciation of the rise of digital nomadism quotes Peters (1999: 19–20) who suggests that:

> nomadism dispenses altogether with the idea of a fixed home or center. Whereas exile often occurs in relation to some looming authority figure who wields power over life and death, nomadism can involve active defiance of or furtive avoidance of the sedentary authority of the state and society (often to the peril of actual nomadic societies). If

> diaspora suggest a geographically dispersed network, the concept of nomadism suggests a face-to-face community, usually linked by ties of kinship stemming from a real or imagined common ancestor that travels as a unit… For nomads, home is always mobile. Hence there is a subtle doubleness here: being at home everywhere but lacking any fixed ground.

D'Andrea (2006: 108) puts it more simply: 'traditional nomadism can be defined as mobile household communities that carry their means of production within a single ecological niche' (Cribb 1991: 20; Rao 1987; Khazanov 1984). He goes on to emphasise that they can be divided into two main groups—those that move animals to better pastures and 'neo' nomads (hippie traders, smugglers, DJs, alternative therapists, tattoo artists, etc.) who exercise their skills along the way and use impermanence as a source of learning and charisma to their professional advantage (D'Andrea 2004; Rao 1987; McKay 1996). Nomadism is defined by both culture and economics, and 'wandering' is no part of their lives. Barfield (1993: 12) confirms that they both know where they are going and why, and possess both goods and tools. They merely remain unattached to land.

Zembylas and Vrasidas (2007: 66) look at how the metaphor of the nomad is useful in showing how ICT can be inclusive for marginalized people. They define a nomad as:

> constantly on the move, connecting with others, assuming heterogeneous identities, and celebrating plurality in contradiction to unitary models of Western thought that exclude certain populations. Nomads learn to live with the discomfort of uncertainty and multiplicity, and do not allow themselves to collapse their identities into that of 'global villagers' who are assumed to have identical and universal needs and desires.

As such, their value in understanding the problems and requirements of maritime governance are clear where the issues of plurality are inadequately addressed through the minimal acknowledgement of stakeholders and the focus on a unitary and inflexible approach to policy-making.

The origins of nomadology are outlined by De Lange (2009: 2) who stresses the significance of the work of Deleuze and Guattari (1988) and their discussion of the characteristics of nomadic life and how it can provide a metaphor for the relationship between the state and the 'war machine'. In so-doing, they hope to liberate thinking about identity. Their emphasis on the state and its fixed, static role is reminiscent of our earlier discussion with specific reference to the maritime sector, its narrowly defined stakeholders, institutionalism and governance that lacks dynamism. Consequently, the characteristics of nomadism provide a framework for the interpretation of the move that maritime governance needs to make from its current somnolence to becoming more zoetic. Lancaster and Lancaster (1998: 32) make this clear:

> Nomadism, with its flexible multi-resource economic strategy is ideally suited to the unpredictable environment… External factors such as trade routes, governments or states are grist to the mill; they are the necessary substrate. (Nomads) change because the economic and political climate changes, and nomadism is still the best method of adapting and surviving.

Peters (1999: 25) notes that nomad is of Greek origin relating to seemingly contrasting concepts of pasturing of animals (*nomas*) and law (*nomos*) and reflected the

Athenian view that community (*polis*) was an essential part of humanity (and consequently nomadism was inhuman). The current use of the word nomadism dates from 1841 and the work of Emerson and its use in connection with how the 'soul may exist in the world' (Emerson 1981; Peters 1999: 30). It retains its 'inhuman' characteristics reflected in the difficulties faced by modern-day travellers across Europe. Emerson rejects this view seeing homelessness as to be celebrated. His attitude is that of 'the tent-dweller for whom home is always already here. As his disciple Thoreau put it, the art is to travel without ever leaving home' (Peters 1999: 31).

Speaking of tents, in themselves serious contenders for a nomadic icon, they are also commonly seen to be a major part of the nomadic metaphor. Peters (1999: 24) provides some clarity. For example, the Bible puts it forward as a 'symbol of rule, a cosmic link with the heavens, and a redoubt against harm (Nibley 1966)'. Peters sees the tent as an 'enduring sign of both worldly authority and the flaunting thereof, and of mobility as part of the human estate' and as such it plays a central part in nomadic sociology.

> Abraham, Isaac and Jacob, the patriarchs who became strangers in a strange land, all dwelt in tents. The tabernacle bearing the Ark of the Covenant was clearly a type of tent, a mobile abode for God's presence. Long after the Jewish people traded a nomadic existence for the settled life of farming and cities, the pastoral life retained an ideological privilege. Even at the pinnacle of the Kingdom of Israel, King David boasted of his childhood as a shepherd… Isaiah compares the people of Israel to a tent (Isaiah 54:2 compare Isaiah 33:20). The sedentary emulation of mobile others is an enduring pattern: it continues in both social theory and social life today. Peters (1999: 24–25).

Peters goes on to elaborate upon the Jewish Festival of Tents and its relationship to space, place, process, change and movement and there may be something here that has relevance for maritime governance and its failure.

Nomadology relies upon the use of nomadic metaphor and plenty of examples exist. Peters (1999: 18) suggests a range of personalities:

> Abraham, the sojourner and stranger, never to return to his home; Odysseus, who finally returns to Penelope after his odyssey; Oedipus, an outcast from his city; the legend of the wandering Jew; *flâneurs*, loafers and Bohemians; gypsies, gypsy scholars, sea gypsies, and gypsy truckers; hoboes, tramps, drifters, vagabonds and flimflam artists; sociologists, private eyes, journalists, men and women of the street; sailors, soldiers of fortune, adventurers and explorers; border crossers of all sorts; gauchos, cowboys, and guerrilla fighters; pioneers, pilgrims and crusaders; knights errant, troubadours, minstrels, charlatans, and journeymen; Huns, Vandals, Goths, Mongols, Berbers, and Bedouins; tourists, travellers, *haji*, refugees, immigrants, the stateless and the homeless; commuters, telecommuters, jet-setters, migrant workers, and *Gastarbeiter*; automobilists, bikers, and circus people.

There is disagreement about those defined as nomadic although generally three types have been widely identified. Pastoral nomads: 'societies specializing in animal husbandry requiring periodic movements' (Barfield 1993: 4). Hunter-gatherers: whose mode of existence sets them apart from traditional pastoralists and sedentary farmers (Bogue 2004: 172). And gypsies, tinkers, weavers, mime artists, magicians, musicians, horse-dealers, circus performers, etc. Salo and Salo (1982: 276) describe these as 'spatially mobile peoples who exploit resources primarily in the social environment'.

Turning specifically to the use of nomads as metaphors and their relationship to nomadology, process and change we find a considerable body of opinion with some contrasting views. This includes Peters (1999: 18) who considers their role in Western metaphorical culture, and Wolff (1993: 224) who notes their proliferation during the early 1990s partly as a postmodern response but also their inherent sexism and favourable characterisation of male nomadism. Cresswell (2006: 43) approaches nomadology from the point of view of sociology but his comments are still helpful as they focus upon the contrast between 'movement, mobility, and contingent ordering, rather than upon stasis, structure and social order', something that sounds very familiar in our consideration of maritime governance. He goes on; this involves looking at 'corporeal, imagined and virtual mobilities of people' (Urry 2000a: 18) and the interactions between people and objects and increasing importance of 'transnational, global, forms of governance' (Cresswell 2006: 43).

Bauman (1998: 240) is critical of the use of nomads as a metaphor for contemporary society in Postmodern times. He considered nomads as circling 'around a well-structured territory with long invested and stable meaning assigned to each fragment'. Thus is far from the conventional image of Postmodernism (see for example Roe 2013 in a maritime context) although he did accept that in this 'fluid stage of modernity' the settled majority is in some ways 'ruled by the nomadic and extraterritorial elite' (Bauman 2000: 13). They lack a final destination which guides their itinerary, not specific and privileged places compared with which all other places and movements are subservient. They 'move from place to place in a strictly regular succession following the order of things rather than composing that order as they move in and dismantling it again as they move out'. As such, their use as a metaphor for dynamism in a Postmodern world can be questioned as inappropriate.

Bogue (2004: 170) meanwhile notes Miller's (1993: 11–12) criticism of nomadology and the contribution of Deleuze and Guattari (1986). Miller suggests that the realities of nomadism are ignored, that selected elements from 'scattered anthropological sources' have tended to romanticise their image creating a process of 'pseudo-colonial subjugation'. Bogue refutes much of Miller's arguments suggesting that Deleuze and Guattari actually made none of these claims which have derived only from their many and varied interpreters; however, the debate remains unresolved.

Bauman (1998: 240), cited in Urry (2001: 240), is also critical of nomadic metaphors asserting that nomads actually follow regular and predictable patterns. From a postmodern perspective particularly appropriate to the times, vagabonds and tourists were much more plausible since they lacked regularity. The vagabond is a 'pilgrim without a destination, a nomad without an itinerary' (Urry 2000a: 29), whilst the tourist:

> pay(s) for their freedom; the right to disregard native concerns and feelings, the right to spin their own web of meanings… The world is the tourist's oyster… to be lived pleasurably – and thus give meaning. Bauman (1998: 243).

Wolff (1993) continues the criticism in particular of the masculinist character of many nomadic metaphors whilst Clifford (1997: 377) suggests that the postmodern nomad is 'unmarked by the traces of class, gender, ethnicity, sexuality and geography' and thus clearly unrealistic. This view is backed by Jokinen and Veijola (1994) who also suggest that maleness could be overcome at least in part by further use of nomads but in the form of prostitute, babysitter and au pair.

Meanwhile, Budd et al. (1990: 176) suggest that:

> Unless it is reflexive and critical, nomadic subjectivity is unlikely to organize meaningful political thought or activity, especially against elites whose thinking is more organized and purposeful. People who are nomads cannot settle down.

Cresswell (1997: 364) places nomadology in the context of state versus individual and stasis in comparison with dynamism, something that is familiar in an appreciation of maritime governance. The nomad is:

> never re-territorialized, unlike the migrant who slips back into the ordered spaces of arrival. The metaphorical space of the nomad is the desert, flat, smooth, curiously isotropic. The nomad shifts across this tactile space making the most of circumstance, not unlike the rhizomic vegetation that shifts location with changes in the weather.

This individualism, dynamism, flexibility contrasts with that of the state, which acts in Cresswell's terms as 'the metaphorical enemy of the nomad, attempting to take the tactile space and enclose and bound it' something re-emphasised by Miller (1993: 13). The state is dedicated to controlling flows, evidenced repeatedly in its failure to address dynamism in governance, to make them 'run in conduits'. In Deleuze and Guattari's view, the nomad represents all that is mobile, free, flowing, dynamic—rioting, revolution, guerrilla warfare:

> Nomads provide new models for existence and struggle. The nomad-self breaks from all molar segments and cautiously disorganizes itself. Nomad life is an experiment in creativity and becoming, and is anti-traditional and anti-conformist in character. The postmodern nomad attempts to free itself of all roots, bonds and identities and thereby resist the state and all normalizing powers. Best and Kellner (1991: 103).

Urry (2001: 239) contrasts the nomad with the desires of the state whose main task is to 'striate the space over which it reigns… not only to vanquish nomadism, but to control migrations and, more generally, to establish a zone of rights over an entire exterior, over all flows. Deleuze and Guattari (1986: 59)'. All things can be perceived as a journey, not least the process of policy-making and the challenges this presents can also be woven into this journey characterised by the actions and desires of the major stakeholders, characteristically in our case shipowners and their allies.

Gilroy (1993: 4) takes this further using the metaphor of the ship in his discussion of the dispersion of Afro-Caribbean diaspora. He sees it as a 'living, micro-cultural, micro-political system in motion'.

> Ships were the living means by which the points within that Atlantic world were joined. They were mobile elements that stood for the shifting spaces in between the fixed spaces that they connected. Accordingly they need to be thought of as cultural and political units… (ships) were… a distinct mode of cultural production. Gilroy (1993: 16–17).

Nomads have a clear relationship with territory explored by a number of authors and this also has a close connection with our discussion of nation-states and governance. Malkki (1992: 31) agrees quoting Deleuze and Guattari (1988: 23) who suggest:

> History is always written from a sedentary point of view and in the name of a unitary State apparatus, at least a possible one, even when the topic is nomads. What is lacking is a Nomadology, the opposite of a history.

They also go on:

> We know about the problems States have always had with journeymen's associations, or compagnonnages, the nomadic or itinerant bodies of the type formed by masons, carpenters, smiths etc. Settling, sendetarising labour power, regulating the movement of the flow of labour, assigning it channels and conduits, forming corporations in the sense of organisms, and, for the rest, relying on forced manpower recruited on the spot (corvee) or among indigents (charity workshops) – this has always been one or the principal affairs of the State, which undertook to conquer both a band vagabondage and a body nomadism. Deleuze and Guattari (1988: 368).

Continuing in this vein, they emphasise that nomads have their own territories moving from one point to another, aware of points with value—water, dwelling, assembly, etc. However, what makes them nomadic is that the territory is always subordinate to the paths between them. They are thus focussed on the process rather than the form; the movement rather than the stationary; dynamism rather than stasis; and as such, their configuration is very much a mirror of how governance should be.

Noyes (2004: 159) stresses the links between nomads and territory in that although society increasingly focuses upon individuals, shopping and surfing the Internet like nomads, with technologies increasingly releasing us from vocational ties to territory, we remain confronted by territorial demands and constraints. Thus, on the one hand, society desires a more nomadic existence and on the other the importance of territory (land and property ownership, a national focus, a home) remains a contradiction which is undeniable. The maritime industry is much the same but exhibiting perhaps a more deliberate trade-off of the territorialised/deterritorialised dialectic (national flag, domestic subsidy versus global markets and regulations) taking the best from each and rejecting the less palatable. Braidotti (1994: 5) suggests:

> Not all nomads are world travellers; some of the greatest trips can take place without physically moving from one's habitat. It is the subversion of set conventions that defined the nomadic state, not the literal act of travelling.

Urry (2000a: 28) considers the relationship between nomad and territory to be fundamental, if negative. Thus, he talks of nomad territorialisation which aims to 'challenge disciplinary limits and hegemonic cultural practices, to *marginalise the centre* and especially the masculinist, imperial, white and academic culture of the *west*' (italics original).

Cresswell (1997: 365) locates nomadic territory in urban space, understandable if considering a true and unadulterated nomad rather than the nomadic metaphor we have adopted. Utilising Deleuze and Guattari's notion of smooth and striated space, he locates the nomad in the former, threatening the power which is found in the striated space of the city. This smooth space is:

> sprawling, temporary, shifting shantytowns of nomads and cave dwellers, scrap metal and fabric, patchwork, of which the striations of money, work, or housing are no longer even relevant. Deleuze and Guattari (1988: 481).

Consequently, one of the main tasks of the modern state is to striate the space which it controls—something relatively easy in urban areas but which is almost impossible when we come to the sea which despite its proximity to national power, remains elusive and consequently of great value to the smooth space searching shipowner who acts almost as if nomadic. Despite addressing the urban nomad, this all sounds eerily familiar in a maritime world where vessels are moved around both physically and virtually, between temporary and shifting registers, characterised by a patchwork of regulations and institutions. However, the regulated striations of housing, work and money, the product of government regulators of all jurisdictions are increasingly irrelevant. True nomads are never accommodated or incorporated within the striated spaces of power and the nomadic maritime equivalent is much the same.

All this can be seen from a postmodern viewpoint (see Roe 2013 once again for a maritime perspective). Mobility as a concept is essentially postmodern in that it emphasises change, movement, dynamism and flux in opposition to stability and stasis—the latter in direct contrast to much of the postmodern movement. We return to this later with a discussion of time geography and its attempts to accommodate place not as statically rooted but as an arena for people as they act out 'place ballets'—the 'collective effect of individual bodies moving through space' (Seamon 1980). Seamon and Nordin (1980: 35–36) use the example of a Swedish marketplace to illustrate the value of such an approach in understanding the structure of society, an arena where institutions and individuals move together to generate the daily activity of a particular location or event. The maritime sector falls easily into place here and the use of metaphors of this type to understand what is actually going on beneath the labyrinth of attitude, power, emotion and the associated subtleties can be valuable.

Cresswell (2006: 29) sees movement as essentially dysfunctional as the principle of least effort is always more rational—generally speaking nothing moves unless it has to. Thus, not only nomads, but all forms of transport (including shipping) are inefficient and policy-making (at least conventionally) should be trying to minimise this.

Not everyone agrees of course—this is postmodernism. Morley (1999: 158) for example suggests that the 'idea that somehow we all experience some new form of postmodern nomadology… appears little more than a cruel nonsense', particularly there are in effect many different relations of space and place and none of us can escape from at least some of these relationships. Meanwhile, Noyes (2004: 160) discusses the duality of 'nomad capitalism' first identified by Williams (1989: 124) which in turn led to the emergence of Palan's (2003) 'nomad millionaires'. This new breed had lifestyles 'increasingly dependent on a virtual world-space, a technological negation of both physical space and solar time'. This generates an increasing divide between the mobile rich and poor; between the traditional and the postmodern nomad; between the disembodied wanderer and the brute reality of the vagabond.

However, despite the obvious differences between two groups characterised in particular by the same feature (excessive mobility), they co-exist and the same can be seen throughout postmodern society; across differing communities, societies, cultures, industries, professions and so on. The maritime sector is far from exempt, and the contrast between the rich and poor shipping interests, between the sophistication of a new Maersk vessel and that of an ageing tramp, between third and first world port facilities, and between supply chains in Africa and Europe is clear, and yet all characterised by a desire for mobility.

Peters (1999: 32–33) brings it all together seeing nomads as central to a postmodern conception of life with 'passing' far from traumatic but rather a 'characteristic motion of subjectivity through signs, otherness and time. Most of them see in the nation-state no promised land, just another Pharaoh to challenge'. In Deleuze and Guattari's minds; a postmodern hero.

Cresswell (2006: 26) sees all this in metaphysical terms contrasting the sedentarist with the nomad requiring an appreciation of the relationships between mobility, spatial order and place. The sedentarist places mobility into the realm of place and rootedness making mobility 'morally and ideologically suspect, a by-product of a world arranged through… spatial order'. The nomad places themselves first revelling in 'notions of flow, flux and dynamism'. Place is an 'irrelevance, trapped in the past, both confining and reactionary as well as practice and material culture'. Thus, sedentarists see mobility as a threat, a disorder that needs controlling, reflecting much of the existing framework for maritime governance with its strict structures and focus on stasis.

> The drifter, the shiftless, the refugee and the asylum seeker have been inscribed with immoral intent. So too, the travelling salesman, the gypsy traveller, and the so-called wandering Jew. These have all been portrayed as figures of mobile threat in need of straightening out and discipline.

Existing maritime governance is equally as dismissive of flux and change, opting for a structure to policy-making that dates from at best the early twentieth century and at worst some centuries before this. This sits uncomfortably with the shipping industry and its enthusiastic attitude for embracing some of the impacts of globalisation (for example, flag-hopping; tonnage tax; global markets for seafarers) which are inherently mobile and rely on the ephemeral characteristics of mobility to prosper. This in turn is more a reflection of the desire of the maritime sector to compete within a globalised world than some sort of moral or metaphysical position.

Malkki (1992: 32) can help us to understand this considering sedentarist attitudes reaffirming a:

> common-sense segmentation of the world into things like nations, states, countries and places. This process is so ingrained as to be invisible… such thoughts actively territorialize identities in property, in region, in nation – in place. They simultaneously produce discourse and practice that treats mobility and displacement as pathological.

As Hannam (2009: 103) indicates, 'mobility, then, as a concept, has circulated metaphorically', and the nomad can help to 'fill contemporary discourses of social and spatial mobility' reflecting a need to reverse the obsession in western society for order whether for society, the organisation of space or how it is all governed. In reaction to this, the ideas of complexity and disorder, with an appreciation that perhaps there is not always an order or even a truth that is objective and that all such interpretations are expressions of power that need to be understood, reflected or resisted in the development of governance itself.

In the words of Zembylas and Vrasidas (2007: 71), the nomad therefore is someone who 'learns to live with the discomfort of uncertainty and the complexity of change', and as a consequence forms an ideal contrast to the fixity of current maritime governance and a reflection of the direction in which it needs to move.

Global Fluids

> Neither boundaries nor relations mark the difference between one place and another. Instead, sometimes boundaries come and go, allow leakage or disappear altogether, while relations transform themselves without fracture. Sometimes then social space behaves like a fluid. (Mol and Law 1994: 643).

We can now return to the ideas of fluids, an interpretation of the impact and mechanism of globalisation that has become increasingly popular (see for example Peters 2002: 11–12). Hier and Greenberg (2007: 321–322) consider the significance of fluidity to mobility, privacy and resistance utilising the concepts of liquidity and solidity as a basis for understanding changes in human mobility. Taking Bauman's (2000) approach, they see liquidity as a metaphor for modernity, something which may have value for maritime governance. Mol and Law (1994) took blood and more specifically anaemia as a vehicle to illustrate issues of networks, regions and power. However, this time we will take the concepts of movement and flux and their metaphorical interpretation and use the fluid analogy to clarify the need for flexibility, change and progress in maritime governance.

To get a clear definition of what is meant by the term global fluid, we can turn to the originator of the concept. Urry (2000a: 38, 2010) moves away from Bauman's 'gardener' interpretation of policy-making, turning to 'game-keeping' metaphors (Spaargaren and Mol 2008: 352) suggesting they are 'remarkably uneven and fragmented flows of people, information, objects, money, images and risks across regions in strikingly faster and unpredictable shapes'. The emphasis is away from networks and upon 'heterogeneous, uneven and unpredictable mobilities'. Mol (2007: 302) sees them as 'spatial patterns structured neither by boundaries nor by more or less stable relations, but by large flexibility, liquidity, gel-like movement and permissible boundaries'. Others have contributed to this debate even in some cases, if they do not all realise it—see for example Deleuze and Guattari (1986, 1988), Lefebvre (1991), Mol and Law (1994), Waters (1995), Albrow (1996), Kaplan (1996), Eade (1997), Shields (1997), Mol and Spaargaren (2005: 97), Morgan (2005: 28), Hannam et al. (2006), and Kemp (2009: 90–91).

Hannigan (2002: 281–282) suggests that the significance of global fluids in globalisation has been captured particularly well by Appadurai (1990) outlining five dimensions that in turn characterise globalisation.

- Ethnoscape: this is the 'landscape of persons who constitute the shifting world in which we live—tourists, immigrants, refugees, exiles, guest workers and other groups and individuals on the move'.
- Technoscape: this is the configuration of global technology moving information at high speed across national boundaries almost at will.
- Finanscape: typified by the flow of 'megamonies' through currency markets, commodity exchanges and stock exchanges.
- Mediascapes: 'image-centred, narrative-based accounts of reality and the infrastructure required to produce and disseminate them'.
- Ideoscapes: these are also image-based but relate to the 'ideologies of states and the counter-ideologies of social movements'. They are inherently political.

Each of the five 'scapes' is interrelated. Hannigan suggests that modern cinema provides excellent examples citing Mira Nair's *Monsoon Wedding*.

> Set in the week before the wedding of the daughter of a well-to-do Punjabi family in New Delhi.... The groom is a computer engineer living in Houston and the bride works in a television station where she is having an affair with her boss, a talk show host. Others in the wedding party have flown in from Australia. Even as the routines of a traditional arranged marriage unfold, everyone is constantly talking on their cell phones, including the wedding coordinator who calls himself an 'event planner' and distributes business cards with his new email address. The bride's younger brother frequently skips school and spends most afternoons watching cooking shows on television. There is a sense here that members of these two Indian families are constantly in motion, actively re-inventing their lives, even as they continue to embrace tradition.

There are a number of key features upon which most agree and summarised by Urry (2000a: 38–39):

- They demonstrate no clear point of departure or arrival, just de-territorialised movements or mobility. It is here that the concept of rhizomatic in contrast to arboreal raises its head;
- They are channelled along particular territorial 'scapes' or 'routeways' which can wall them in;
- They are relational in that they productively affect relations between the spatially varying features of a scape which would otherwise remain functionless;
- The move in a particular direction at certain speeds but with no necessary end-state or purpose;
- They possess different properties of viscosity and, as with blood, can be thicker or thinner and hence move in different shapes at different speeds;
- They move according to certain temporalities, over each minute, day, week, year and so on;
- They do not always keep within walls—they may move outside or escape like white blood corpuscles through the wall of the scape into tinier and tinier capillaries;

- Their power is diffused through these various fluids into very many often minute capillary-like relations of domination/subordination;
- Their power is exercised through the intersection of various fluids working on diverse senses;
- Different fluids spatially intersect in the empty meeting grounds of the non-places of modernity, such as motels, airports, service stations, the Internet, international hotels, cable television and expense account restaurants.

Urry (2005a: 245–246) provides a number of examples: world money (Eatwell and Taylor 2000), automobility (Urry 2004a), social movements (Sheller 2000), digitized information (Brand 1999), the Internet (Plant 1997), international terrorism (Gunaratna 2002), to which could be added shipping. Urry stresses the ambiguity in all this with one of the best examples of a global fluid (itself an excellent example of globalisation) being the anti-globalization movement:

> Like a virus, uncontrollable and untameable, this inspiration flowed from city to city, country to country, spreading at the same speed as the trillions of dollars involved in the reckless unsustainable money game of international capital… Capital's dream of super fast networks… was turned on its head. Aingers et al. (2003: 65).

Hannigan (2002: 278) provides an interpretation of Urry's definition placing global fluids within his understanding of the relationship between culture, social cohesion and globalisation and using three metaphors of space—regions, networks and global fluids. The regions can be represented by the bounded nation-state, the continental trade and the global economy; the networks are the flows and their interconnections that bind these states together (financial, physical, communications, etc. exemplified by stock exchange markets, commodity trades, street gangs and drug traffickers); and global fluids which are a much more chaotic and flexible 'de-territorialized movement(s) of people, information, objects, money, images and risks across regions in undirected and nonlinear fashion and at variable speeds' (Hannigan 2002: 278). They have purpose, speed and definition but not necessarily any particular end-state.

Hannam (2009: 106) also provides a link between the flexibility and chaos of the nomad with the use of fluids as metaphors and in so-doing provides a framework for understanding the needs of effective maritime governance, the latter lying as it does, within a chaotic yet patterned globalised marketplace, exhibiting the flexibility and mobility associated with fluids. Hassan cites Urry (2003: 101) who suggests that society is not wholly organised according to 'globally integrated networks', characterised by structures such as transnational corporations with, in Hassan's words, 'the ability to nullify (and exploit) space-time constraints'. Instead, there are global fluids, defined by Urry as highly mobile and viscous formations whose shapes are uneven, contingent and unpredictable and which 'create over time their context of action'. Can we recognise the international shipping industry here acting as a global fluid? Adaptable, moveable, coherent yet multifaceted, taking advantages of every possibility in the space-time encyclopaedia of options.

Before we go on to look at global fluids in more detail and their contribution to the debate on maritime governance, we need firstly to spend some time looking at the use of fluids as metaphors more generally. Padgett (1980: 584) is emphatic about the failure of networks as a metaphor for governance systems because most organisations do not utilise a process of fluid participation and are constrained by much more rigid structures.

Kennedy (2007: 272) emphasises the increasing reliance upon movement metaphors which includes those referring to fluids:

> Thus, the globalization literature is replete with concepts and metaphors attempting to clarify… altered or intensified spatial (and related temporal) realities including the experience of movement; de-spatialization and de-territorialization (Appadurai 1990; Scholte 2000); time-space compression (Harvey 1989); scapes (Appadurai 1990); the space of flows (Castells 1996); diverse mobilities, global fluids, globally integrated networks and complex human and inhuman hybrids (Urry 2003); the ubiquity of nomadic lives and place polygamy (Beck 2000); and the possibility of leading bi-focal and multi-focal lives in several locations simultaneously through the new trans-national migration spanning borders (Basch et al. 1994; Vertovec 1999).

Thus, the increasingly global governance demanded by the maritime sector has to reflect increasing place-bound and mobile societies, based upon the impact of proximity and distance but also the new human inter-relationships operating through machines, objects, texts, symbols and images (Urry 2000a, 2003, and cited in Kennedy 2007: 273). Sheller and Urry (2000: 11–12) see the impact of global fluids in particular on public institutions, diluting their national characteristics. Instead, we have a range of global institutions which they categorise into three types:

- Those associated with orchestrating consumption (Olympic movement, World Football Cups, CNN, MTV, etc.) (Roche 2000);
- Global economic publics, such as stockholders in multinationals, the World Bank, IMF and WTO
- Global political publics (at state level such as the EU, UN, IMO and IATA); international NGOs (Amnesty International, Greenpeace, Amnesty International) and international social movements (anti-WTO; Zapatistas of Mexico, anti-globalists).

They also see the emergence of a global public, with a strong sense of communication and mobile opportunities, living within cultures that are mobile and flexible. This is exemplified by the changes that have occurred in what they term the 'staging of publicity'. Communication has always been with us although its extent, reach and speed have changed markedly in recent years. What has also changed are the characteristics of this 'publicity' as national citizenship manifesting itself through the public broadcasting of the twentieth century, has given way to the broadcast of opportunities and the prime use of mass communication as a marketing and sales opportunity. This has been accompanied by the extension of flows of fluids of information, images, sounds and the like, reinforcing the development of global fluids which necessitate new mechanisms for effective governance. The maritime sector reflects all the best and worst in these changes, with clear examples of global fluids in the effectively free movement of information, money, people, goods and the like across national boundaries and with increasing speed and simplicity. What makes it perhaps more interesting is its ability to take advantage of these features of fluid globalisation to enhance its position both within its own bounds and across other global activities, for example within major global or supranational political organisations where its influence can appear at times to be over-stated. Classic examples of the industry's influence at the European Commission enabling it to 'bend' the normally fiercely upheld principles of the Treaty of Rome (tonnage tax, liner shipping exemptions, etc.) are notable. Shipping is fluid and consequently the globalisation of this fluidity is more marked.

Not everyone is so enamoured by the concept. By definition, global fluids are chaotic and unpredictable and consequentially commonly attract criticism for these very features. Hannigan (2002: 281) for example suggests that they are indisciplined because they are neither hierarchically nor territorially bound, lacking the formality and predictability that such features would inject.

Kemp (2009: 91) suggests that global fluids have their place but commonly lack analytical rigour. Urry's tendency to categorise fluids generally—for example regarding all travelling people together as one fluid—inevitably means that the fluid is unpredictable, heterogeneous and variable. Kemp goes on to cite McLennan (2003: 555) who suggests that a more finely devised separation of the constituent parts of fluids would make more sense and be more useful. 'The *flows* of refugees and transnational capitalists are likely to have different characteristics and logics'.

McLennan (2003: 555) suggests that Urry's (2000b: 186) concept of global fluid also demands acceptance of another 'implausible and polarisation' of society rather than accepting 'successive styles of understanding and enablement/control'. Urry suggests that the idea of societal fluids only applies to the present era and that not long ago movements of people, things, money, ideas, images and wastes were inconsequential. McLennan suggests that mobilities have been tracked by sociologists for many years and certainly since the Industrial Revolution from the eighteenth century. 'Fluids and their associated networks are not new, but changed'. 'Commerce, ideas, fashions, machinery, cuisine, cultures of fellowship and association have always been on the move, filtered and stylized through particular networks and associations, passing through many channels of transmission'. Urry (2000b: 192) quotes Law (1994: 23): 'left to their own devices, human actions and words do not spread very far at all'. McLennan describes this as 'extraordinarily shallow' as human actions and words are never left to their own devices, and what seems far today will seem close tomorrow.

Angelides and Caiden (1994: 227) provide us with a link between the concept of global fluids and policy-making that helps to illustrate the benefits that might be realised from taking such an approach to governance in any sector, maritime included. They start by identifying those features of current policy-making that we have noted earlier, which in turn make effective maritime governance unlikely:

- definition of problems abstracted from their environment and believed to be solvable;
- searches for a best solution and quantification of information;

- reliance on data and models as modes of enquiry;
- assumption that the scientist is an unbiased observer outside the system being studied;
- tendency to ignore the individual in the search for generality;
- time is viewed as linear in which the past is separate from the present and future;
- assumption of linearity—with no discontinuities, no critical thresholds;
- supremacy of rationality over intuition in decision-making;
- incrementalism; notion of absolute control; reliance on scapegoating when things go wrong;
- a tendency to exclude the interests of future generations.

This range of what he describes as 'global pragmatics and future problematics' present 'difficult challenges' to contemporary maritime policy-making. How to:

- Tolerate and live with uncertainty, ambiguity and complexity without resorting to simplification and reductionism;
- How to acquire freedom, develop and embrace creative and innovative learning;
- How to develop new limits of knowledge and a new kind of 'thick knowing' to replace the existing 'thin' form of detached and reductionist reasoning;
- How to get rid of old and heavy ideological baggage and move beyond the many dichotomies in order to redesign institutions that will better function in a fluid and uncertain manner;
- How to embrace complementarity and acknowledge that conceptual pluralism is necessary to provide a full account of reality;
- How to start thinking in terms of both/and rather than either/or.

Global fluids present one opportunity to address some of these fundamental issues that may well in some cases be unachievable but which in Angelides and Caiden's terms need to be 'articulated'. Peters (2002: 13) agrees seeing policy-making as addressing 'streams of problems, solutions, opportunities and actors', acting with fluidity.

Complexity

> Hegel said that true thinking is thinking that looks death straight in the face. We could add that true thinking is thinking that looks disorder and uncertainty straight in the face. Morin (2002: 329).

> Facets of public policy are more difficult to study systematically than most other phenomena investigated empirically by political scientists. Our attempt to test hypotheses with some rigor demonstrated that public policy becomes troublesome as a research focus because of inherent complexity – specifically because of the temporal nature of the process, the multiplicity of participants and policy provisions, and the contingent nature of theoretical effects. Greenberg et al. (1977: 1532).

> The protestors are winning. They are winning on the streets. Before too long they will be winning the arguments. Globalisation is fast becoming a cause without credible arguments. Financial Times (17th August, 2001).

The comment from the Financial Times in particular provides a vital link between globalisation, governance and complexity in that the process of change (here exemplified by globalisation and the effect of protest) is typically and almost inevitably complex. Taking this idea further, McLennan (2003: 555) provides a link between the global fluids of the last chapter and notions of complexity which are central to the difficulties of designing and implementing meaningful maritime governance. He follows the line of Urry (2000b) in focussing upon:

> the loose ends, by-products and chaotic spillovers of systems, phenomena and their interaction. If we are compelled to conclude that spiralling global disequilibrium and constant disorder are prevalent today... then we need a greater armoury of notions that register this non-linearity.

The convention of establishing conceptual and social ordering for the processes that drive governance needs to be replaced by systematic pluralisation. The existing orderly conventions have a long history. Thomas More in *Utopia* (1965: 102) for example suggests that there is a type of person who: 'rather than live in wretched poverty at home, volunteers for slavery in Utopia'. Parker (2009: 1297) suggests that this:

> is what organization means to them. A steady job, shops with food in them, and a police force that enforces the law: this has its attractions, and anyone who studies organization will understand the importance of certain sorts of predictability. Lucifer would rather 'reign in Hell than serve in Heaven' (More 1965: 263).

And this is central to a demand that the role of metaphors be accepted to replace the questionable 'deeper, leaner and more revelatory vocabulary'. He cites global fluids as a move in this direction and later (Urry 2004b: 18) confirms the relationship between fluids, globalisation, complexity and order seeing the fluid model reflecting the organisation that exists within the disorder. Law and Urry (2004: 401) follow this up suggesting that complexity theory is based on three assumptions which in turn provide a definition:

1. There is no necessary proportionality between causes and effects
2. Individual and statistical levels of analysis are not equivalent
3. System effects do not result from the simple addition of individual components.

Complexity theory has many contributors including Thorn and Welford (1994: 667), Medd and Marvin (2005: 44–45) and Celek and Er (2006: 879–887). Thrift (1999: 35) treats complexity theory as a 'set of metaphors concerning holistic emergent order', something that harmonises with recent discussions of governance and holism, particularly in the EU maritime sector. The metaphors of complexity theory are 'able to travel and gradually become a *commonplace* structure of intelligibility'. Thus, the metaphor provides the mechanism for complexity to be understood within the context of governance and its inherently diverse, nonlinear and dynamic characteristics. And governance is inherently complex:

> If you see the whole thing - it seems that it's always beautiful. Planets, lives… But close up a world's all dirt and rocks. And day to day, life's a hard job, you get tired, you lose the pattern. Le Guin (1974) quoted in Marston et al. (2005: 416).

> Or as Thrift believes, knowledge is no longer based in eternal 'truths' but is simply 'an archipelago of islands of epistemic stability in a sea of disorder, fluctuations, noise, randomness and chaos. Thrift (1998: 32) in Herod et al. (1998: 4).

Urry also questions the relationship between our earlier discussion of metaphor and complexity (2004b: 15) asking 'whether complexity can generate productive metaphors' for the analysis of the post-societal world? He goes on to take a positive view of the relationship as social scientists are 'in the business of formulating the metaphors for this new science, metaphors that with luck, will guide the way these sciences are done over the next fifty years or so' (Arthur 1994: 680). Well we are at least 20 years into that period and although the value of complexity and the metaphor has been recognised, there remains much to be done in their application particularly to governance and even more so in the maritime sector. Urry (2005a: 249) continues in a later paper to quote Gray (2001) who suggests that the current state of the globe is 'an intractably disordered world' but one which Urry feels complexity can provide metaphors for analysing the disorderliness.

Artigiani (1987: 251–252) notes Prigogine and Stengers's (1984) contribution to the development of complexity theory and in particular the 'dissipative structures' model that can be applied to both scientific and social disciplines (see also Gemmill and Smith 1985: 708–709; Allen et al. 1985: 85). Although he looks at its application to political revolution, the value and applicability of the approach is clear. The model itself focuses upon bifurcation points (similarly stressed by Wilson 1981) and the significance of change:

> Locatable moments at which a destabilized structure can make an unpredictable leap to one of several alternative stable states. There, randomly joined elements can form a nucleation with a privileged relationship to the system, determined by feedback loops that enable it to become dominant, producing a new structure. The stable state chosen is selected as a result of local conditions that are completely aleatory. But bifurcation points are reached through historical processes that can be fully described because they are essentially deterministic. Thus history, the record of its experiences as a system in time, defines a dissipative structure, while the randomly generated leaps at bifurcation points which generate its evolutionary growth are inherently non-deterministic. Artigiani (1987: 251).

Despite attacks on philosophical grounds (see for example Edens 2001; and Bishop 2004), the principles of the model have become widely accepted and the issue of organisational system change has become established in organisational learning (Argyris and Schon 1978; Argyris 1982; Golombiewski et al. 1975; Sheldon 1980; Davis 1982; Miller and Friesen 1982). Later, Nicolis and Prigogine (1989: ix) see two disciplines as essential to the early development of complexity theory—non-equilibrium physics and dynamical systems, the latter particularly characterised by instability, and as both problems of equilibrium and instability characterise governance this makes its application beyond the physical sciences that much more relevant.

Dillon (2000: 4) quotes Stengers (1997: 4): ‘the theme of complexity has played an ambiguous role in discourses on science’, deriving from the physical sciences, nonlinear mathematics and microbiology. Its application to the social sciences, and therefore its applicability to the issues of governance, has been a long and tortuous one that now has applications to international politics, strategic thinking, military science and all aspects of globalisation (including shipping) (Alberts and Czerwinski 1997; Cebrowski and Gartska 1998; de Landa 1991; Jervis 1997; Rosenau 1992). This has not been achieved without a struggle:

> We usually opt for one level of analysis exclusively, without considering the range of other alternatives. To judge from the literature this choice is a private act of faith, not to be reported publically. Watson (1978) quoted in Meentemeyer (1989: 163).

And this is despite the recognition of complexity that prevails; that:

> logic and philosophy are messy, that language is messy, that chemical kinetics is messy, that physics is messy and finally that the economy is naturally messy. And it’s not that this is a mess created by the dirt that’s on the microscope glass. It’s that this mess is inherent in the systems themselves. You can’t capture any of them and confine them to a neat box of logic. Arthur (1994), cited in Waldrop (1993: 329) and quoted in Thrift (1999: 32).

Benko and Strohmeyer (1997: 291) add their view that a process of *reflexive accumulation* is going on that is inherently complex in nature, linking together culture and the economy bringing together the decline of the ‘national’ along with the rise of the ‘regional’ and ‘local’. Thus, globalisation itself is a concept where it is helpful to take a complex approach.

Smith and Jenks (2006: 4) suggest that the origins of complexity theory are as complex as the issue itself although Urry (2005b: 3) is insistent that ‘complexity’ does not necessarily mean ‘complicated’:

> Complex systems analysis investigates the very many systems that have the ability to adapt and co-evolve as they organize through time. Such complex social interactions are likened to walking through a maze whose walls rearrange themselves as one walks through… Complexity investigates emergent, dynamic and self-organizing systems that interact in ways that heavily influence the probabilities of later events.

Some of the earliest moves came with Lorenz, noted in Gleik (1987: 11–31) whilst constructing computer models of weather systems. Meanwhile, many others have noted the development of complexity theory over time including Hayles (1991, 1999), Eve (1997), Price (1997) who relates it to Postmodernism, Cilliers (1998), Byrne (1998) and Schnitman and Schnitman (2002).

Urry (2005b: 1) notes the origins of the ‘incursion’ of complexity into the social sciences from the 1980s derived from developments in the formal sciences along with economics and emerging from what Fraser et al. (2005) consider neo-vitalism in social thought. This was accompanied by a more general increase in complexity in the ‘structure of feeling’ which challenged the existing social order.

Urry continues identifying 'transformations' such as chaos, complexity, nonlinearity and dynamical systems analysis. There was a 'shift from reductionist analyses to those that involve the study of complex adaptive (*vital*) matter that shows ordering but which remains on *the edge of chaos*'. Urry goes on to identify a range of applications where complexity has now become a significant intellectual contributor including alternative healing, architecture, consultancy, consumer design, economics, defence, fiction, garden design, geography, history, literary theory, management, New Age, organisational studies, philosophy, politics, post-structuralism, small world analyses, sociology, stock car racing and town planning (Thrift 1999).

He also provides a reading list to die for (or perhaps to die as a consequence of attempting to read them all). They include Stewart (1989), Kauffman (1993), Cohen and Stewart (1994), Casti (1994), Arthur (1994), Nicolis (1995), Luhmann (1995), Krugman (1996), Capra (1996), Prigogine (1997), Jervis (1997), Rescher (1998), Holland (1998), Byrne (1998), Kelly (1998), Cilliers (1998), Hayles (1999), Watts (1999), Rycroft and Kash (1999), Rasch and Wolfe (2000), Capra (2001), Gladwell (2002), Buchanan (2002), Wolfram (2002), De Landa (2002), Barabasi (2002), Taylor (2003), Watts (2003), Surowiecki (2004), Ball (2004). And that is only up to 2004. These all followed the earliest developments noted in Keil and Elliott (1996) and to which can be added Buchanan (2002).

Thrift (1999: 39) adds to the list of complexity contributions suggesting contributions by de Lillo (1990), Argyros (1991), Mirowski (1994), Jones (1994), Barnett et al. (1996), Isard (1996), Benitez-Rojo (1996), Jencks (1996), Khalil and Boulding (1996), Livingston (1997), and Ferguson (1997), in areas additional to those cited by Urry including town planning, regional science, literary theory, anthropology, art, film and drama.

However, an understanding of complexity theory is far from complete and continues to develop. McLennan (2003: 4) for example notes that:

> complexity and emergence are still in the mode of general ideas rather than modelled solutions and so whilst the trail of suggestive notions – phase transitions, constrained generating procedures, state trajectories, time irreversibility, self-organizing criticality, increasing returns, positive feedback, strange attractors, networks, nodes, co-evolution, membranes of reorganization, broken symmetries and the rest - is exciting to follow, it is less clear that these leads are *applicable* as such.

The definition of complexity is as complex as the concept itself although as Pollitt (2009: 213) remarks it is part of a very abstract and generalised theory 'about almost everything, rather than a theory about some specific sector, process or problem'. Consequently, it is ambitious and underneath it has the aim of understanding the modern condition of governance amongst many other things. As such, it has much to offer an understanding of maritime governance and policy-making.

Mihata (1997: 31) notes that complexity takes much from how global structures arise from both local interactions as well as a series of 'simple rules' emerging from a process of central coordination. Law and Urry (2004: 400–401) see complexity as 'a wide array of innovative notions that… take social investigation a long way from conventional linear analysis of structure or action/agency'. It is characterised by 'unpredictable' yet patterned results can be generated, with 'small causes on occasions producing large effects and vice versa'. Derived from this notion, complexity theory is based upon three assumptions (Law and Urry 2004: 401):

1. There is no necessary proportionality between cause and effect;
2. Individual and statistical levels of analysis are not the same;
3. System effects do not result from the simple addition of individual components.

Lee (1997: 20–21) also contributed to the complex task of defining complexity, suggesting it:

> has to do with the interrelatedness and interdependence of components as well as their freedom to interact, align, and organize into related configurations. The more components and the more ways in which the components can possibly interact align and organize, the higher the complexity.

Elliott and Kiel (1997: 65) suggest that complexity remained a relatively new concept at the time they were writing and as such there was no generally agreed definition. Mayer-Kress (1994) for example proposed that definitions were an annoyance as they hindered understanding the very complex nature of complexity. Gell-Mann (1994) considered that complexity was a function of interactions between elements in a system rather than behaviour, the latter preferred by Nicolis and Prigogine (1989). However, all commentators seem agreed that complex systems exhibit nonlinear behaviour (see for example Lewin 1992; Waldrop 1992 and Morgan 1998: 222, 2005: 27). Coveney and Highfield (1991: 7) for example suggest that complexity theory comprises:

> ...the study of the behaviour of macroscopic collections of (interacting) units that are endowed with the potential to evolve in time. Their interactions lead to coherent collective phenomena, so-called emergent properties that can be classified only at higher levels than those of individual units.

Or to put it another way:

> From the interaction of the individual components (of a system)... emerges some kind of property... something you couldn't have predicted from what you know of the component parts... And the global property, this emergent behaviour, feeds back to influence the behaviour... of the individuals that produced it. (Langton, quoted in Thrift 1999: 33–34).

Urry (2005a: 237) taking the work of Mitleton-Kelly (2003) suggests that complexity 'science' is designed to 'investigate systems that adapt and evolve as they self-organize through time'. Consequently, it tends to focus upon 'emergent, dynamic and self-organizing systems that interact in ways that heavily influence the probabilities of later events' (Prigogine 1997: 35). As such, it is particularly pertinent for the development of meaningful governance where the issue of later events is central to policy-making and the concepts of emergence and dynamism need to be a focus. Urry cites Axelrod and Cohen's (1999: 14) use of the term 'dynamical zoo', 'wildly unlike the smoothly additive changes of their simpler cousins'. This is once again familiar in our consideration of existing maritime governance and its placid acceptance of unchanging governance, stable policies and amendments to existing policy which are clearly additive. He contrasts this simplistic view to the real world, full of 'avalanches, of founder effects, self-restoring patterns, apparently stable regimes that suddenly collapse, punctuated equilibria, butterfly effects and thresholds as systems tip from one state to another'. In the same year (Urry 2005b: 5), he goes on to stress the importance of emergence in complexity, the spontaneous development of collective properties or patterns which are not implicit in individual components, non-reducible and nonlinear (Nicolis 1995).

Urry rejects what he calls the 'dichotomies of determinism and chance' as well as stasis and change suggesting that there is always both order and disorder in any system including maritime governance, commonly in balance but close to chaos—something we return to later. Prigogine (cited in Capra 1996: 184) agrees—'islands of order' located within increasing turbulence and disorder. 'Very small perturbations or fluctuations can become amplified into gigantic, structure-breaking waves' (Prigogine and Stengers 1984: xvii):

> Elements at one location have significant time-space effects elsewhere through multiple connections and trajectories, such as individual local decisions to drive by car (rather than to use slower modes, take public transport or live closer to work or family) resulting in extensive emergent 'far from equilibrium' effects of an 'out-of-control' global car system (see Cilliers 1998; Urry 2004a). Urry (2005a: 238).

Thrift (1999: 34) notes how there have been innumerable claims of the appropriateness of applying complexity theory to many disciplines—chaos theory, fractal modelling, artificial life, cellular automata, neural nets—and an associated vocabulary that has grown up—chaos, attractors, fractals, emergent orders, self-organisation, implicate order, autopoiesis, etc. suggesting an ill-defined but expansive discipline. Given the wide-reaching nature of maritime governance, this may be a hopeful sign.

Buijs et al. (2009: 37–38) provide another perspective suggesting 'complex causation and trajectories in and between complex cases' (Byrne 2005). They also see as essential an understanding of patterning which may be simple or less so but which always underlies complex systems. Two types of complexity are identified. General complexity assumes a set of general rules from which emergent complexity follows. They suggest that this forms the basis from which 'rules of pattern and order' follow (Holland 1998). Situated complexity assumes that reality is 'deeply complex and inherently contingent' and that general rules do not always apply. However, both the general and situated complexities have in common that they deal with patterns. These patterns may be simple and linear, distinct and obvious, indistinct and light, oscillating or continuous. But patterning there is.

And finally, Law and Urry (2004: 402) see 'criss-crossing societies (as) diverse systems in complex interconnections with their environments', that order and chaos are always found together, and that self-organisation in society is predominant, and consequently the application of complexity theory to the social sciences is obvious. Kemp (2009: 84) stresses the value of taking a complexity perspective on things and the social sciences in particular, bringing its own 'vocabulary and explanatory repertoire to account for the social world'. It 'offers concepts that are held to apply to both the social and natural worlds, rather than arguing of a clear cut division between the two in the manner of interpretive social thought'—something also to be found in the school of critical realism (Archer 1998). Kemp (2009: 85) links this to the work of Urry (particularly *Global Complexity*, 2003), in bringing together the issues of globalisation and complexity and consequently providing empiricism to what was increasingly seen as a theoretical distraction.

So in terms of defining this rather difficult to define concept perhaps Thompson's attempt is clearest (2004: 412) and also benefits from being one of the very latest, taking its roots from the work of many others. The reader is asked to think the 'maritime governance and the maritime industry' all the whilst in considering Thompson's definition:

> Complexity is the term used to refer to self-reinforcing dynamic systems with many feedback mechanisms. In these complex systems operating in a social context, behaviour is modified as a reaction to what other agents do. The non-linearity of these systems means that small amounts of a change in inputs can have dramatic and unexpected effects on outputs. Formally, complexity is equated with the number of different items or elements that must be dealt with simultaneously by the organism or organization. But its distinctive feature is to stress the world as a system in construction, a dynamic formulation encouraging the notion of a continual process of spontaneous emergence. Multiple possible outcomes are typically associated with the mathematically inscribed non-linear modelling techniques used to isolate the network topologies. Turbulence and uncertainty abound in this environment, often further described as 'open system ecologies', where perpetual novelty results. Filling one niche simply provides new niches, and small perturbations can affect the future of multiple combinations of events.

This focus upon nonlinearity is seen by Elliott and Kiel (1997: 69) as central to the value of complexity theory in understanding policy-making and governance. Complexity should be recognised and accommodated whenever an analyst is 'aware of changing relationships between variables… and when a study involves a highly multivariate phenomenon in which multiple interactions also present the potential for non-linear behaviour'. Maritime governance is a clear candidate.

The whole issue is made more confusing by the variation in terminology that has been used. Complex systems are probably the most commonly found deviant. Dennis and Urry (2009: 59–60) use them to analyse the role of the private motor car whilst Smith and Jenks (2006: 13) provide an extensive definition which is remarkably similar to those for complexity we have already covered. Others who refer to them directly include Buijs et al. (2009) in considering public management, Engelen (1988) looking at urban areas, Mihata (1997: 31) and Lansing (2003) who takes it all on a step further to examine Complex Adaptive Systems (CAS). The latter also represent a minor deviation to core complexity theory along with Choi et al. (2001: 352–365) in particular who consider a range of issues that they believe are central to CAS—logistics, supply chains, governance and dimensionality included and illustrated with a range of examples from supply chain management with its obvious relevance to the maritime sector. Pel (2009) also provides examples of complexity and its application to transport and more specifically traffic management.

Urry (2005a) identifies a range of applications for complexity theory, reflecting Thrift's earlier (1999) contribution. He suggests that much has been borrowed from physical science and is progressively being applied in the social sciences something further emphasised by Buijs et al. (2009: 40–43) in their consideration of positivism and post-positivism and their relationship to complexity—and incidentally also the reverse where physicists are turning to the sociology of social networks (for example Watts 1999, 2003; Barabasi 2002; Buchanan 2002). All this activity is based on much earlier work by Giddens (1990) who implicitly applied complexity concepts in conceiving the modern world as a 'driverless out-of-control juggernaut which has set in motion irreversible processes across the globe' (Urry 2005a: 235). Clear divisions between the social and physical sciences are consequently unrealistic (Wallerstein 1996: 61, 63). Harvey's interpretation of time-space compression is also clearly orientated towards the complex (Harvey 1989); Bauman's 'liquid modernity' similarly so (Bauman 2000); and Hardt and Negri's (2000: 136) discussion of nation-states and globalisation rings particularly true in the light of our discussion of governance. They also note the contribution of Marx (Urry 2005a: 240–243) to the discussion of change throughout The Communist Manifesto and in particular its relevance to globalisation:

> the need for a constantly changing market chases the bourgeoisie over the whole surface of the globe. It must settle everywhere, establish connexions everywhere. (Marx and Engels 1952: 46–47. Originally published 1848).

Law and Urry (2004: 400) provide justification for considering complexity as some sort of bridge between the physical and social sciences. Both are dominated by complexity and quoting the Gulbenkian Commission on the Restructuring of the Social Sciences of the 1990s, one should not be conceiving of humanity as mechanical, but rather instead conceiving nature as active and creative, to ensure that 'the laws of nature (are) compatible with the idea of events, of novelty, and of creativity' (quoted in Wallerstein 1996: 61, 63). Thus, application of complexity theory to the social sciences (and consequently policy-making and governance) makes sense.

> Criss-crossing societies may be seen as diverse systems in complex interconnections with their environments; that there are many chaotic effects distant in time and space from their location of origin; that there are positive feedback mechanisms that mean that order and chaos are always intertwined; that there are self-organizing global networks and global fluids moving systems far from equilibrium; and that a social order is never accounted for by purified social processes.... Material worlds (are) unpredictable, unstable, sensitive to initial conditions, irreversible and rarely societally organized. Law and Urry (2004: 401).

There is much in maritime governance that needs to accommodate these complex issues; and much that currently does not.

Urry (2010: 358) later also points out the contribution of Byrne (1998), Cilliers (1998) and Wallerstein (1998) but particularly emphasises complexity's roots in science where mathematical formulae and computer algorithms can be applied to the large number of iterative events that can occur (Prigogine and Stengers 1984). However, he also suggests that sociology can be similarly as complex and thus susceptible to theoretical applications of complexity. Although limited examples exist (see for example Mingers 1995), much more could be done. Governance and policy-making are notably complex and hence may have much to offer.

Complexity, Space and Time

> Scientific analysis based on the dynamics of the non-equilibria, with its emphasis on multiple futures, bifurcation and choice, historical dependence, and… intrinsic and inherent uncertainty should be the model for the social sciences. Ilya Prigogine (in Wallerstein 1996: 61).

Kiel (1991: 431) outlines the emergence of what he calls a new 'paradigm in the natural sciences' during the twentieth century with various strands including nonlinear dynamics, chaos theory and the science of complexity, citing Gleik (1987) and Pagels (1988) to support his claim. This new paradigm is centred around nonlinearity, instability and uncertainty and moves away from Newtonian simplicity, linearity and certainty to that of nonlinearity, asymmetry and the unexpected.

Nicolis and Prigogine (1989: 2) see complexity everywhere, across all spaces, places and times although they also see it as complex behaviour rather than a complex system (1989: 8). Teisman et al. (2009: 7–13) note three characteristics of governance complexity that relate to space and time. Firstly, nonlinear dynamics is a clear function of governance, albeit along with linear dynamics as well. The nonlinearity refers to the erratic nature of governance and the issues to which it has to be applied in all sectors, maritime included. Thus, an 'incentive given to a certain sub-system may result in a certain kind of behaviour at a certain time, but a repetition of this incentive may result in a different response'. One need only think of countless differing responses to environmental stimuli by maritime stakeholders, all attempting to be defined by the same governance framework. Thus, governance, nonlinearity and complexity come together.

Secondly, complexity is often found along with self-organisation. Thus, shipping companies will continuously adapt to remain successful within a marketplace or in response to changing global economic environments. Routes and ports may be changed, vessels adapted, cargoes changed, crews recruited elsewhere, flags hopped—all common features of the shipping sector and examples of self-organisation that governance needs to embrace and which are inherently complex.

Thirdly co-evolution, where actors come together in a process of polycentric adaption and mutual recognition and adaptation (Norgaard 1984, 1994; McKelvey 2002; Gerrits 2008). This notion of co-evolution is easily recognisable in the shipping sector where professional bodies and industrial representatives bring participants together in a complex process of discussion, development and reformation as well as the subtle and indefinable process of political lobbying. Teisman et al.

(2009: 12) stress the irregularity of co-evolution, how it is interspersed with patterns of competition and resistance rather than collaboration and this intensifies the complexity. However, overall the industry develops because of co-evolution and the mutual interaction it represents with all the complexity this suggests.

Law and Urry (2004: 401) make a strong case for the role of complexity and its relationship to space and time. Complexity they feel helps to explore how:

> components of a system can, through dynamic interaction, spontaneously develop collective properties or patterns… that are not implicit in the same way within its components… Complexity argues against reductionism, against reducing the whole to the parts.

They relate this to the issues of space and time suggesting complexity 'transforms scientific understanding of far-from-equilibrium structures, of irreversible times and of non-Euclidean mobile spaces. Space and time are not containers of bodies that move along various dimensions (Capra 1996). They are internal to the processes by which the physical world operates helping to constitute the very power of objects' (Law and Urry 2004: 401), something particularly pertinent to shipping where its very activity makes space and commodities accessible and can compress time between them. There are multiple times and spaces partly dependent in the maritime case on the type and usage of vessel, the weather, the administrative arrangements in ports, the flexibility of the financial system and so on. Complexity abounds.

The value of a holistic rather than a reductionist approach has been emphasised by many commentators. These include Freeman (1991: 34) with respect to neuro-science, Zohar (1997: 43) who refers back to 'old paradigm science' and the ancient Greek philosophers who focussed upon separation and fragmentation (with four atomic elements—earth, air, fire and water) in contrast to modern holism, and Angelides and Caiden (1994: 226) who consider that 'the properties of the parts can only be understood through an understanding of the dynamics of the whole' and this is exacerbated because the 'dynamics of the whole may be greater than the summation of the parts'. They agree with Mintzberg (1989: 344) who considers that the powerful hold that reductionism has held on policy-making needs to be removed enabling the connections between the effects and implementation of policies to be realised. The alternative is a series of ad hoc policies that address specific and individualised problems in a world where the interconnections between them are far more significant (if more complex). In Angelides and Caiden's terms, 'governance systems thus learn how to be objective in their superficiality'. Or as William Blake suggested in contemplating how man could gain clarity of perception through reducing all things to their elements, 'if the doors of perception were cleansed every thing would appear to man as it is, infinite. For man has closed himself up, till he sees all things thro' narrow chinks of his cavern' (Blake 2000).

White (1992: 132–133) is not quite so convinced by the holistic argument, suggesting that what he calls the structuralist argument is extreme, 'the antipodes to individualism', claiming that all action is shaped by the overall. He accepts that context does shape actions and in particular in the social where culture and perception are so important not least throughout governance, but he sees it as a diversion within locality and boundary issues rather than a fundamental need to tend towards reductionism or holism:

> Augustine's theology... conceives of an (priestly) order almost exclusively with reference to the individual who exercises it... Such a view would have been impossible to hold but for the virtual breakdown of the old, jealously corporate notion of the local Church and its local ministry during the two preceding generations. Augustine's theory is in itself a proof that by circa AD 400 that idea was extinct as a living force. Dix (1957).

White continues his campaign against holism and supporting the reductionist position with another quote this time referring to an earlier period:

> Earlier Christianity began as a renewal movement within Judaism brought into being through Jesus... after AD 70 Pharisaism gained the upper hand in Judaism, and the Christians were excommunicated... Wandering charismatics were the decisive spiritual authorities in the local communities, and local communities were the indispensable material and social basis for the wandering charismatics. Both owed their legitimation and existence to their relationship to the transcendental bearer of revelation. It was the homeless wandering charismatics who handed on what was later to take independent form as Christianity... the local groups of sympathizers remained within the framework of Judaism... entangled in the old situation... Wandering prophets and teachers were still the decisive authorities at the time of the Didache (in the first half of the second century)... Their superiors were still the 'apostles' who were all allowed to stay no more than three days in one place. All these wandering charismatics had a higher reputation than local ministers. Theissen (1978: 7).

Urry (2004b: 15) is also clear about the importance of complexity to space and time. He notes the failures of the social sciences to understand the new relationships in space and time generated by globalisation and suggests that complexity theory has a significant role to play:

> Complexity maintains that there is no 'structure' and no 'agency', no 'macro' and no 'micro' levels, no 'societies' and no 'individuals', and no 'system-world' and no 'life-world' – in that each of these is presumed to be separate and distinct essences brought into external juxtaposition with its other. Overall the argument here is one of 'relationality', a position not only central to complexity but also to actor-network theory and various post-structuralist formulations... and relationality is effected through a wide array of networked or circulating relationships implicated within different overlapping and increasingly convergent mobile, material worlds.

Time and space are drawn to complexity by the very nature of their characters and consequently globalisation dominated as it is by time and space is also embroiled. And we know the relationship between shipping and globalisation. Therefore, governance of the maritime sector itself is dominated by time and space issues, and consequently is complex. The elements of globalisation 'interact physically and because of de-materialising transformations' (or otherwise referred to as time-space compression), 'informationally over multiple time-spaces'. Interactions are commonly nonlinear (as we have seen), determined by a mix of the local and global, and operate through a variety of temporal and spatial 'distanced effects... through multiple connections and mobile trajectories'. The complexities identifiable in globalisation—and consequently its governance and by association that of the maritime sector—are frighteningly large. But to ignore them is to condemn maritime governance to an irrelevant impotence.

To quote Urry again (2004b: 17):

> Thus criss-crossing societies are diverse systems in complex interconnections with their environments, there are many chaotic effects time-space distanced from where they originate, there are positive as well as negative feedback mechanisms that mean that order and chaos are always intertwined, there are self-organising global networks and global fluids moving systems far from equilibrium, and there is not social order accounted for by purified social processes. Such complexity-thinking enables the transcendence of the dichotomies of determinism and free will, especially through seeing material worlds as unpredictable, unstable, sensitive to initial conditions, irreversible and rarely societally organised.

Urry continues the theme (2005b: 4) suggesting that Pre-twentieth-century science (and even later it could be claimed) viewed time as Newtonian—'invariant, infinitely divisible into space-like units, measurable in length, expressible as a number and reversible…' Objects are viewed as being contained within such boundaries of absolute time and space (Coveney 2000). However, since then things have changed stemming as much as anywhere from Einstein's view that time is never fixed or absolute, independent of the system to which it refers and inherently complex:

> Time is a local, internal feature of any system of observation and measurement. It varies on where and how it is measured. It can stretch and shrink. Further time and space are not separate from each other but are fused into a four-dimensional time-space curved under the influence of mass. Time and space are internal to the processes by which the physical and social worlds themselves operate, helping to constitute their powers… Space and time are dynamic qualities; when a body moves, or a force acts, it affects the curvature of space and time, and in turn the structure of time-space affects the way in which bodies move and forces act. Urry (2005b: 4).

Complexity and Governance

> One shouldn't complicate things for the pleasure of complexity, but one should also never simplify or pretend to be sure of such simplicity where there is none. If things were simple, word would have gotten round. Derrida (1988: 119).

> If Russia can be destroyed, the United States can also be beheaded. They are like little mice. (Osama Bin Laden, quoted in Reeve and Foden 2001 and Urry 2002).

White (1992: 116) links the role of institutions in policy-making with increasing complexity in society whilst Lee (1997: 20) suggests that complexity is a key concept in modelling 'change processes' and in so-doing hints at the importance of complexity to governance. Blom-Hansen (1997: 670) says much the same emphasising that political systems were getting increasingly fragmented and specialised. As a consequence, policy formulation involved an increasing number of institutions and the role of interdependencies had increased alongside. Societal problems were getting more and more complex and the need for governance to match this degree of complexity was clear. Meanwhile, much later, both Van Gils et al. (2009: 76) and Boons et al. (2009: 231) provide examples of where complexity and governance collide in particular in public management and decision-making (for example, Kaufman 1991; Allen 1997; Haynes 1999; Morcul 2003; Teisman 2005; Gerrits 2008; Dennard et al. 2008). Boons et al. (2009: 233) go on to stress the links between the dynamics inherent in complexity to the characteristics of governance to reflect upon how the relationship is so close. They cite a number of reasons for this:

- (Maritime) Governance processes normally develop in dynamic ways that cannot be predicted from previous processes and/or from the initial conditions at the beginning of the process.
- Due to self-organisation within (maritime) processes, the coevolution between processes, the occurrence of change events and interactions within a multiple and ever-changing context, nonlinearity in processes and outcomes often occurs.
- The desire for change may initiate (maritime) governance processes and new processes and dynamics, but the initiator does not control the dynamics that occur after initiation. A specific governance initiative is just one manifestation of self-organizing capacity within societies.
- Initiators of (maritime) governance processes facing complex (maritime) governance systems develop and apply simplified pictures of these systems, often over-estimating their knowledge as well as their ability to change and control the system. Due to these limited boundary judgements, their actions often generate more nonlinear dynamics.

McLennan (2003: 554) notes Urry's (2000b: 186) acceptance that there is a need to move from a vision of post-nationalism to one of post-societalism as globalisation progresses. To achieve this, there is a requirement to accommodate both complexity and chaos. The new rules of globalised governance make a simplistic approach inadequate. The pre-eminence of a national dimension has been difficult to shrug off (and continues to be) but McLennan's view is that society is now most definitely non-national and consequently approaches to governance, including that for the maritime sector, must accept this or fail.

Angelides and Caiden (1994: 225) had earlier said much the same thing emphasising how the:

> interaction between global pragmatics and future problematic imposes great difficulties on the formation of accurate, up-to-date and complete images or present and future realities. One in particular is the gap between the phenomena with which governance is faced and those it perceives and with which it is prepared to deal.

Policy-makers simplify in the light of the complexity that surrounds them, much to the detriment of governance. The maritime sector is a good example where the complexities of the market, the global environment and the ambitions of the stakeholders that make up the industry present a dizzying mélange that seems almost impossible to manage. However, addressing the issues of complexity head-on may present an opportunity to accommodate and overcome some of these problems. They go on to suggest how to do this:

- Recognise uncertainty and distinguish between what can and cannot be known, thereby admitting areas of ignorance and separating substantive from speculative analysis.
- Guard against oversimplification, especially by over-relying on common sense, by recognising that complexity is the only means of handling complexity and only variety can handle variety.
- Adopt advanced notions of policy reasoning by merging intuition and reason, logic and feeling, that is the two sides of the brain (Trist 1980).
- Allow for emerging strategies, which call for abilities to change one's mind, improvise and engage in constant rejection and self-evaluation, i.e. policies need to be allowed to 'form without necessarily being formulated' (Mintzberg and Jorgenson 1987: 219).
- View decisions as fuzzy gambles (Dror 1990) and bets (Godet 1980).
- Tolerate value trade-offs, contradictions, paradoxes and dilemmas as being inherent to complexity and uncertainty.

Many of these points can be seen to be clearly relevant to the maritime sector with issues such as common sense, the failure to admit ignorance and the significance of speculation often predominant in policy-making decisions.

Getting right up to date, Klijn and Snellen (2009: 17) point out that although issues of complexity have been recognised in public administration and consequently governance and policy-making by all public institutions including government ministries, the UN and IMO, the European Union and more locally at local and regional level, there remains a shortage of application. 'The history of the field of public administration could be viewed as an on-going attempt to search for concepts to grasp the complexity of day-to-day practices in policy-making and decision-making' (Klijn et al. 2009: 17). Policy-makers meanwhile have been evident in their ability to avoid such complexity, commonly adopting 'policy escapism', focussing more easily upon what they wish to avoid rather than what they hope to achieve. The need to address complexity is reduced consequently. In Lindblom's words (1965: 147): 'they deal more confidently with what is wrong than with what in the future may or may not be right'.

Complexity and Chaos

> Moral condemnation, the sacred alliance against terrorism are in direct proportion to the prodigious jubilation at seeing this global superpower destroyed. Jean Baudrillard (*Le Monde*, November 3rd, 2001).

Elliott and Kiel (1997: 73) suggest that the link between complexity theory and chaos is so close that knowledge of the former could help to control the latter. This would be achieved by identifying the 'small perturbations in chaotic systems in an effort to alter or smooth its dynamics'. Taking the idea that very unimpressive and easily overlooked changes in a complex environment can have major chaotic effects, then minimal effort might be employed to have the maximum effect—in the words of Peat (1991), this 'gentle action' could be used to stimulate adjusting feedback to alter the temporal dynamics' of society. Kiel (1991: 431) notes much the same suggesting that minimal change taking place in a complex environment is akin to nonlinear systems where dynamic relationships occur between variables in which the relationship between cause and effect may not be proportionate. The significance of nonlinearity is well supported (see for example Loye and Eisler 1987; Baumol and Benhabib 1989; Saperstein and Mayer-Kreiss 1988 amongst many others). The result can be chaos. We can see much the same across large swathes of maritime governance where, for example, a minor environmental piece of legislation may have serious ramifications for the prosperity of shipowners, ferry operators or fish farmers in remote islands, or calamitous implications ultimately for climate change. The invention of the sea container and its impact on globalisation can be viewed similarly.

Arrighi and Silver (1999: 21–22) and Morgan (2005: 27) emphasise the significance of chaos whilst Thietart and Forgues (1995: 20) provide an introduction to chaos theory and its origins in the natural sciences (see for example Ruelle and Takens 1971; Swinney 1983) where it was defined as a 'system behaviour which is apparently random even though it is driven by deterministic rules'. Angelides and Caiden (1994: 224) took this up and were part of a number of researchers who applied it to the social sciences noting that increasingly policies have to reflect 'greater adversity, turbulence, conflict and possibly chaos', consequences of a future that is more 'complex, novel and uncertain', and a view supported by Back (1997: 49–50) who suggests that chaos theory has its roots in complex structures. However, Allen et al. (1985: 65) emphasise that there is often order in chaos—a 'hidden constancy underlying processes of apparent change'.

Eve (1997: 270) stresses that chaos and complexity are not the same thing at all, although there is plenty of literature that seems to suggest that this is the case, and by definition chaos always tends towards the chaotic and sometimes the unpredictable, characteristics with which the shipping industry has much affinity. Eve also believes that there is always order in chaos (1997: 271) and that the concept is universally applicable if not always dominant. Complex phenomena commonly exhibit what Smith and Jenks (2006: 5) term patterned limits, pathways and recurrences, and chaos characterised by upheaval and randomness is not an appropriate term. The poststructuralist interpretation of chaos—more randomness, more conventionality and a presumed emphasis upon deconstruction, interpretation and reconstruction—confirms this. Complex systems can and do exhibit 'different degrees of complexity, interdependence and robustness of self-organisation. They can also die' (Smith and Jenks 2006: 5–6). Meanwhile, McLennan (2003: 556–557) links complexity and chaos together in the social sciences, reflecting the increased dynamism we have noted that is sadly lacking in maritime governance.

Turning to chaos and away from its direct links to complexity, Brock (1986) provides a mathematical interpretation whilst Lorenz (1993) looks in detail in excess of the analysis needed here. At the same time, Thietart and Forgues (1995: 20) suggest a simple explanation for chaos theory:

> Chaos theory takes its roots in the study of nonlinear dynamic systems. Nonlinear dynamic systems have special properties that mathematicians have studied for more than a century (Poincare, 1892-1899)... interest in these systems has grown among researchers of different scientific fields such as physics, chemistry and economics. Interest has been mainly stimulated by these systems' capabilities in representing what were perceived, in the past, as noise and randomness.

Thietart and Forgues (1995: 21) go on to indicate areas of the social sciences where, at the time of writing, attention had been drawn to the possibilities of using chaos theory including economics (Grandmont 1986; Anderson et al. 1988); organization science (Gemmill and Smith 1985); social systems transformation (Smith 1986); firm self-renewal (Nonaka 1988); organizational chaotic behaviour (Priesmeyer and Baik 1989); management systems (Rasmussen and Mosekilde 1988); management (Stacey 1992, 1993); and strategic processes (Zimmerman 1990).

Dooley and Van de Ven (1999: 358) are highly critical of the approaches to chaos up to that date suggesting that the focus had been far too much upon the implications of chaos and needed to be looking much more at how organisational chaos comes about. This would require a more dynamic view (something that rings true in the maritime sector) looking for patterns of events (Abbott 1990). Using models developed in this way, chaotic events and processes could be better understood and predicted and they suggest 'interventionist strategies' (Glick et al. 1990; Pettigrew 1990; Van de Ven and Poole 1990). They also call for a closer examination of alternative patterns of organisational behaviour that can be identified—for example, periodic, coloured noise and white noise.

Dennis and Urry (2009: 52–53), along with Dooley and Van de Ven (1999: 358) citing Dooley et al. (1995) and Goldstein (1994), bring this discussion of chaos theory back to some of our earlier considerations of time whilst Collier and Esteban (1999: 179) did the same for process supporting the idea that chaos and complexity have a role to play in understanding the needs of governance. Chaotic systems can be viewed over time and should always be seen as processes where change is a permanent feature, characterised by both stability and instability. The chaotic model should be seen as a mechanism for holding in 'check both the tendency to emphasise purpose at the expense of responsiveness and the impulse to respond to the external environment in a way which disregards purpose' referring also to the relationship between minor events and their major consequences which has also been noted (Stacey 1996: 284).

Chaos also has a direct relationship to governance. Collier and Esteban (1999: 178) point out that 'understanding of governance in the participative organization can most usefully be based on non-linear dynamics and chaos theory, or more properly *complexity theory*' (Black 1962: 240) with a foundation in open systems theory (Scott 1992; Flood and Jackson 1991). Cheng and Van de Ven (1996:

594–595) provide a link between chaos, governance and innovation which can be traced through change and dynamism:

> Chaos requires a dynamic model. That is, the variables at any given time are a function, at least in part, of the same variables at an earlier time. Also the functional form of the model must be nonlinear in the variables. It need not be very complicated (May 1976). Nonlinearity simply requires that there be at least two not-entirely-compatible underlying forces or sources of demands. Stated differently, this means that there must be both positive and negative feedback loops. With this type of system, irregular and unpredictable behaviour can arise endogenously – that is without any exogenous, truly random inputs. This occurs when the balance between the positive and negative feedbacks is especially severe. (Koput 1992: 20–21).

Much of this we can identify in maritime governance. The maritime sector reflects earlier events at all times with both tradition and history vital components. There is at times severe nonlinearity (think of climate change and the environmental impact of shipping; the cataclysmic effect of containerisation, IT and communication changes). The demands that underlie shipping commonly contradict and both negative and positive feedback loops are present—the former exemplified by the relationship between cheap labour, flag-hopping and safety, the latter by environmental control and improvement of ships and market demand for quality.

Thus, maritime governance is positioned directly within a framework determined by process, in itself largely described by characteristics of complexity and chaos, fluid in its significance and ever-changing—all of which need to be incorporated in its design and application if it is to be effective. Almost none have even been considered despite widespread recognition of the need to do so and the failure within the maritime sector that results.

Two characteristics of this idealised dynamic governance for the maritime sector remain to be considered in the next two chapters—flow and speed, and it is to these issues that we now turn.

References

Abbott, A. (1990). A primer on sequence methods. *Organizational Science, 1*(4), 375–392.

Aingers, K., Chesters, G., Credland, T., Jordan, J., Stern, A., & Whitney, J. (2003). *We are everywhere: The irresistible rise of global anti-capitalism*. London: Verso.

Alberts, D. S., & Czerwinski, T. (1997). *Complexity, global politics and national security*. Washington, DC: National Defense University.

Albrow, M. (1996). *The Global Age*. Cambridge: Polity Press.

Allen, P. (1997). *Cities and regions as self-organizing systems*. London: Taylor and Francis.

Allen, P. M., Sanglier, M., Engelen, G., & Boon, F. (1985). Towards a new synthesis on the modelling of evolving complex systems. *Environment and Planning B, 12*, 65–84.

Alvesson, M. (1993). The play of metaphors. In J. Hassard & M. Parker (Eds.), *Postmodernism and organisations*. London: Sage.

Anderson, P. W., Arrow, K. J., & Pines, D. (Eds.) (1988). *Economy as an evolving complex system* (Vol. V), Santa-Fe Institute Studies in the Science of Complexity. Redwood City, CA: Addison-Wesley.

Angelides, C., & Caiden, G. (1994). Adjusting policy-thinking to global pragmatics and future problematic. *Public Administration and Development, 14*, 223–239.

Appadurai, A. (1990). Disjuncture and difference in the global cultural economy. In M. Featherstone (Ed.), *Global culture: Nationalism, globalization and modernity* (pp. 295–310). London: Sage.
Archer, M. (1998). Introduction: Realism in the social sciences. In M. S. Archer, R. Bhaskar, A. Collier, T. Lawson, & A. Norrie (Eds.), *Critical realism: Essential readings* (pp. 189–205). Abingdon: Psychology Press.
Argyris, C. (1982). *Reasoning, learning and action*. San Francisco: Jossey-Bass Publishing.
Argyris, C., & Schon, D. (1978). *Organizational learning: A theory of action perspective*. Reading, MA: Addison-Wesley.
Argyros, A. J. (1991). *A blessed rage for order: Deconstruction, evolution and chaos*. Ann Arbor, MI: University of Michigan Press.
Arrighi, G., & Silver, B. J. (1999). *Chaos and governance in the modern world system*. Minneapolis, MN: University of Minnesota Press.
Arthur, B. (1994). *Increasing returns and path dependency in the economy*. Ann Arbor, MI: University of Michigan Press.
Artigiani, R. (1987). Revolution and evolution: Applying Prigogine's dissipative structures model. *Journal of Social and Biological Structures, 10*, 249–264.
Axelrod, R., & Cohen, M. (1999). *Harnessing complexity*. New York: Free Press.
Bachelard, G. (1942). *Water and dreams: An essay on the imagination of matter*. Farrell, TX: Pegasus.
Back, W. W. (1997). Chaos and complexity. Necessary myths. In R. Eve, S. Horsfall, & M. Lee (Eds.), *Chaos, complexity and sociology* (pp. 39–51). Thousand Oaks, CA: Sage.
Bailey, N. T. J. (1957). *The mathematical theory of epidemics*. New York: Hafner.
Ball, P. (2004). *Critical mass: How one thing leads to another*. London: Heinemann.
Barabasi, A.-L. (2002). *Linked: The new science of networks*. Cambridge, MA: Perseus.
Barfield, T. (1993). *The nomadic alternative*. Englewood Cliffs, NJ: Prentice Hall.
Barnes, T. J., & Duncan, J. S. (Eds.). (1992). *Writing worlds: Discourse, text and metaphor in the representation of landscape*. London: Routledge.
Barnett, W. A., Kirman, A. P., & Salmon, M. (Eds.). (1996). *Nonlinear dynamics and economics*. Cambridge: Cambridge University Press.
Basch, L., Schiller, N. G., & Blanc, C. S. (1994). *Nations unbound: Transnational projects, post-colonial predicaments and deterritorialized nation states*. New York: Gordon and Breach.
Bauman, Z. (1998). *Globalization: The human consequences*. Cambridge: Polity Press.
Bauman, Z. (1999). *Postmodern ethics*. London: Blackwell.
Bauman, Z. (2000). *Liquid modernity*. Cambridge: Polity Press.
Baumol, W. J., & Benhabib, J. (1989). Chaos, significance, mechanism and economic applications. *Journal of Economic Perspectives, 3*, 77–105.
Beck, U. (2000). *What is globalization?*. Cambridge: Polity Press.
Bendor, J., Moe, T. M., & Shotts, K. W. (2001). Recycling the garbage can: an assessment of the research program. *American Political Science Review, 95*(1), 169–190.
Benko, G., & Strohmeyer, U. (1997). *Space and social theory. Interpreting modernity and post-modernity*. Oxford: Blackwell.
Benitez-Rojo, A. (1996). *The repeating island*. Durham, NC: Duke University Press.
Benjamin, A. (1995). Complexity: Architecture/art/philosophy. *Journal of Philosophy and the Visual Arts, 6*, 6–23.
Bennett, C. (1991). What is policy convergence and what causes it? *British Journal of Political Science, 21*, 215–233.
Bennett, C. J. (1997). Understanding ripple effects: The cross-national adoption of policy instruments for bureaucratic accountability. *Governance, 10*(3), 213–233.
Bennett, C. J., & Howlett, M. (1991). *When states learn do they change? American lessons, Canadian learning and the conceptualization of policy change*. Annual Meeting of the American Political Science Association.
Bergson, H.-L., Ansell-Pearson, K., & Mullarkey, J. (2002) *Henri Bergson: Key writings*. London: Continuum-3PL.

Berry, B. J. L. (1973). A paradigm for modern geography. In R. J. Chorley (Ed.), *Directions in geography* (pp. 3–21). London: Methuen.

Berry, F. S., & Berry, W. D. (1990). State lottery adoptions as policy innovations. *American Political Science Review, 84*(2), 395–416.

Best, S., & Kellner, D. (1991). *Postmodern theory: Critical interrogations*. New York: Guilford Press.

Bicchieri, C. (1988). Should a scientist abstain from metaphor? In A. Klamer, D. McCloskey, & R. Solow (Eds.), *The consequences of economic rhetoric*. Cambridge: Cambridge University Press.

Bishop, R. C. (2004). Non-equilibrium statistical mechanics Brussels-Austin style. *Studies in the History and Philosophy of Modern Physics, 35*, 1–3.

Black, M. (1962). *Models and metaphors*. Ithaca, NY: Cornell University Press.

Blake, W. (2000). *The marriage of Heaven or Hell*. New York: Dover Publications.

Blom-Hansen, J. (1997). A 'new institutional' perspective on policy networks. *Public Administration, 75*, 669–693.

Bogue, R. (2004). Apology for nomadology. *Interventions, 6*(2), 169–179.

Boons, F., Van Buuren, A., Gerrits, L., & Teisman, R. (2009). Towards an approach of evolutionary public management. In G. R. Teisman, A. van Buuren, & L. Gerrits (Eds.), *Managing complex governance systems: Dynamics, self-organization and coevolution in public investments* (pp. 231–249). New York: Routledge.

Borzel, T. (1998). Organizing Babylon—On the different conceptions of policy networks. *Public Administration, 76*, 253–273.

Braidotti, R. (1994). *Nomadic Subjects*. New York: Columbia University Press.

Brand, S. (1999). *The clock of the long now*. New York: Basic Books.

Brock, W. A. (1986). Distinguishing random and deterministic systems: Abridged version. *Journal of Economic Theory, 40*, 168–195.

Browning, D. (1965a). Preface. In D. Browning (Ed.), *Philosophers of process* (pp. xxiii–xxv). New York: Random House.

Browning, D. (1965b). Process. In D. Browning (Ed.), *Philosophers of process* (pp. 305–312). New York: Random House.

Buchanan, M. (2002). *Small world: Uncovering nature's hidden networks*. London: Weidenfeld and Nicolson.

Budd, M., Entman, R. M., & Steinmann, C. (1990). The affirmative character of US cultural studies. *Critical Studies in Mass Communication, 7*(2), 169–184.

Buijs, J.-M., Eshuis, J., & Byrne, D. (2009). Approaches to researching complexity in public management. In G. R. Teisman, A. van Buuren, & L. Gerrits (Eds.), *Managing complex governance systems: Dynamics, self-organization and coevolution in public investments* (pp. 37–55). New York: Routledge.

Buttimer, A. (1982). Musing on helicon: Root metaphors and geography. *Geografiska Annaler B, 64*(2), 89–96.

Byrne, D. S. (1998). *Complexity theory and the social sciences*. London: Routledge.

Byrne, D. S. (2005). Complexity, configuration and cases. *Theory, Culture and Society, 22*(5), 95–111.

Calhoun, C. (1995). *Critical social theory*. Oxford: Blackwell.

Cameron, L., & Low, G. (Eds.). (1999). *Researching and applying metaphor*. Cambridge: Cambridge University Press.

Capra, F. (1996). *The web of life*. London: HarperCollins.

Capra, F. (2001). *The hidden connections: A science for sustainable living*. London: HarperCollins.

Carlson, B. L., Koenig, J., & Reid, G. L. (1986). *Lessons from Europe: The role of the employment security system*. Washington, DC: National Governors' Association.

Castells, M. (1996). *The rise of the network society*. Oxford: Blackwell.

Casti, J. (1994). *Complexification*. London: Abacus.

Cebrowski, A., & Gartska, J. (1998). *Network-centric warfare: Its origins and future*. Annapolis MD: United States Naval Institute Proceedings.
Celek, M., & Er, I. D. (2006). *Application requirements of catastrophe theory in maritime transportation industry* (pp. 879–887). 3rd International Conference on Maritime Transport, Barcelona.
Cheng, Y.-T., & Van de Ven, A. H. (1996). Learning the innovation journey: Order out of chaos? *Organization Science, 7*(6), 593–614.
Choi, T. Y., Dooley, K. J., & Rungtusanatham, M. (2001). Supply networks and complex adaptive systems: Control versus emergence. *Journal of Operations Management, 19*, 351–366.
Chorley, R. J., & Kennedy, B. A. (1971). *Physical geography: A systems approach*. London: Prentice Hall.
Cilliers, P. (1998). *Complexity and postmodernism*. London: Routledge.
Clifford, J. (1992). Travelling cultures. In I. Grossberg, C. Nelson, & P. A. Treichler (Eds.), *Cultural studies*. New York: Routledge.
Clifford, J. (1997). *Routes. Travel and translation in the late twentieth century*. Cambridge, MA: Harvard University Press.
Cohen, J., & Stewart, I. (1994). *The collapse of chaos*. Harmondsworth: Penguin.
Coleman, J. S., Katz, E., & Menzel, H. (1966). *Medical innovation: A diffusion study*. Indianapolis, IN: Bobbs-Merrill Co.
Coleman, W. D. (1994). Policy convergence in banking: A comparative study. *Political Studies, XLII*, 274–292.
Collier, J., & Esteban, R. (1999) Governance in the participative organisation; Freedom, creativity and ethics. *Journal of Business Ethics, 21*, 2/3, 173–188.
Collier, D., & Messick, R. E. (1975). Prerequisites versus diffusion: Testing alternative explanations of social security adoption. *American Political Science Review, 69*(4), 1299–1315.
Congreve, W. (1923). In M. Summers (Eds.), *The complete works* (4 Vols.). London: Nonesuch Press.
Coveney, P. (2000). A clash of doctrines: the arrow of time in modern physics. In P. Baert (Ed.), *Time in contemporary intellectual thought*. Amsterdam: Elsevier.
Coveney, P., & Highfield, R. (1991). *The arrow of time*. New York: Fawcett Columbine.
Cox, R. (1999). *Ideas, policy borrowing and welfare reform*. Conference on global trajectories: Ideas, international policy transfer and models of welfare reform, Florence, Italy, March 25–26.
Crain, R. L. (1966). Fluoridation: The diffusion of an innovation among cities. *Social Forces, 44*(4), 467–476.
Cresswell, T. (1997). Imagining the nomad: Mobility and the postmodern primitive. In G. B. Benko & U. Strohmeyer (Eds.), *Space and social theory* (pp. 360–382). London: Wiley-Blackwell.
Cresswell, T. (2006). *On the move*. London: Routledge.
Cribb, R. (1991). *Nomads in archaeology*. Cambridge: Cambridge University Press.
D'Andrea, A. (2004). Global nomads: Techno and new age as transnational countercultures in Ibiza and Goa. In G. Saint-John (Ed.), *Rave Culture and Religion* (pp. 236–255). London: Routledge.
D'Andrea, A. (2006). Neo-nomadism: A theory of post-identarian mobility in the global age. *Mobilities, 1*(1), 95–119.
Da Vinci, L. quoted in Richter (2008). *The Notebooks of Leonardo da Vinci*. CreateSpace, pp. 402–403.
Davis, S. (1982). Transforming organizations. The key to strategy is context. *Organizational Dynamics, 10*, 64–80.
De Landa, M. (1991). *War in the age of intelligent machines*. New York: Zone Books.
De Landa, M. (2002). *Intensive science and virtual philosophy*. London: Continuum.
De Lange, M. (2009). *Digital nomadism: A critique*. Draft version Ph.D. http://blog.bijt.org.
Deleuze, G., & Guattari, F. (1986). *Nomadology*. New York: Semiotext(e).
Deleuze, G., & Guattari, F. (1988). *A thousand plateaus. Capitalism and schizophrenia*. Minneapolis, MI: University of Minnesota Press.

De Lillo, D. (1990). *White noise*. New York: Picador.
Dennard, L., Richardson, K., & Morcul, G. (Eds.), (2008) *Complexity and policy analysis: Tools and concepts for designing robust policies in a complex world* (*Exploring Organizational Complexity Volume 2*). Goodyear, AZ: ISCE Publishing.
Dennis, K., & Urry, J. (2009). *After the car*. Cambridge: Polity Press.
Derrida, J. (1987). *Positions*. London: Athlone Press.
Derrida, J. (1988). *Limited Inc*. Evanston, IL: Northwestern University Press.
Dillon, M. (2000). Restructuralism, complexity and poetics. *Theory, Culture and Society, 17*(5), 1–26.
Dix, G. (1957). The ministry of the early church. In K. E. Kirk (Ed.), *The apostolic ministry* (pp. 183–303). London: Hodder and Stoughton.
Dolowitz, D. (1997). *Where's the State? The political process of globalisation*. Globalisation Critical Perspectives Conference, University of Birmingham, UK.
Dolowitz, D., & Marsh, D. (1996). Who learns from whom: A review of the policy transfer literature. *Political Studies, 44*, 343–357.
Dommergues, P., Sibille, H., & Wurzburg, E. (1989). *Mechanisms for job creation: Lessons from the United States*. Paris: OECD.
Dooley, K., Johnson, T., & Bush, D. (1995). TQM, chaos and complexity. *Human Systems Management, 14*, 1–16.
Dooley, K. J., & Van de Ven, A. H. (1999). Explaining complex organizational dynamics. *Organization Science, 10*(3), 358–372.
Dror, Y. (1990). Fateful decisions as fuzzy gambles with history. *The Jerusalem Journal of International Relations, 12*(3), 1–12.
Eade, J. (Ed.). (1997). *Living the global city*. London: Routledge.
Eatwell, J., & Taylor, L. (2000). *Global finance at risk*. New York: The New Press.
Edens, B. (2001). *Semigroups and symmetry: An investigation of Prigogine's theories*. Institute for History and Foundations of Science, Utrecht University. Ph.D. Thesis. http://philsci-archive.pitt.edu/documents/disk0/00/00/04/36/.
Edwards, R., Nicoll, K., Solomon, N., & Usher, R. (2004). *Rhetoric and educational discourse*. London: Routledge.
Eichenbaum, J., & Gale, S. (1971). Form, function and process: A methodological inquiry. *Economic Geography, 47*(4), 525–544.
Eisen, A. M. (1986). *Galut: Modern Jewish reflection on homelessness and homecoming*. Bloomington, IN: Indiana University Press.
Eisenstein, S. (1969). Through theatre to cinema, in S. Eisenstein. In *Film form*: *Essays in film theory* (p. 3). Boston MA: Harcourt.
Elliott, E., & Kiel, L. D. (1997). Nonlinear dynamics, complexity and public policy. Use, misuse and applicability. In R. A. Eve, S. Horsfall, & M. Lee (Eds.), *Chaos, complexity and sociology* (pp. 64–78). Thousand Oaks, CA: Sage.
Emerson, R. W. (1981). *Selected writings of Ralph Waldo Emerson, edited by Donald McQuade*. New York: Modern Library.
Engelen, G. (1988). The theory of self-organization and modelling complex urban systems. *European Journal of Operational Research, 37*, 42–57.
Etheredge, L. S. (1981). Government learning. An overview. In S. L. Long (Ed.), *The handbook of political behavior* (Vol. 2). New York: Pergamon.
Evans, M., & Davies, J. (1999). Understanding policy transfer: A multi-level, multi-disciplinary perspective. *Public Administration, 77*(2), 361–385.
Eve, R. A. (1997). Afterword. So where are we now? A final word. In R. A. Eve, S. Horsfall, & M. Lee (Eds.), *Chaos, complexity and sociology* (pp. 269–279). Thousand Oaks, CA: Sage.
Eyestone, R. (1977). Confusion, diffusion and innovation. *Political Science Review, 71*, 441–447.
Farina, C., & Kelly, M. (1983). Innovation policy and the social sciences. *Policy Studies Review, 3*, 21–28.
Ferguson, N. (Ed.). (1997). *Virtual history: Alternatives and counterfactuals*. London: Picador.

Fischer, F. (1993). Burger, Experten und Politik nach dem 'Nimby' – Prinzip: Einn Pladoyer fur die Partizipatorische Policy-Analyse. *PVS-Sonderheft, 24*, 451–470.
Fischer, F., & Forester, J. (Eds.). (1993). *The argumentative turn in policy analysis and planning*. Durham, NC: Duke University Press.
Flood, R. L., & Jackson, M. C. (1991). *Critical systems thinking*. Chichester: Wiley.
Forster, E. M. (1931). *Howard's End*. London: Penguin.
Fraser, B. (1979). The interpretation of novel metaphors. In A. Ortony (Ed.), *Metaphor and thought*. Cambridge: Cambridge University Press.
Fraser, M., Kember, S., & Lury, C. (Eds.) (2005). Inventive life: Approaches to the new vitalism. *Theory, Culture and Society*, *22*(1), 1–14.
Freeman, R. (1999) *Policy transfer in the health sector*. School of Social and Political Science, University of Edinburgh, Working Paper.
Freeman, W. J. (1991). The physiology of perception. *Scientific American*, February 34–41.
Game, A., & Metcalfe, D. (1996). *Passionate sociology*. London: Sage.
Gay, P., Hall, S., Janes, L., Madsen, A. K., Mackay, H., & Negus, K. (1996). *Doing cultural studies: The story of the Sony Walkman*. London: Sage.
Gell-Mann, M. (1994). *The quark and the jaguar: Adventures in the simple and the complex*. New York: Freeman.
Gemmill, G., & Smith, C. (1985). A dissipative structure model of organization transforming. *Human Relations, 38*(8), 751–766.
Gerrits, L. M. (2008). *The gentle art of coevolution managing and developing estuaries in Germany, Belgium and the Netherlands*. Ph.D. Thesis, Erasmus University, Rotterdam.
Ghosh, A. (1986). The Imam and the Indian. *Granta, 20*(Winter), 135–146.
Giddens, A. (1990). *The consequences of modernity*. Stanford, CA: Stanford University Press.
Gilroy, P. (1993). *The black Atlantic: Modernity and double consciousness*. London: Verso.
Gladwell, M. (2002). *Tipping points: How little things can make a big difference*. Boston, MA: Little Brown and Co.
Gleik, J. (1987). *Chaos: Making a new science*. New York: Viking.
Glick, W., Huber, G., Miller, C., Doty, D. H., & Sutcliffe, K. (1990). Studying changes in organizational design and effectiveness. *Organizational Science, 1*(3), 293–312.
Godet, M. (1980). Europe facing its futures. *Technological Forecasting and Social Change, 18*(2), 161–173.
Goldstein, J. (1994). *The unshackled organization*. Portland, OR: Productivity Press.
Golombiewski, R., Billingsley, K., & Yeager, S. (1975). Measuring change and resistance in human affairs: Types of change generated by OD designs. *Journal of Applied Behavior Science, 12*, 133–157.
Grandmont, J. M. (Ed.). (1986). Symposium on nonlinear economic dynamics. *Journal of Economic Theory*, *40*(1), 1–196.
Gray, J. (2001). The era of globalisation is over. *New Statesman*, 24th September.
Gray, V. (1973). Innovation in states; A diffusion study. *American Political Science Review, 67*(4), 1174–1185.
Gregor, J. (1971). *An introduction to metapolitics*. New York: The Free Press.
Greenberg, G. D., Miller, J. A., Mohr, L. B., & Vladeck, B. C. (1977). Developing public policy theory: Perspectives from empirical research. *American Political Science Review, 71*(4), 1532–1543.
Griffin, D. R. (1998). Process philosophy. In E. Craig (Ed.), *Routledge encyclopedia of philosophy*. London: Routledge.
Gunaratna, R. (2002). *Inside Al-Qaeda: Global networks of terror*. New York: Columbia University Press.
Haas, E. B. (1990). *When knowledge is power: Three models of change in international organizations*. Berkeley, CA: University of California Press.
Hagerstrand, T. (2004). The two vistas. *Geografiska Annaler B, 86*(4), 315–323.
Hancher, L., & Moran, M. (1989). Organizing regulatory space. In L. Hancher & M. Moran (Eds.), *Capitalism, culture and economic regulation* (pp. 271–299). Oxford: Clarendon Press.

Hannam, K. (2009). The end of tourism? Nomadology and the mobilities paradigm. In J. Tribe (Ed.), *Philosophical issues in tourism* (pp. 101–134). Bristol: Channel View.
Hannam, K., Sheller, M., & Urry, J. (2006). Editorial. Mobilities, immobilities and moorings. *Mobilities, 1*, 1–22.
Hannigan, J. (2002). Culture, globalization, and social cohesion; Toward a de-territorialized, global fluids model. *Canadian Journal of Communication, 27*, 277–287.
Hanson, S., & Pratt, G. (1995). *Gender, work and space*. London: Routledge.
Hardt, M., & Negri, T. (2000). *Empire*. Cambridge, MA: Harvard University Press.
Hartshorne, C. (1965). Introduction. The development of process philosophy. In D. Browning (Ed.), *Philosophers of Process* (pp. v–xxii). New York: Random House.
Harvey, D. (1989). *The condition of postmodernity*. Oxford: Blackwell.
Hayles, D. (Ed.). (1991). *Chaos and order*. London: University of Chicago Press.
Hayles, N. K. (1999). *How we became posthuman*. Chicago, IL: University of Chicago Press.
Haynes, P. (1999). *Complex policy planning: The government strategic management of the social care market*. Aldershot: Ashgate.
Heclo, H. (1974). *Modern social politics in Britain and Sweden*. New Haven, CT: Yale University Press.
Hernes, G. (1976). Structural change in social processes. *American Journal of Sociology, 82*, 513–547.
Herod, A., O'Tuathail, G., & Roberts, S. M. (Eds.). (1998). *An unruly world? Globalization, governance and geography*. London: Routledge.
Hier, S. P., & Greenberg, J. (2007). *The surveillance studies reader*. New York: McGraw Hill.
Holland, J. A. (1998). *Emergence*. Reading, MA: Addison-Wesley.
Hogwood, B. W., & Peters, B. G. (1983). *Policy dynamics*. Brighton: Wheatsheaf Books.
Huggett, R., & Perkins, C. (2004). Landscape as form, process and meaning. In J. A. Matthews & D. T. Herbert (Eds.), *Unifying geography: Common heritage, shared future* (pp. 224–239). London: Routledge.
Ikenberry, G. J. (1990). The international spread of privatization policies; inducements, learning and 'policy band wagoning'. In E. Suleiman & J. Waterbury (Eds.), *The political economy of public sector reform and privatization*. Boulder, CO: Westview Press.
Inkeles, A. (1999). *One world emerging? Convergence and divergence in industrial societies*. Boulder, CO: Westview Press.
Isard, W. (1996). *Commonalities in art, science and religion*. London: Avebury.
Jencks, C. (1996). *The architecture of the jumping universe, a polemic: How complexity science is changing architecture and culture*. London: Academy Editions.
Jervis, R. (1997). *System effects: Complexity in political and social life*. Princeton, NJ: Princeton University Press.
Jokinen, E., & Veijola, S. (1994). *The death of the tourist. Seven improvisations*. XIIIth World Congress of Sociology, Bielefeld, Germany.
Jones, S. (1994). Demonology: Some thoughts towards a science of chaos in recent British theatre. *Contemporary Theatre Review, 11*, 49–59.
Kaplan, C. (1996). *Questions of travel*. Raleigh, NC: Duke University Press.
Kaufamn, H. (1991). *Time chance and organizations: Natural selection in a perilous environment*. Chatham, NJ: Chatham House.
Kaufmann, S. (1993). *The origins of order*. New York: Oxford University Press.
Kelly, K. (1998). *New rules for the new economy*. London: Fourth Estate.
Kemp, S. (2009). Unpredictability and nonlinearity in complexity theory. *A critical appraisal, Emergence: Complexity and Organization, 11*(1), 84–93.
Kennedy, P. (2007). Global transformations but local 'bubble' lives: Taking a reality check on some globalization concepts. *Globalizations, 4*(2), 267–282.
Kerr, C. (1983). *The future of industrial societies: Convergence or continuing diversity?*. Cambridge, MA: Harvard University Press.
Khalil, E. L., & Boulding, K. E. (Eds.). (1996). *Evolution, order and complexity*. London: Routledge.

Khazanov, A. (1984). *Nomads and the outside world*. Cambridge: Cambridge University Press.
Kiel, L. D. (1991). Lessons from the nonlinear paradigm: Applications of the theory of dissipative structures in the social sciences. *Social Science Quarterly, 72*(3), 431–442.
Kiel, L., & Elliott, E. (Eds.). (1996). *Chaos theory in the social sciences*. Ann Arbor, MI: University of Michigan.
Klijn, E.-H., & Snellen, I. (2009). Complexity theory and public administration. In G. Teisman, A. Van Buuren, & L. Gerrits (Eds.), *Managing complex governance systems* (pp. 17–36). New York: Routledge.
Knoepfel, P., & Kissling-Naf, I. (1998). Social learning in policy networks. *Policy and Politics, 26*(3), 343–367.
Kolodny, A. (1973). *The lay of the land: Metaphor as experience and history in American life and letters*. Chapel Hill, NC: University of North Carolina Press.
Koput, K. (1992). *Dynamics of innovation idea generation in organizations: Randomness and chaos in the development of a new medical device*. Unpublished Ph.D. Dissertation, School of Business, University of California at Berkeley, CA.
Kristeva, J. (1982). *The powers of horror: An essay on abjection*. New York: Columbia University Press.
Krugman, P. (1996). *The self-organizing economy*. Cambridge, MA: Blackwell.
Lakoff, G., & Johnson, M. (1980). *Metaphors we live by*. Chicago, IL: University of Chicago Press.
Lancaster, W., & Lancaster, F. (1998). Who are these nomads? What do they do? Continuous change or changing continuities. In J. Ginat & A. Khazanov (Eds.), *Changing nomads in a changing society*. Portland, OR: Sussex.
Landau, M. (1972). *Political theory and political science*. New York: Macmillan.
Lansing, J. S. (2003). Complex adaptive systems. *Annual Review of Anthropology, 32*, 183–204.
Law, J. (1994). *Organizing modernity*. Oxford: Blackwell Publishers.
Law, J., & Urry, J. (2004). Enacting the social. *Economy and Society, 33*(3), 390–410.
Law, R. (1999). Beyond 'women and transport': Towards new geographies of gender and daily mobility. *Progress in Human Geography, 23*(4), 567–588.
Lee, M. E. (1997). From enlightenment to chaos: toward nonmodern social theory. In R. A. Eve, S. Horsfall, & M. E. Lee (Eds.), *Chaos, complexity and sociology* (pp. 15–29). Thousand Oaks, CA: Sage.
Lefebvre, H. (1991). *The production of space*. Oxford: Blackwell.
Le Guin, U. (1974). *The dispossessed: An ambiguous utopia*. New York: Harper and Row.
Lewin, R. (1992). *Complexity. Life at the edge of chaos*. New York: Macmillan.
Lindblom, C. E. (1965). *The intelligence of democracy*. New York: Free Press.
Lindblom, C. E., & Cohen, D. K. (1979). *Usable knowledge: Social science and social problem-solving*. New Haven, CT: Yale University Press.
Livingston, I. (1997). *Arrow of chaos: Romanticism and postmodernity*. Minneapolis, MN: University of Minnesota Press.
Livingstone, D. N., & Harrison, R. T. (1980). The frontier: Metaphor, myth and model. *The Professional Geographer, 32*, 127–132.
Livingstone, D. N., & Harrison, R. T. (1981). Meaning through metaphor: Analogy as epistemology. *Annals of the Association of American Geographers, 71*(1), 95–107.
Locke, J. (1996). *An essay concerning human understanding*. London: Penguin.
Locke, R., & Jacoby, W. (1997). The dilemmas of diffusion. *Politics and Society, 25*(1), 34–65.
Lorenz, E. N. (1993). *The essence of chaos*. London: UCL Press.
Loye, D., & Eisler, R. (1987). Chaos and transformation: Implications of non-equilibrium theory for social science and society. *Behavioral Science, 32*, 53–65.
Luhmann, N. (1995). *Social systems*. Stanford, CA: Stanford University Press.
Lury, C. (1987). The objects of travel. In C. Rojek & J. Urry (Eds.), *Touring cultures* (pp. 75–95). London: Routledge.
Lynn, L. (Ed.). (1978). *Knowledge and policy: The uncertain connection*. Washington, DC: National Academy of Sciences.

Majone, G. (1993). Wann ist poicy-deliberation wichtig? *PVS Sonderheft, 24*, 97–115.

Malkki, L. (1992). National geographic: The rooting of peoples and the territorialisation of national identity among scholars and refugees. *Cultural Anthropology, 7*(1), 24–44.

Marston, S. A., Jones, J. P, I. I. I., & Woodward, K. (2005). Human geography without scale. *Transactions of the Institute of British Geographers NS, 30*, 416–432.

Marx, K., & Engels, F. (1952). *The manifesto of the communist party*. Moscow: Foreign Languages Publishing House.

Matless, D. (1995). Culture run riot? Work in social and cultural geography, 1994. *Progress in Human Geography, 19*, 395–403.

May, R. M. (1976). Simple mathematical models with very complicated dynamics. *Nature, 261*, 459–467.

Mayer-Kress, G. (1994). *Presentation at chaos and complexity: Their meaning for business, economics and society*. Dallas, TX: University of Texas.

McCloskey, M. A. (1964). Metaphors. *Mind. New Series, 73*(290), 215–233.

McDowell, L. (1996). Off the road. Alternative views of rebellion, resistance and 'the beats'. *Transactions of the Institute of British Geographers NS, 21*, 412–419.

McKay, G. (1996). *Senseless acts of beauty: Cultures of resistance since the sixties*. London: Verso.

McKelvey, B. (2002) *Managing coevolutionary dynamics*. 18th FGOS Conference, Barcelona.

McLennan, G. (2003). Sociology's complexity. *Sociology, 37*(3), 547–564.

McVoy, E. C. (1940). Patterns of diffusion in the United States. *American Sociological Review, 5*, 219–227.

Medd, W., & Marvin, S. (2005). From the politics of urgency to the governance of preparedness: A research agenda on urban vulnerability. *Journal of Contingencies and Crisis Management, 13*(2), 44–49.

Meentemeyer, V. (1989). Geographical perspectives of space, time and scale. *Landscape Ecology, 2*(3/4), 163–173.

Mihata, K. (1997). The persistence of 'emergence'. In R. Eve, S. Horsfall, & M. Lee (Eds.), *Chaos, complexity and sociology* (pp. 30–38). Thousand Oaks, CA: Sage.

Miller, C. L. (1993). The postidentitarian predicament in the footnotes of *A Thousand Plateaus*: Nomadology, anthropology and authority. *Diacritics, 23*(3), 6–35.

Miller, E., & Friesen, P. (1982). Structural change and performance: Quantum vs piecemeal-incremental approaches. *Academy of Management Journal, 25*(4), 867–892.

Mills, W. J. (1982). Metaphorical vision: Changes in western attitudes to the environment. *Annals of the Association of American Geographers, 72*(2), 237–253.

Mingers, J. (1995). *Self-producing systems*. New York: Plenum.

Mintzberg, H. (1989). *Mintzberg on management: Inside our strange world of organizations*. New York: The Free Press.

Mintzberg, H., & Jorgenson, J. (1987). Emergent strategy for public policy. *Canadian Public Administration, 30*(2), 214–229.

Mirowski, P. (1994). *Natural images in economic thought*. Cambridge: Cambridge University Press.

Mitleton-Kelly, E. (2003). *Complex systems and evolutionary perspectives of organisations*. London: Elsevier.

Mol, A. P. J. (2007). Boundless biofuels? Between environmental sustainability and vulnerability. *Sociologica Ruralis, 47*(4), 297–315.

Mol, A. P. J., & Law, J. (1994). Regions, networks and fluids: Anaemia and social topology. *Social Studies of Science, 24*, 641–671.

Mol, A. P. J., & Spaargaren, G. (2005). From additions and withdrawals to environmental flows. *Organization and Environment, 18*(1), 91–107.

Morcul, G. (2003). *Complexity and public administration*. Public Administration Theory Network Annual Conference, Anchorage, AK.

More, T. (1965). *Utopia*. London: Penguin.

Morgan, A. (2005). The global dimension. Contexts within contexts. *Zeitschrift fur Internationale Bildungsforschung und Entwicklungspadagogik, 28*(2), 26–28.
Morgan, G. (1998). *Images of organization*. Thousand Oaks, CA: Sage.
Morin, E. (2002). The notion of the subject. In D. F. Schnitman & J. Schnitman (Eds.), *New paradigms, culture and subjectivity*. New Jersey, NJ: Hampton Press.
Morley, D. (1999). Bounded realms: Household, family, community and nation. In H. Naficy (Ed.), *Home, exile, homeland* (pp. 151–168). New York: Routledge.
Morris, M. (1988). At Henry Parkes Motel. *Cultural Studies, 2*, 1–47.
Nash, W. (1989). *Rhetoric: The wit of persuasion*. Oxford: Blackwell.
Nedley, A. (1999). *Policy transfer and the developing country experience gap: Taking a southern perspective*. York: University of York.
Nibley, H. (1966). Tenting, toll and taxing. *Western Political Quarterly, 19*, 599–630.
Nicolis, G. (1995). *Introduction to non-linear science*. Cambridge: Cambridge University Press.
Nicolis, G., & Prigogine, I. (1989). *Exploring complexity: An introduction*. New York: Freeman.
Nonaka, I. (1988). Creating organizational order out of chaos: Self-renewal in Japanese firms. *California Management Review, 30*(3), 57–73.
Nordlinger, E. (1981). *On the autonomy of the democratic state*. Cambridge, MA: Harvard University Press.
Norgaard, R. B. (1984). Coevolutionary development potential. *Land Economics, 60*(2), 160–173.
Norgaard, R. B. (1994). *Development betrayed: The end of progress and a coevolutionary revisioning of the future*. London: Routledge.
Noyes, J. K. (2004). Nomadism, nomadology, postcolonialism. *Interventions, 6*(2), 159–168.
Nullmeier, F. (1993). Wissen und Policy Forschung, Wissen spolitologie und rhetorisch-dialektisches Handlungsmodell. *PVS-Sonderheft, 24*, 175–196.
O'Loughlin, J. O., Ward, M. D., Lofdahl, C. L., Cohen, J. S., Brown, D. S., Reilly, D., et al. (1998). The diffusion of democracy 1946–1994. *Annals of the Association of American Geographers, 88*(4), 545–574.
Ortony, A. (Ed.). (1979). *Metaphor and thought*. Cambridge: Cambridge University Press.
O'Toole, L. J. (1993). Multiorganizational policy implementation: Some limitations and possibilities for rational-choice contributions. In F. W. Scharpf (Ed.), *Games in hierarchies and networks*. Campus: Frankfurt.
Padgett, J. F. (1980). Managing garbage can hierarchies. *Administrative Science Quarterly, 25*(4), 583–602.
Paechter, C. (2004). Metaphors of space in educational theory and practice. *Pedagogy, Culture, Society, 12*(3), 449–464.
Pagels, H. (1988). *The dream of reason*. New York: Simon and Schuster.
Palan, R. (2003). *The offshore world: Sovereign markets, virtual places and nomad millionaires*. Ithaca, NY: Cornell University Press.
Parker, M. (2009). Angelic organization: Hierarchy and the tyranny of heaven. *Organization Studies, 30*(11), 1281–1299.
Peat, F. D. (1991). *The philosopher's stone. Chaos, synchronicity and the hidden order of the world*. New York: Bantam.
Pel, B. (2009). The complexity of self organisation. Boundary judgements in traffic management. In G. Teisman, A. Van Buuren, & L. Gerrits (Eds.), *Managing complex governance systems* (pp. 116–133). New York: Routledge.
Peters, B. G. (2002). *Governance: A garbage can perspective*. Institut fur Hohere Studien (HIS), Wien, Austria.
Peters, J. D. (1999). Exile, nomadism and diaspora. In H. Naficy (Ed.), *Home, exile, homeland* (pp. 17–41). New York: Routledge.
Pettigrew, A. (1990). Longitudinal field research on change. *Organizational Science, 1*(3), 267–292.
Plant, S. (1997). *Zeros and ones*. London: Fourth Estate.

Platt, J. (1970). Hierarchical growth. *Bulletin of the Atomic Scientists, 26*(9), 2–4, 46–48.

Pollitt, C. (2009). Complexity theory and evolutionary public administration. In G. R. Teisman, A. van Buuren, & L. Gerrits (Eds.), *Managing complex governance systems: Dynamics, self-organization and coevolution in public investments* (pp. 213–230). New York: Routledge.

Polsby, N. W. (1984). *Political innovation in America: The politics of policy initiation*. New Haven, CT: Yale University Press.

Porteus, J. (1985). Smellscape. *Progress in Human Geography, 9*, 356–378.

Pratchett, T. (1989). *Guards! Guards!*. London: Victor Gollancz.

Price, B. (1997). The myth of postmodern science. In R. Eve, S. Horsfall, & M. Lee (Eds.), *Chaos, complexity and sociology* (pp. 3–14). Thousand Oaks, CA: Sage.

Priesmeyer, H. R., & Baik, K. (1989). Discovering the patterns of chaos. *Planning Review, 17*(6), 14–21, 47.

Prigogine, I. (1997). *The end of certainty*. New York: Free Press.

Prigogine, I., & Stengers, I. (1984). *Order out of chaos*. London: Heinemann.

Punter, D. (2007). *Metaphor*. London: Routledge.

Radaelli, C. M. (2000). Policy transfer in the European Union: Institutional isomorphism as a source of legitimacy. *Governance, 13*(1), 25–43.

Rao, A. (Ed.). (1987). *The other nomads: Peripatetic minorities in cross-cultural perspective*. Cologne: Bohlau.

Rasmussen, D. R., & Mosekilde, E. (1988). Bifurcations and chaos in a generic management model. *European Journal of Operational Research, 35*, 80–88.

Rasch, W., & Wolfe, C. (Eds.). (2000). *Observing complexity*. Minneapolis, MN: University of Minnesota Press.

Reeve, S. and Foden, G. (2001) A new breed of terror, *The Guardian*, September 12th.

Rescher, N. (1998). *Complexity*. New Brunswick, NJ: Transaction.

Rescher, N. (2000). *Process philosophy*. Pittsburgh, PA: University of Pittsburgh Press.

Roche, M. (2000). *Mega events and modernity*. London: Routledge.

Roe, M. S. (2013). *Maritime governance and policy-making*. London: Springer.

Rogers, E. M. (2003). *Diffusion of innovations*. London: Simon and Schuster.

Rogers, E. M., & Shoemaker, F. (1971). *Communications of innovations: A cross-cultural approach*. New York: Free Press of Glencoe.

Rojek, C., & Urry, J. (Eds.). (1997a). *Touring cultures. Transformations of travel and theory*. London: Routledge.

Rojek, C., & Urry, J. (1997b). Transformations of travel and theory. In C. Rojek & J. Urry (Eds.), *Touring cultures. Transformations of travel and theory* (pp. 1–22). London: Routledge.

Rose, R. (1988). Comparative policy analysis: The program approach. In M. Dogan (Ed.), *Comparing pluralist democracies* (pp. 219–241). Boulder, CO: Westview.

Rose, R. (1991). What is lesson drawing? *Journal of Public Policy, 11*(1), 3–30.

Rose, R. (1993). *Lesson-drawing in public policy*. Chatham, NJ: Chatham House.

Rosenau, J. (1992). *Governance without consent: Order and change in world politics*. Cambridge: Cambridge University Press.

Ruelle, D., & Takens, F. (1971). On the nature of turbulence. *Communications in Mathematical Physics, 20*, 167–192.

Russell, B. (1912). The philosophy of logical atomism. *The Monist, 28*(4), 495–527.

Ryan, B., & Gross, N. C. (1943). The diffusion of hybrid seed corn in two Iowa communities. *Rural Sociology, 13*, 15–24.

Rycroft, R., & Kash, D. (1999). *The Complexity challenge*. London: Pinter.

Sabatier, P. A. (1978). The acquisition and utilization of technical information by administrative agencies. *Administrative Science Quarterly, 6*, 386–411.

Sabatier, P. A. (1993). Policy change over a decade or more. In P. A. Sabatier & H. Jenkins-Smith (Eds.), *Policy change and learning: An advocacy coalition approach*. Boulder, CO: Westview.

Sachs, S. (1978). *On metaphor*. Chicago, IL: University of Chicago Press.
Said, E. (1994). *Culture and imperialism*. London: Vintage.
Salo, M. T. (1986). *Peripatetic adaptation in historical perspective* (pp. 7–36). Commission on Nomadic Peoples, Osaka, Japan, 21/22.
Salo, M. T., & Salo, S. (1982). Romnichel economic and social organisation in urban New England 1850–1930. *Urban Anthropology, 11*(3–4), 273–314.
Saperstein, A. M., & Mayer-Kreiss, G. (1988). A non-linear dynamic model of the impact of SDI on the arms race. *Journal of Conflict Resolution, 32*, 636–671.
Savage, R. L. (1985). Diffusion: Research traditions and the spread of policy innovation in a Federal system. *Publius, 15*(4), 1–28.
Scharpf, F. W. (1993). Positive und negative Koordination in Verhandlungssystemen. *PVS-Sonderheft, 24*, 57–83.
Schneider, A., & Ingram, H. (1988). Systematically pinching ideas: A comparative approach to policy design. *Journal of Public Policy, 8*(1), 61–80.
Schnitman, D. F., & Schnitman, J. (Eds.). (2002). *New paradigms, culture and subjectivity*. Cresskill, NJ: Hampton Press.
Scholte, J. A. (1996). Beyond the buzzword: Towards a critical theory of globalization. In E. Korfman & G. Youngs (Eds.), *Globalization, theory and practice* (pp. 50–51). London: Cassel.
Scholte, J. A. (2000). *Globalization: A critical introduction*. Basingstoke: Macmillan.
Schon, D., & Rein, M. (1994). *Frame reflection: Toward the resolution of intractable policy controversies*. New York: Basic Books.
Scott, W. R. (1992). *Organizations: rational, natural and open systems*. Englewood Cliffs, NJ: Prentice Hall.
Seamon, D. (1980). Body-subject, time-space routines and place ballets. In A. Buttimer & D. Seamon (Eds.), *The human experience of space and place* (pp. 148–165). London: Croom Helm.
Seamon, D., & Nordin, C. (1980). Marketplace as place ballet. A Swedish example. *Geografiska Institutonem, Goteborg, Series B*, 35–41.
Sfard, A. (1998). On two metaphors for learning and the dangers of losing just one. *Educational Researcher, 27*(2), 4–13.
Sheldon, A. (1980). Organizational paradigms: A theory of organizational change. *Organizational Dynamics, 8*, 61–80.
Sheller, M. (2000). *The mechanisms of mobility and liquidity. Re-thinking the movement in social movements*, Sociology Department, Lancaster University.
Sheller, M., & Urry, J. (2000). *Mobile transformations of 'public' and 'private' life*. American Sociological Association Annual Conference, Washington, DC.
Sheller, M., & Urry, J. (2006). The new mobilities paradigm. *Environment and Planning A, 38*, 207–226.
Shields, R. (1997). Flow as a new paradigm. *Space and Culture, 1*, 1–4.
Shrivastava, P. (1983). A typology of organizational learning systems. *Journal of Management Studies, 20*, 7–28.
Simmons, B. A., & Elkins, Z. (2004). The globalization of liberalization: Policy diffusion in the international political economy. *American Political Science Review, 98*(1), 171–189.
Singer, O. (1993). Policy Communities und Diskurs-Koalitionen: Experten und Expertse in der Wirtschaftspolitik. *PVS-Sonderheft, 24*, 149–174.
Sitwell, O. F. G. (1981). Elements of the cultural landscape as figures of speech. *Canadian Geographer, 25*, 167–180.
Smith, C. (1986). Transformation and regeneration in social systems: A dissipative structure perspective. *Systems Research, 3*(4), 203–213.
Smith, J., & Jenks, C. (2006). *Qualitative complexity*. London: Routledge.
Spaargaren, G., & Mol, A. P. J. (2008). Greening global consumption: Redefining politics and authority. *Global Environmental Change*, 350–359.

Stacey, R. D. (1992). *Managing chaos: Dynamic business strategies in an unpredictable world.* London: Kogan Page.

Stacey, R. D. (1993). *Strategic managment and organizational dynamics.* London: Pitman.

Stacey, R. D. (1996). *Complexity and creativity in organizations.* San Fransisco: Berrett-Koehler Publications.

Stallybrass, P., & White, A. (1986). *The politics and poetics of transgression.* London: Methuen.

Stengers, I. (1997). *Power and invention: Situating science.* Minneapolis, MI: Minnesota University Press.

Stewart, J. (1989). *Does god play dice? The mathematics of chaos.* Oxford: Basil Blackwell.

Stone, D. (1999). Learning lessons and transferring policy across time, space and disciplines. *Politics, 19*(1), 51–59.

Stone, D. (2001). *Learning lessons. Policy transfer and the international diffusion of ideas,* CSGR Working Paper 69/01, Centre for the Study of Globalisation and Regionalisation, University of Warwick, Coventry.

Surowiecki, J. (2004). *The wisdom of crowds.* London: Little Brown.

Swinney, H. (1983). Observations of order and chaos in nonlinear systems. *Physica, 7D*, 3–15.

Taylor, M. (2003). *The moment of complexity: Emerging network culture.* Chicago, IL: University of Chicago Press.

Taylor, W. (1984). Metaphors of educational discourse. In W. Taylor (Ed.), *Metaphors of education.* London: Heinemann.

Teisman, G. R. (2005). *Public management on the edge of chaos and order.* The Hague: Academic Services (in Dutch).

Teisman, G. R., Gerrits, L., & Van Buuren, A. (2009). An introduction to understanding and managing complex process systems. In G. Teisman, A. Van Buuren, & L. Gerrits (Eds.), *Managing complex governance systems* (pp. 1–16). New York: Routledge.

Theissen, G. (1978). *The social setting of Pauline christianity.* Philadelphia, PA: Fortress Press.

Thietart, R. A., & Forgues, B. (1995). Chaos theory and organization. *Organizational Science, 6*(1), 19–31.

Thompson, G. F. (2004). Is all the world a complex network? *Economy and Society, 33*(3), 411–424.

Thoreau, H. D. (2007). *Walking.* Rockville, MD: Arc Manor.

Thorn, C. E., & Welford, M. R. (1994). The equilibrium concept in geomorphology. *Annals of the Association of American Geographers, 84*(4), 666–696.

Thrift, N. (1998). The rise of soft capitalism. In A. Herod, G. O'Tuathail, & S. M. Roberts (Eds.), *An unruly world? Globalization, governance and geography* (pp. 25–71). London: Routledge.

Thrift, N. (1999). The place of complexity. *Theory, Culture and Society, 16*(3), 31–69.

Toulmin, S., & Goodfield, J. (1962). *The architecture of matter.* London: Hutchinson.

Trist, E. (1980). The environment and system-response capability. *Futures, 12*(2), 119–120.

Tuan, Y. (1978). Sign and metaphor. *Annals of the Association of American Geographers, 68*, 362–372.

Urry, J. (2000a). *Sociology beyond societies.* London: Routledge.

Urry, J. (2000b). Mobile sociology. *British Journal of Sociology, 51*(1), 185–203.

Urry, J. (2001). Transports of delight. *Leisure Studies, 20*, 237–245.

Urry, J. (2002). The global complexities of September 11th. *Theory, Culture and Society, 19*(4), 57–69.

Urry, J. (2003). *Global complexity.* Cambridge: Polity Press.

Urry, J. (2004a). The 'system' of automobility. *Theory, Culture and Society, 21*(4–5), 25–39.

Urry, J. (2004b). The complex spaces of scandal. In J. O. Baerenholdt & K. Simonsen (Eds.), *Space odysseys, spatiality and social relations in the 21st century* (pp. 15–25). Aldershot: Ashgate.

Urry, J. (2005a). The complexities of the global. *Theory, Culture and Society, 22*(5), 235–254.

Urry, J. (2005b). The complexity turn. *Theory, Culture and Society, 22*(5), 1–14.

Urry, J. (2010). Mobile sociology. *The British Journal of Sociology, 61*, 347–366.
Van de Ven, A., & Poole, A. S. (1990). Methods for studying innovation development in the Minnesota innovation research program. *Organizational Science, 1*(3), 313–335.
Van Gils, M., Gerrits, L., & Teisman, G. R. (2009). Non-linear dynamics in port systems, change events at work. In G. Teisman, A. Van Buuren, & L. Gerrits (Eds.), *Managing complex governance systems* (pp. 76–96). New York: Routledge.
Van Potter, R. (1971). *Bioethics*. Englewood Cliffs, NJ: Prentice-Hall.
Vertovec, S. (1999). Conceiving and researching transnationalism. *Ethnic and Racial Studies, 22*, 447–462.
Vico, G. (1968). *The new science of Giambattista Vico*. Ithaca, NY: Cornell University Press.
Waldrop, M. M. (1992). *Complexity: The emerging science at the edge of order and chaos*. New York: Simon and Schuster.
Waldrop, M. M. (1993). *Complexity*. New York: Viking.
Walker, J. L. (1969). The diffusion of innovations among the American states. *American Political Science Review, 63*, 880–899.
Walker, J. L. (1974). *The diffusion of knowledge and policy change: Toward a theory of agenda setting*. Annual Meeting of the American Political Science Association.
Wallerstein, I. (1996). *Open the social sciences. Report of the Gulbenkian Commission on the Restructuring of the Social Sciences*. Stanford, CA: Stanford University Press.
Wallerstein, I. (1998). *The heritage of sociology; The promise of social science*. Presidential Address, 14th World Congress of Sociology, Montreal.
Waters, M. (1995). *Globalization*. London: Routledge.
Watson, M. K. (1978). The scale problem in human geography. *Geografiska Annaler B, 60*, 36–47.
Watts, D. (1999). *Small worlds*. Princeton, NJ: Princeton University Press.
Watts, D. (2003). *Six degrees: The science of a connected age*. London: Heinemann.
Weimar, D. R. (1966). *The city as metaphor*. New York: Random House.
Weiss, C. (1977a). Research for policy's sake; The enlightenment function of social research. *Policy Analysis, 3*, 531–545.
Weiss, C. (1977b). *Using social research in public policy-making*. Lexington, KY: D.C. Heath.
White, H. C. (1992). *Identity and control. A structural theory of social action*. Princeton, NJ: Princeton University Press.
Whitehead, A. N. (1929). *Process and reality*. Cambridge: Cambridge University Press.
Whitehead, A. N. (1967). *Adventures of ideas*. New York: Free Press.
Whitehead, A. N. (1978). In D. R. Griffin & D. W. Sherburne (Eds.), *Process and reality: An essay in cosmology, corrected edition*. New York: Free Press.
Whyte, L. L. (1944). *The next development in man*. London: The Cresset Press.
Williams, R. (1989). Mining the meaning: Key words in the miners' strike. In R. Gable (Ed.), *Resources of hope*. London: Verso.
Wilson, A. G. (1981). *Catastrophe theory and bifurcation*. London: Croom Helm.
Wolff, J. (1993). On the road again: Metaphors of travel in cultural criticism. *Cultural Studies, 7*, 224–239.
Wolff, J. (1995). On the road again: Metaphors of travel in cultural criticism. In J. Wolff (Ed.), *Resident alien: Feminist cultural criticism* (pp. 115–134). Cambridge: Polity Press.
Wolfram, S. (2002). *A new kind of science*. Champaign, IL: Walfram Media Inc.
Wolman, H. (1992). Understanding cross national policy transfers. *The case of Britain and the US, Governance, 5*(1), 27–45.
Zembylas, M., & Vrasidas, C. (2007). Globalization, information and communication technologies, and the prospect of a 'global village': promises of inclusion or electronic colonization? *Journal of Curriculum Studies, 37*(1), 65–83.
Zimmerman, B. J. (1990) *Nonequilibrium: The flipside of strategic processes*. Working Paper, Faculty of Administrative Studies, York University, North York, Canada.
Zohar, D. (1997). *Rewiring the corporate brain*. San Francisco, CA: Berrett-Koehler Publishers.

Chapter 6
Flow

Abstract Money laundering, drugs, asylum-seeking, arms trades, people smuggling, slave trades and let us not forget the movement of terrorist materials around the world are all common features of the maritime sector and facilitated by the heady mix of globalisation and shipping. Central to the concept of change and flux in policies and the governance of the maritime sector is the concept of flow, representing the movement that needs to be repeated within maritime policy if it is to become dynamic and to reflect the constantly changing world of the shipowner, port manager, maritime lawyer and banker, and other stakeholders within the industry. It is a concept that is also needed to accommodate maritime policies within a governance framework in which they have to operate outside the maritime sector and within the wider spread of stakeholders that exists but which are frequently overlooked. In this chapter, we shall consider the work of Manuel Castells and the *Space of Flows,* but although highly significant, this is not the only aspect of flow that needs to be debated from a governance perspective and within a maritime context. Other issues include the relationship of flow to process and how although closely related, they are not the same thing. Flows, hierarchies and fluids will also be central to the discussion along with the relationship of flow to our earlier consideration of space, territories and boundaries. And after all this, we draw it together looking at the development of flows in governance and policy-making and the issue of speed.

> While the actual location of high-level centers in each period is critical for the distribution of wealth and power in the world, for the perspective of the spatial logic of the new system what matters is the versatility of its networks. The global city is not a place, but a process. A process by which centers of production and consumption of advanced services, and their ancillary local societies, are connected in a global network, whilst simultaneously downplaying the linkages with their hinterlands, on the basis of information flows. Castells (1996: 417) quoted in Ballve (2011).

> Standing by a river we see the perpetual flowing of the water. But to grasp it conceptually, and to communicate it to others, we do not think and say 'Look at the perpetual flowing of the water', we say, 'Look at how fast the river is flowing'. We say, 'the wind is blowing', as if the wind were actually a thing at rest which at a given point in time, begins to move and blow. We speak as if the wind were separate from its blowing, as if a wind could exist which did not blow. Elias (1978: 112, quoted in Moore 2008: 219).

M. Roe, *Maritime Governance*, DOI 10.1007/978-3-319-21747-5_6

> The flows from the wild zones of people, risks, substances, images, Kalashnikovs… increasingly slip under, over and through the safe gates, suddenly and chaotically eliminating the indivisibilities that had kept the zones apart. Through money laundering, the drug trade, urban crime, asylum-seeking, arms trading, people smuggling, slave trading and urban terrorism, the spaces of the wild and the safe are chaotically juxtaposed, time and space is being curved into new complex configurations. Urry (2002: 63–64).

Urry's comments on flows and security ring frighteningly true in the context of maritime governance—money laundering, drugs, asylum-seeking, arms trades, people smuggling, slave trades and let us not forget the movement of terrorist materials around the world—are all common features of the maritime sector and facilitated by the heady mix of globalisation and shipping. Central to the concept of change and flux in policies and the governance of the maritime sector is the concept of flow, representing the movement that needs to be repeated within maritime policy if it is to become dynamic and to reflect the constantly changing world of the shipowner, port manager, maritime lawyer and banker, and other stakeholders within the industry. It is a concept that is also needed to facilitate maritime policies within a governance framework which has to operate outside the maritime sector, within the wider spread of stakeholders that exists but which is frequently overlooked.

In this chapter, we shall consider the work of Manuel Castells and the *Space of Flows*, but although highly significant, this is not the only aspect of flow considered from a governance perspective and within a maritime context. Other issues include the relationship of flow to process and how although closely related, they are not the same thing. Flows, hierarchies and fluids will also be central to the discussion along with the relationship of flow to our earlier consideration of space, territories and boundaries.

And after all this, we draw it together looking at the development of flows in governance and policy-making and the issue of speed. This is taken up in some detail in the next chapter where the central features of process and governance are considered. At that point, we can begin to consider what needs to be done next to refresh policy-making and governance in the maritime sector to accommodate the movement, change and dynamism that is so significant and yet sadly lacking and which makes the current situation untenable. But first to a clarification of flow before the more specific issues that need to be considered can be introduced.

Flow: Definition, Significance and Context

> We live in mobile societies. Although most people continue to think and talk about social life as though it can be easily explained using well-defined, clearly established parameters, domains or spheres – for example the nation-state, family, culture, work, love, publicity, privacy and citizenship – the reality is that our lives are constructed around flows of information, capital, waste, technology, people, images sounds etc. Castells (1996) quoted in Hier and Greenberg (2007).

> Ancient atomism is inseparable from flows, and flux is reality itself, or consistency. Deleuze and Guattari (1988: 361).

Sidaway (1995: 487) points out the rise in interest in motion and flow that has occurred, describing this as unprecedented using the movement of information and consequently power by senior Saudi families as an example of its diverse and significant impact. Stalder (summarised in 2006: 152, and detailed from 154 on) could not be clearer in his support. He sees nearly all 'strategic dominant activities—generating most of the financial wealth and administering the most powerful institutions' operating as flows, including not only those associated with the Internet but also what he terms 'private and closed networks'. Meanwhile, Shields (1997: 1) stresses the importance of transfers, movement, speed but particularly flows as cultural images and economic factors. He suggests that flow is a new paradigm.

> The notion of flow, most widely known from the work of Deleuze and Irigaray, (see Lorraine 1999), occurs repeatedly in social theory. Associated with a paradigm shift within cultural studies and sociology from the analysis of objects to processes, it is also linked by geographers to the notion of nomadism and the breakdown of the fixity of boundaries and barriers. More poignantly, it is the lived experience of the global mass migrations and movements of refugees. In effect, the dominant metaphors for discussions of sociality have swung from models of affinity to those of viscosity. Shields (1997).

Shields also notes a range of research focussing upon the significance of flows that was active in the late twentieth century. This includes de Courville-Nicol (1997), Dykrton (1997), O'Connor (1997), Ironstone-Catterall (1997) and McCarthy (1997) reflecting a wide spectrum of opinion on the need to place flow at the centre of human perspective.

Allmendinger (2001: 34) looks at the relationship between processes and flows and consequently helps to draw closer our discussion in the last two chapters. He sees flows with 'no locus or closure' (2001: 34). He links the two through the work of Foucault on postmodernism (McNay 1994) rejecting the 'unidirectional state or class-orientated power struggle of Marxist or Enlightenment thinking'. Power is seen as a flow that forms a central and fluid process characteristic of postmodern society (Allendinger 2001: 226). Webster (2002: 107) meanwhile links processes to Manuel's *Space of Flows* suggesting that the 'global city is not a place but a process' (Castells 1989: 386). Castells himself views the new 'informational technological paradigm' that he has identified as founded upon the combination of a well-educated labour force, an appropriate social organisation and an institutional framework that maximises 'information flows and connects them to development tasks' (Castells 1989: 15–17). This is all underlined with a network of processes along which self-regulating flows of information ensure efficiency and efficacy.

Shields also describes the characteristics of flow (1996: 3) noting that it normally has a tempo and rhythm as well as direction. Rather than just another way of expressing process, they always possess content, whilst process:

> generally indicates the transformative gap between states or dispositions (and is) strongly defined on the basis of origin and terminus as a definite line or path between two points… flows signal pure movement, without suggesting a point of origin or a destination, only a certain character of movement, fluidity and direction. They are relational… they have a Fate but no destiny.

He goes on to use Virilio's (1999) example of ice flows which under pressure become plastic and thus exhibit intensity and motion. Viscosity measures the degree of intensity as a relative motion—a tendency to flow. Viscosity is therefore time and material together whilst flow combines viscosity and direction. Flows have direction but no purpose, and they do not flow to any specifically defined place but are controlled by topology (hills, valleys, communication channels, railways, etc.) and also by laws, taboos, and many social regulations and inhibitions. This in globalisation terms leads to *deterritorialisation* and *reterritorialisation* with flows of people, money, data and commodities crossing from territory to territory, across boundaries something also identified by Brenner (1999: 60) in considering new geographic networks and increased mobility and flexibility. Along with Castells' *Space of Flows* (1996), he points out Jameson's postmodern hyperspace (1991), Ruggie's (1993) interpretation of the EU as a postmodern political form, Appadurai's (1996) ethnoscapes (Sidaway 1995: 493), Ohmae's (1995) borderless world and O'Brien's (1992) end of geography. In addition, Rosow (2005) says much the same.

Meanwhile, Lefebvre (1991: 206) identifies a rhythmic quality to flows:

> The way in which rhythms may be said to both embrace both cyclical and linear is illustrated by music, where the measure and the beat are linear in character, whilst motifs, melody and particularly harmony are cyclical… Much the same may be said of dance, a gestural system whose organisation combines two codes, that of the dancer and that of the spectator (who keeps time by clapping or with other body movements): thus, as evocative (paradigmatic) gestures recur, they are integrated into a ritually linked gestural chain.
>
> What do we know about rhythms, as sequential relationships in space, as objective relationships? The notion of flows (of energy, matter etc) is self-sufficient only in political economy. It is in any case always subordinate to the notion of space. As for drive, this idea is a transposition onto the psychic level of the fundamental, but at the same time disassociated, idea of rhythm. What we live are rhythms – rhythms experienced subjectively. Which means that, here at least, lived and conceived are close – the laws of nature and the laws governing our bodies tend to overlap with each other – as perhaps too with the laws of so-called social reality.

Both Shields and Lefebvre are not the only ones to note the relationship between rhythm and flow. Deleuze and Guattari (1988: 363–364) had suggested that there were important connections at a much earlier time even placing it within a maritime context:

> The sea as a smooth space is a specific problem of the war machine. As Virilio shows, it is at sea that the problem of the *fleet in being* is posed, in other words the task of occupying an open space with a vertical movement that can rise up at any point. In this respect, the recent studies on rhythm, on the origin of that notion, do not seem entirely convincing. For we are told that rhythm has nothing to do with the movement of waves but rather that it designates *form* in general, and more specifically the form of a *measured, cadenced movement*. However, rhythm is never the same as measure. And though the atomist Democritus is one of the authors who speak of rhythm in the sense of form, it should be borne in mind that he does so under very precise conditions of fluctuation and that the forms made by atoms are primarily large, non-metric aggregates, smooth spaces such as

> the air, the sea or even the earth... There is indeed such a thing as measured, cadenced rhythm, relating to the course of a river between its banks or to the form of a striated space; but there is also a rhythm without measure, which relates to the upswell of a flow, in other words, to the manner in which a fluid occupies a smooth space.

Miller (1993: 29) continues the theme commenting on the contribution of Deleuze and Guattari and suggesting that they have their own 'ethic of flow' and relating this directly to nomadology. Flows are seen as 'abstract but real' and can only be represented as indexes on a segmented line. However, 'that line and those indexes exist only by virtue of the flow suffusing them' (Deleuze and Guattari 1988: 218 and 266).

We saw Urry's (2010: 353–354) use of fluids in the last chapter to emphasise the significance of process and its metaphorical relationship in understanding globalisation. This in turn helps to understand the changes that are needed in the design of maritime governance and policy-making. Taking this further, he continues by emphasising how the fluid metaphor can draw us into an appreciation of flows and goes on to suggest that globalisation is actually characterised by a wide variety of machines and technology that:

> dramatically compress or shrink time-space. These technologies carry people, information, money, images and risks and flow within and across national societies in increasingly brief moments of time.

Castells (1996: 469) sees this as a network, but networks in themselves do nothing, only providing the framework upon which these movements can flow like fluids. Cresswell (2006:1) emphasises the importance of flow in discussing mobility suggesting that it is *everywhere*, playing a:

> central role in the discussions of body and society (Bale and Sang 1996; Cresswell 1999; Young 1990). It courses through contemporary theorizations of the city (Graham and Marvin 2001). Culture we are told, no longer sits in places but is hybrid, dynamic—more about routes than roots (Appadurai 1996; Augé 1995; Chambers 1994; Clifford 1997; Grossberg 1993; Gupta and Ferguson 1992).

Flows are seen as a representation (evidence even) of progress, freedom, opportunity and modernity although contrasting as we have seen with shiftlessness, deviance and resistance (cf. nomadism). Flows and the mobility they represent reflect society as it has developed, the intensity of globalisation, and as such have to be represented in maritime governance and policy-making. Flow, mobility, process and change are central to the human experience and in many ways lie behind the more obvious manifestation of place. Cresswell (2006: 25) agrees:

> If something can be said to be fluid, dynamic, in flux, or simply mobile, then it is seen to be progressive, exciting, contemporary. If on the other hand, something is said to be rooted, based on foundations, static, or bounded, then it is seen to be reactionary, dull, and of the past.

Serres (1993) sees it all in terms of angels who as messengers act as representatives of movement (8–12). He sees flux and change everywhere, news of which at some significant times (e.g. most famously at Christmas) is brought by angels (25–35):

> *The Reader*: Why should we be interested in angels nowadays?
>
> *The Author*: Because our universe is organized around message-bearing systems, and because, as message-bearers, they are more numerous, complex and sophisticated than Hermes, who was only one person, and a cheat and a thief to boot.
>
> Each angel is a bearer of one or more relationships; today they exist in myriad forms, and every day we invent billions of new ones. However we lack a philosophy of such relationships.
>
> Instead of weaving networks of things or of beings, let us therefore map some of the interlacings of paths. The angels are increasingly drawing up the maps of our new universe. Serres (1993: 293).

Pred (1984: 280) even interpreted place in terms of flow suggesting it can be 'conceptualised partly in terms of the unbroken flow of local events'. Any social event is actually the 'passing manifestation of a complex process' (Pred 1984: 292).

Cresswell goes further in analysing the positive and optimistic attitude that exists towards flows and suggests that this is at times unfounded. Taking the human reproductive system as his example, he suggests that 'textbook descriptions… are remarkable for the way they give meaning to bodily processes' in a way that academic literature would normally avoid. Martin (1990: 76) provides support with evidence that menstruation is commonly associated with language such as 'degenerate, decline, lack and deteriorate'. Meanwhile, popular textbooks on male physiology are somewhat different:

> The mechanisms which guide the remarkable cellular transformation from spermatoid to mature sperm remain uncertain… Perhaps the most amazing characteristic of spermatogenesis is its sheer magnitude: the normal male may manufacture several hundred million sperm per day. Martin (1990: 76).

Language of this sort remains common today with a 'mobile sperm cell penetrating an immobile egg', and this mobility is equated with 'agency'. Martin continues: eggs 'drift' and are 'transported'; sperm 'delivers' with 'velocity', propelled by 'strong' tails. Ejaculation 'propels the semen into the deepest recesses of the vagina' where the sperm are aided by 'energy' so that with a 'whiplashlike movement and strong lurches' they 'burrow through the egg coat' and 'penetrate' the egg. The flow of sperm is active; the egg is a passive recipient. However, more recently, Schatten and Schatten (1996) have started to reflect a new attitude:

> The classic account, current for centuries, has emphasised the sperm's performance and relegated to the egg the supportive role of sleeping beauty. The egg is central to this drama to be sure, but it is as passive a character as the Brother Grimm's Princess. Now it is becoming clear that the egg is not merely a yolk-filled sphere into which the sperm burrows to endow new life. Rather, recent research suggests the almost heretical view that sperm and egg are mutually active partners.

Flows can be understood by what Cresswell (2006: 3) suggest are three relational movements:

- Physical movement—the target of modellers, the movement of ships, cargoes, people, money and data.
- Representational flows—ideological in nature and conveying through metaphors other meanings of flows and movement—freedom, transgression, creativity and the like.
- Flows as a representation of mobility in the way we exist in the world. Our way of walking, our inability to sleep on a flight and our response to wind in our faces are all reactions to flows. Delaney (2003) suggests that 'human mobility implicates both physical bodies moving through material landscapes and categorical figures moving through representational spaces'.

However, earlier this had been taken much further by Knox (1995: 244–245) who identified six principal categories (also based in part on the work of Appadurai 1990):

- *Technoscapes* produced by flows of technology, software and machinery disseminated by transnational corporations, supranational organisations and government agencies.
- *Finanscapes* produced by rapid flows of capital, currency and securities and made visible not only through teleports and concentrations of financial service workers but also through the rapidly changing geography of investments and disinvestments.
- *Ethnoscapes* produced by flows of business personnel, guest workers, tourists, immigrants, refugees, etc.
- *Mediascapes* produced by flows of images and information through print media, television, social media, smartphones, film, etc.
- *Ideoscapes* produced by the diffusion of ideological constructs, mostly derived from Western world views—including democracy, sovereignty, citizenship and welfare rights.
- *Commodityscapes* produced by flows of high-end consumer products and services including clothes, interior design, food, personal and household objects which signify taste and distinction.

To them can be added flows of raw materials and manufactured goods which, whilst less fashionable in the debate on globalisation and modern society, remain the most significant of all in terms of quantity and in the maritime sector are dominant both in the bulk and liner sectors. Whatever categories of flow dominate, they have always been on the move, 'filtered and stylised through particular networks and associations, passing through many channels of transmission' (McLennan 2003: 555). Flows are nothing new, and as such the seeming inability satisfactorily to accommodate them into maritime governance is inexcusable.

It is already apparent that flows have a close relationship to globalisation. Taylor (2000a, 2005: 705) considers that a network of city-states (e.g. Singapore, Hong Kong) and suprastates (e.g. the EU) now operates over and above nation-states. In some cases, this can even be state-like cities within nation-states that have assumed so much power and influence that they operate as intra-states (e.g. in the maritime sector Rotterdam, Piraeus, London, Oslo). The flows of information, money and influence that characterise these network processes undermine the traditional role of the nation-state generating a globalised society exemplified by shipping and associated port communities. Globalisation thrives (in fact both needs and generates) flows of many types, and effective governance must take explicit note of this in its design and application.

Paasi (1998: 72) places it in the context of territoriality using Sack (1986: 1) as his inspiration. Territoriality is a 'spatial strategy which can be employed to affect, influence or control resources and people by controlling area'. This globalisation strategy is part of the new 'world of flows' themselves becoming increasingly complex characterised by overlap and conflict. Following postmodern trends, the spatiality of the national state has become associated much more with flow rather than territory and across all scales—local, regional, national, supranational and global.

> Power is diffused in global networks of wealth, information and images 'which circulate and transmute in a system of variable geometry and dematerialized geography' (Castells 1997: 359). Power flows in the codes of information and in the 'images of representation around which societies organize their institutions and people build their lives and decide their behaviour' (Castells 1997: 359). Paasi (1998: 82).

Storper (1997: 177–180) takes all this much further. He imagines the emergence of a 'global supply oligopoly' which would operate within a hierarchical structure and replace all semblance of a territorial society as we know it. Resources would have no specific place dependence and would flow between locations contingent upon issues like scale economies. Production could take place almost anywhere facilitated by the smoothness of flows between locations which would virtually eliminate costs. Although it may seem unrealistic at the moment, one day in the future it may well be possible to move materials without transport (i.e. at nil or virtually nil cost) which would free up almost the entire planet for locational choices. This would of course destroy the shipping industry overnight but leaving that aside for the moment, just think of the nineteenth-century postal service—and what those using it at that time would think of emails, faxes, texts, television, radio and so on. All virtually instantaneous and almost free (certainly by comparison); times do change.

This flexibility in location is what the current moves in globalisation are moving towards. Production is moving towards infinite flexibility as transport friction declines, technology makes resources more widely available, and training and education make workforces more locally suitable (consider the likelihood of the use of North Korean or Chinese labour 50 years ago). The result is what Storper terms 'pure globalisation':

> Low wage, low skill, low sunk cost manufacturing processes, certain highly standardised consumer durable manufacturing (where sunk costs are higher, but modular and widely available equipment is used) and certain consumer services where centralised production can be combined with local delivery, come to mind.

Rodrigue et al. (2009) (Fig. 6.1) suggest that there are three main types of globalised flows, and although it is rather simplistic (there is no mention of money for example), it provides an interesting basis for appreciating the close relationship that exists.

And finally, Taylor (1996: 10) sees the new world of flows focussing upon cities rather than flows reflecting to a certain extent the feelings of Marston et al. (2005: 423) citing Smith (2003: 570) as an example of the feitishisation of 'overzealous flow enthusiasts'.

> In contrast to Sassen's (2006) interest in scales, boundaries and territories, any ontology of globalisation fluidifies such solidified thinking revolving around such motifs as fluidity and flow, movement and mobility, folds and networks. A consequence of that ontology – where all that is solid melts into air – is a rejection of scales and boundaries altogether as globalisation and world cities are too intermingled through scattered lines of humans and non-humans to be delimited in any meaningful sense.

Reverting to the principles of hierarchies, Taylor suggests that global governance may well need to be based upon a political organisation that 'recognises leagues of cities at different levels of that hierarchy. Such a framework would cut across states and nations and provide the architecture for a relatively egalitarian, decentralised, non-territorial world'. He goes on in a later paper (Taylor 2000c: 1111–1112) to place the discussion in the context of globalisation suggesting that statism and the belief in the inevitability and value of nation-states are working against the true reality of the domination of flows over artificial territoriality. The traditional vision of the world as a 'mosaic of states' is not a necessary (or perhaps even desirable) one. Some form of 'metageography' is needed to 'escape from embedded statism' which will place flow central to understanding the structure of

	Trade	**Migration**	**Telecommunications**
Nature	Flows of physical goods	Flows of people	Flows of information
Types	Raw materials, energy, food, parts and consumption goods	Permanent, temporary (migrant workers), tourism, business transactions	Communication, power exchanges, symbolic exchanges
Medium	Transport modes and terminals (freight)	Transport modes and terminals (passenger)	Transport modes and terminals (postal), telecommunication systems
Network	Hub and spoke with interconnections	Hub and spoke	Redundant and diffuse (point to point)
Main Gateways	Ports	Airports	Global cities
Speed	Low to average	Slow to fast	Instantaneous
Capacity	Very large	Large	Almost unlimited

Fig. 6.1 The flows of globalisation. *Source* Rodrigue et al. (2009)

the global economy and society. He quotes Arrighi (1994: 81) who says it better than most; our 'deficiencies in perceptual habits' causing:

> non-territorial spaces-of-flows to have gone unnoticed alongside the national spaces-of-places throughout the history of the modern world system.

Space of Flows

> Confucius: 'Do you think me a learned well-read man?'
>
> 'Certainly' replied Zi-Gong. 'Aren't you?'
>
> 'Not at all,' said Confucius. 'I have simply grasped one thread which links up the rest.'
>
> *Confucius*, quoted in Friedmann (2000: 112) from Castells (1996: 1).

Castells' *Space of Flows* is not the central focus of the debate about maritime governance, but as a major philosophical step in the development of an understanding of modern society and focusing as it does on flows, it cannot be entirely ignored. It is also central to an understanding of globalisation, itself a major force lying behind the need to consider dynamic governance:

> The internationalization of post-war capitalism has produced a lived experience in which 'the space of flows... supersede(s) the space of places' (Henderson and Castells 1987). Watts (1991: 9).

There is a mountain of literature which has looked at the concept. We cannot consider it all here, but it includes Knox (1995: 244–245), Thrift (1995: 18, 20), Waterman (1999), Soja (2000: 212–216), Taylor (2000b: 161), Friedman (2000), Harvey (2000: 195), Yeung (2000: 201), Dicken et al. (2001: 93), Webster (2002: 107), Allen (2003: 63–64), Mol (2007: 301) and Hassan (2009: 11). In addition, there are also the core Castells texts (1989: 169 and 348–353), (1993), (1996) (particularly 378–428 and 469–478), (1999: 295–296), (2000) and Carnoy and Castells (2001).

Stalder (2002b) refers back to Heraclitus and his discussion of flow and the constant transformation of nature. However, he suggests that Castells' contribution is to take this much further referring to *Space of Flows* as a social condition and a concept that became fully understood during the 1970s and was first outlined by Castells in the 1980s (Stalder 2006: 46). He defines it as 'that stage of human action whose dimensions are created by dynamic movement rather than by static location'—and the relationship to our earlier discussion of change, movement and dynamism is clear. Castells (1996) places 'distant elements—things and people—into an interrelationship that is characterised today by being continuous and in real time' (Stalder 2002a: 5). Because this now takes place in almost instantaneous real time, this enables space to contract and expand quickly changing the very nature of human activity, the application of ideas and policies and the mechanisms needed to govern them. Shipping, despite being the slowest of modes, is affected by this as much as any other activity and because of its central position within globalisation, in many ways much more so. *The Space of Flows* as a concept consequently has much relevance for a new interpretation of maritime governance.

Examples abound. Zook (2003: 1263) suggests the global financial system (Sassen 1991) and offshore banking (Roberts 1994) along with Russian criminal mafia networks and narcotic supply chains (Castells 1996: 414, 1998: 166–205). Stalder (2002b) even provides us with a maritime example. Taking Amsterdam as an old port city, he begins by identifying the three main elements of the *Space of Flows*—the medium through which things flow; the products that flow; and the nodes between which the flows circulate. Historically, Dutch long-distance trading used the ocean as the *medium* possessing certain characteristics that affected the flows which might use it—it was unpredictable and enabled only slow movement for example, and consequently, only certain *products* could flow (e.g. spices) and others could not (e.g. fresh fruit). Flow media and their content are always related—thus financial information lends itself to modern communications (and much has changed in the shipping world because of this); written information on prices and currencies transported by sailing boat would be useless. *Nodes* meanwhile are harbours, ports and trading posts, and if the world had only a single harbour, then 'ships are mere entertainment' (Stalder 2002a: 5). Nodes are interfaces which create a membrane of flows. The distance between ports and the maritime space and time between them are relatively fixed and increasingly predictable, as they show signs of stability. In contrast, the modern development of globalised communications shows no such signs with the promise of infinite expansion in area and volume and infinite contraction in time. This contradiction between the traditional and the future is one that particularly affects the maritime sector and illustrates the problems of developing effective maritime governance.

In a later contribution, Stalder (2006: 152) stresses how the *Space of Flows* is supported by empirical reality:

> Nearly all the strategically dominant activities – generating most of the financial wealth and administering the most powerful institutions – operate through the space of flows and their relative power, compared to activities organized on a purely local basis, has only increased. The global elite is still relatively cohesive, dominating fragmented population, even if the resistance and mutual interconnections of the latter have increased. Finally, the dominant activities are still highly clustered in a few central nodes.

As Castells (1989) wrote:

> While organisations are located in places, and their components are place-dependent, the organizational logic is placeless, being fundamentally dependent on the space of flows that characterizes information networks. But such flows are structured, not undetermined. They possess directionality, conferred both by the hierarchical logic of the organization as reflected in instructions given, and by the material characteristics of the information systems infrastructure.

Central to the concept is the independence that the *Space of Flows* allows for geographical location, something that makes maritime governance particularly difficult. Whilst shipping interests will be located at certain selected nodes, these nodes and the relationship between them are located within an indeterminate flexible space made even more complex by the mobility exhibited by the ships themselves. Whilst it may have to be somewhere (and even this can be questioned as capitalist interests increasingly expand into virtual communities and outer space), very little commercially now needs to be specifically anywhere.

Not everyone is enamoured with Castells' ideas. Thrift (1995: 22) was soon on the case accusing the *Space of Flows* of becoming 'hoary with age' almost as soon as it was outlined. He suggests that it was not a new idea with evidence of circulation within the nation-state of ideas and letters dating from the eighteenth century or even earlier. The spread of the railway and telegraph in the nineteenth century pushed the obsession with speed and flow further which continued throughout the twentieth century to the current day and the domination of electronic (and almost instantaneous) communications.

He goes on to question the suitability of Harvey's concept of time-space compression which is closely linked to that of the *Space of Flows*, associating it with historical accounts of the increasing pace of life with societal hysteria. Virginia Woolf provides a comment on fragmented times:

> After twenty minutes the body and the mind were like scraps of torn paper tumbling from a sack and, indeed, the process of motoring fast out of London so much resembles the chopping up small of identity which precedes unconsciousness and perhaps death itself that it is an open question in what sense Orlando can be said to have existed at the present moment. Woolf (1928) cited in Prendergast (1992: 193).

This does not deny the existence of a speeded-up world nor an electronic society, but it does imply that the *Space of Flows* and similar models are neither new nor nothing more than in some ways an expression of the obvious.

Space and Flow

> That's what you see beyond the galvanized steel guardrails. That is the informational city, a land of virtual networks ever more severed from their social context… Check our basilica's view of US Highway 101 gashing through the flat valley in ominous shades of black and white, a vast parking lot to the left, empty fields to the right. Transmission wires are low across the sky and trail into the distance. This is the space of flows. On the horizon sit carceral towers, the seeming prison houses of software engineers and product managers. Latent in the image are layers of spatial data; vestigial scraps of nature; the low defining hills; cars streaking along the highway, their own vectors in the landscape. Jeff Byles commenting on the photographs of Gabriele Basilico at the San Francisco Museum of Modern Art Exhibition of Silicon Valley, *Modern Painters Magazine*, quoted in *Slow Muse*, February 28, 2008 (Fig. 6.2).

Taylor (2000b: 160) summarises it nicely. 'It is quite odd that in a world produced and reproduced through a myriad of flows, connections and linkages, the metageographic emphasis should be on boundaries'. Meanwhile, Luke (1991: 321) is confident that the concept of place is being 'resituated within the hyperreality of flow, understood in terms of iconic/symbolic access to or process through networks of informational circulation'. Or to put it in understandable English, flows are as important as place in understanding the working of society.

Fig. 6.2 San Francisco, Gabriele Basilico. *Source* http://therumpus.net/2009/04/what-you-think-is-sad-gabriele-basilico-and-san-francisco-noir/

In Stalder's opinion (2006: 10 and reiterated in Undated: 5), the move has been from 'place-based conflicts to flow-based forms'. Although using the terminology of 'place', it is fair enough to interpret much of this as 'space':

> Thus people do still live in places. But because the function and power in our societies are organized in the space of flows, the structural domination of its logic essentially alters the meaning and dynamic of places. Experience, by being related to places, becomes abstracted from power, and meaning is increasingly separated from knowledge. There follows a structural schizophrenia between two spatial logics that threatens to break down communication channels in society. Castells (1996: 459) quoted in Ballve (2011: 2).

Santos (1995: 176) interprets this as a new dimension for space giving it density and depth generated by the increasing number and significance of flows that cross it, a construct that is also pursued by Sum (2000: 232). This is confirmed by Thrift (1995: 27) although he considers the debate on Space of Flows and time-space compression and their interaction in space as at least in part 'illegitimate'. This follows from Ruggie's (1993: 172) interpretation of globalisation which centres on the idea of the side-by-side existence of the Space of Flows (decentred and operating in real time) and the nation-state characterised by its essence of place, the latter having as much significance as ever and at least comparable with that of flows. Friedmann (2000: 113), however, reverts to the ideas of Santos, emphasising the significance of flows over the space they occupy, diminish or virtually eliminate.

Blatter (2001: 176) places it all in the context of the European Union contrasting the traditional 'spaces of place' with the emerging Space of Flows and thus presenting immense difficulties for governance. Meanwhile, Cresswell (2006: 2) says much the same thing but interpreting flow as mobility. Mobility is viewed as central to modern society reflecting values of freedom, choice and opportunity and in so doing neoliberalism currently so popular in Western democratic society. Yet this mobility remains largely unspecified and even ignored, 'a kind of blank space that stands as an alternative to place, boundedness, foundations and stability'.

Castells himself does not shy away from the issues of spaces and places in relation to flows (Carnoy and Castells 2001; Castells 2000: 13–14, 2002: 553–554) something that Shields (1997: 6) had noted some time earlier in a rather critical analysis of Castell's limiting the role of flows to the space between places. Castells places his comments specifically in the realm of form, and therefore particularly relevant to this discussion, he sees a bipolar relationship between the Space of Flows and the Space of Places where the former represents the increasingly electronic links between the latter, places where global production and media networking take place. These places form a fundamental part of the flows that characterise modern shipping with the flows providing the connection between them. What he sees as critical and a symbol of the change in globalised society is that these locations are formed in the light of the flows of information, money, data and the like and both flows and spaces in turn feed off one another and are changed by them. Both place and flow are therefore necessary to define the maritime sector, and together, they constitute the form that we can observe. Maritime governance at present provides little accommodation for these relationships and remains institutionalised within a fixed and formalised framework that takes little account of the new 'cultural sociology of space, flows and mobilities' which are a 'key dimension in understanding material practices in society' (Jensen and Richardson 2004: 86). Castells (1996: 412, 423) sees this as a dialectical tension between two forms of spatial logic or forms of rationality—his Spaces of Flow and Place—the relationship between which he presents in some detail (e.g. Castells 1999: 296–297).

Zook (2003: 1262) suggests that Castells' view does not undermine the significance of space or place but instead 'provides the means for the reconstitution and reorganisation of social connections and geographic concentrations at all levels within the economy' referring specifically to the relationship between place and flow. This interaction has been exacerbated by the communications revolution that remains in progress and its relationship to 'existing organisational, economic, political and regulatory structures (Graham and Marvin 2001; Leinbach and Brunn 2001; Wheeler et al. 2000)' (Zook 2003: 1262). Major activities benefitting from physical synergy remain concentrated (one only has to look at maritime clusters in London, Hong Kong, Piraeus, Oslo, etc., for evidence of this) despite the flexibility and freedom inherent in communications (Leaner and Storper 2001; Lo and Grote 2002). At the same time, other maritime activities (e.g. distance training, technical check-ups) have become possible through remote Internet access, whilst maritime consultancies have the opportunity to locate almost anywhere. The result is an industry characterised by complexity in geographical and organisational structures (Leyshon 2001) and by what Gritsenko (2013) and Gritsenko and Yliskyla-Peuralahti (2013) describe as polycentric contextuality.

Place, territory and flow are clearly closely linked, something identified by many including Taylor (1996: 9) suggesting that 'the movement of flow is the opposite of the fixity of territory'. Ruggie (1993) notes that nomads possess sovereignty over flow, something that questions the conventional idea of sovereignty represented by the territory of the nation-state and also hints at the possibility of a governance associated with flow rather than place. Ruggie goes on to suggest that the Space of Flows and its related cyberspace are also an example of a mobile sovereignty with what he terms 'non-territorial regions… undermining territorial absolutism'.

Storper (1997: 188–189) identifies the problems faced by nation-states under pressure from globalisation and the impact that the rise of flows of people, finance and information has had upon their governance and the influence they can have in governing others. Examples abound from the global shipping sector where the 'race to the bottom' and the 'tragedy of the commons' both manifest themselves in flags of convenience, substandard shipping, tonnage taxation and the like. He also goes on to stress the failure to harmonise rules and regulations across nation-states when global competition allows and encourages flows of influence and materials to occur on an unprecedented level. Multinational shipping companies are quick to recognise the opportunities presented by the move from a static formalised industry to one characterised by movement, dynamism and flows.

We noted earlier how Brenner (1999: 60, 61) provides considerable support reflecting on how flows are 'supplanting the inherited geography of state territories that has long preoccupied the sociological imagination', a process of deterritorialisation earlier recognised by Deleuze and Guattari (1988) and re-emphasised by Shields (1997: 4)—'flows tend to appear, in what one might call b-Grade Deleuze-imitators, to be understood as existing only as moments of deterritorialisation and reterritorialisation, passing from one to the next, from here to there from state to state'. Meanwhile, Blatter (2001: 178) notes Agnew's (1998) interpretation of the new world structure where 'cores, peripheries and semi-peripheries are linked together by flows of goods, people and investment'. Thus, territory and flow become one and the same, and regardless of the existence or otherwise of the nation-state and its role, the governance of any sector with the remotest of global characteristics has to be designed to meet this.

The relationship between flows and networks is clear with the latter relying upon the former for their existence and identification. Thus, the maritime sector is characterised by flows of data, goods, people and finance along identifiable networks either predetermined (e.g. electronic communications carrying messages and finance; shipping trade routes carrying containers and people) or generated as a consequence of the industry's needs (e.g. the flow of bunker oil and LNG to provide for shipping movements or the flow of capital between banks to provide financial support to the industry).

Castells' appreciation of a network society has been extensively discussed by Dickens and Ormrod (2007: 105–108) and has already been considered earlier in this chapter. Consequently, it will not be repeated here except to note how it builds on the work of Harvey (e.g. 1989) and is further developed by Barney (2004) and Yeung (1998, 2000: 201) the latter seeing it as a 'space of network relations'. In Friedmann's (2000: 113) words:

> The Castellsian world is polarized between what he calls the Net and the Self. Captured by global capital, the Net is dominant, while the Self, besieged on all sides, retreats into communities of resistance built around primary identities of religion, ethnicity, nation and territory.

Castells emphasises that the network society, dominated by flows, is a consequence of the global deregulation of capitalism, the depersonalisation of ownership and control of capital, the growth of countercultural movements, the resurgence of neo-liberalism, the 'mediafication' of exchange and understanding, the failure of nation-states to understand or appreciate the significance of Urry's (2000) global fluids and the growth of electronic communications (McLennan 2003: 560). The influence of these factors is captured appropriately by Taylor (2005) in his analysis of the relationship between inter-states, suprastates and transstates. Each is characterised by a network of flows that operates on sometimes conflicting and often overlapping levels forming polycentric governance dimensions that are determined by their respective contexts—economic, cultural, social, financial, technical, political, organisational and so on. We return to the idea of polycentric governance in the final chapter.

Governance, Speed and Flow

In anticipation of what comes next, we turn finally in this chapter to issues of speed and attempt to place them in the context of flow. This will conclude our journey through mobilising maritime governance. In so doing, the issues of dynamic governance as considered by both Cashore and Howlett (2007: 532) and Neo and Chen (2007: 1–5, 10–13) and applied to the maritime sector will be drawn together before we begin in the final chapter to look at earlier attempts to develop cultural models of activity which might be applicable to the maritime sector and policy-making and then to place it all within its broad capitalist framework and narrower polycontextual structure; but first to governance and flow.

Friedmann (2000: 113) sees flow at the centre of governance, and this is no less the case for the maritime sector than any other. He identifies a 'world of binaries' combining powerful and weak stakeholders and global and local influences which are linked together by unequal flows of influence and power. In the maritime sector, these binary relationships manifest themselves as shipowners, shipping and international port corporations and global commodity traders battling against (commonly successfully) the traditional nation-states, city port authorities and fragmented international policy-makers exemplified by the IMO, OECD, WTO and ILO.

This is emphasised by Spaargaren and Mol (2008: 351–352) who begin by repeating Friedmann's observation of the binary divide where:

> highly mobile flows and global networks (of capital, production, communication, international institutions, crime, transnational religions) are combined into global structures and processes that brutalise communities and local identities residing in the so-called places of space. What is left for local actors (worker unions, environmental movements, communities etc) are protest actions and the development of protest identities against the overwhelming power of the actors (and their neo-liberalist ideologies) in the space of flows.

The consequence is that the flows have become more important than the nodes between which they flow (the transfer of information of all sorts is superior in influence and impact than the nation-states between which they pass) forcing the nodes into strategic actors playing their part within a governance framework that is dominated by the movement of information rather than its possession, hence the need for maritime governance that focuses upon this movement rather than its specific origin or destination as these are likely to be different by the time decisions are made within a static framework. This constant process of change and disruption makes a mobile, flexible and dynamic form of governance essential if it is to be effective.

Nation-states have been transformed into 'mediators', attempting to improve the competitiveness of their maritime economies by allying their economic interests (flags, registries, financial and legal regimes, taxation incentives, etc.) to those of global shipping interests which results in favourable flows of information, money, influence, goods and people both into and out of their economy. Spaargaren and Mol (2008: 352) cites Sassen (2006) and suggests that:

> the organising and operational logic of markets increasingly infiltrates the organising and operational logic of states. Thus nation-states seem to fall away as 'mediators' between the space of flow and the space of place, which renders the concept of governance problematic.

You can say that again. The very fluid nature of the maritime industry presents some of the more complex difficulties for developing effective governance, but this in turn makes it more important to open the discussion on these issues. In the final chapter, we shall turn to an appreciation of a new design that might move towards some sort of solution but which might as well require some lateral thinking. But remember, at stake are human lives and the global environment. Isn't that enough?

Despite this, all is not lost. Torpey (2007: 52) outlines how Marx had suggested that capitalists had expropriated the means of production from the workers in order to exploit them and make them dependent for wages and survival. Weber turned this somewhat on its head to argue that the modern state had expropriated the right to exercise violence from the individual thus helping to control it and legitimise its usage. Only those other than the state licensed to use violence (e.g. security guards at ports or on-board ships to prevent piracy) were permitted to do so and those that transgressed would be punished. This same argument could be used to explain how nation-states expropriated the legitimate means of movement across international borders of both flows of people and goods; how mechanisms for punishment for violators was devised; and how this has now been discredited by the globalisation of the maritime sector and its associated activities making the need for consideration of a new governance framework essential. Curiously and as we have seen, the original design of movement control continues to rest with the nation-state despite the globalisation of its constituent stimuli. It is this discordance between flow and power manifested in the inadequacies of shipping and ports policy along with the increasing difficulties associated with global migrancy, the transfer of illegal goods (drugs, weapons, etc.) and the increasing illegal (and commonly immoral) financial and legal flows that has to be addressed. As Torpey (2007: 57) suggests:

> In order to extract resources and implement policies, states must be in a position to locate and lay claim to people and goods.

This image of 'penetration' no longer holds true despite Foucault's insistence of extensive 'surveillance' by the state—the latter's grasp no longer embraces societies adequately to achieve this.

Conclusions

Sum (2000: 233) helps to bring all the ideas of time, space and flows together in the context of a multinational, globalised commercial industry which could be shipping as much as anywhere else:

> Commercial time-space flows are influenced by the practices of networks of multinational service firms and their regional/local counterparts located in the 'global-gateway' cities (Sassen 1991). Their operations are permitted on the provision of the producer and distributive services and logistics information (i.e. insurance, legal services, consultancies, logistic management, transportation, retail) that 'facilitate all economic transactions and the driving force that stimulates the production of goods' within the 'regional chain' (Riddle 1986: 26). These networks of service firms co-ordinate and narrate the time and space of global-regional and regional-local next of the production and distributive chains. In temporal terms, service firms in the 'supply-pipeline' manage information flows that balance cost options and lead- and transit- time in time-bound projects. This is increasingly coordinated in 'electronic space' insofar as information is substituted for inventory (i.e. 'virtual' inventory) at the centre so that 'quick responses' can be made directly into the replenishment systems through local outsourcing or procurement (Christopher 1992: 108–124).

In beginning to understand the substantial importance of flow in the future and continuing development of the globalised world including shipping and thus its governance, we can revert to two themes. The first was put forward by Lash and Urry (1994: 323) and their consideration of what they termed disorganised capitalism—a new epoch where 'processes and flows have transformed (the) pattern of a dozen or so of organised capitalist societies within the core of the North Atlantic Rim'. These flows and processes included:

- The flowing of capital and technologies to 170 or so individual 'self-governing' capitalist countries each concerned to defend its territory;

- Time-space compression in financial markets and the development of a system of global cities;
- The growth in importance of internationalised producer services;
- The generalisation of risks which know no national boundaries and the fear of such risks (e.g. Iraq, Afghanistan, Kurdistan, ISIS);
- The putative globalising of culture and communication structures partly breaking free of particular territories;
- The proliferation of forms of reflexivity, individual and institutionalised, cognitive and especially aesthetic;
- Huge increases of personal mobility across the globe, of tourists, migrants and refugees;
- The development of a service class with cosmopolitan tastes especially for endlessly 'fashionable' consumer services provided by one or other category of migrant;
- The declining effectiveness and legitimacy of nation-states which are unable to control such disorganised capitalist flows; and
- The emergence of 'neo-worlds', the kinds of socially and regionally re-engineered cultural spaces which are the typical homelands for cosmopolitan postmodern individuals.

All of which sound remarkably familiar in the light of international shipping and their impact upon maritime governance (Luke 1992; Lodge 1983). Meanwhile from an earlier book, Roe (2013: 422–423) considers the role of outer space and capitalism's need for ever more fixes. Warf (2007: 385) suggests that:

> satellites and earth stations comprise a critical, often overlooked, part of the global telecommunications infrastructure. Castells's space of flows would be impossible without the skein of earth stations and orbital platforms that lie at the heart of the [satellite] industry.

MacDonald (2007: 594) continues the theme of astropolitics and their relationship to flows of information, money and more. Satellites move 'persistently through orbit, structuring the global imaginary, the socioeconomic order and the issue of everyday experience across the planet' (Parks 2005: 7) and as such will have an increasingly important position within the flows of globalised shipping which leads us to speed and someone who has dominated the discussion for some years, Paul Virilio.

We shall look at speed in much more detail in the next chapter, but for the moment, Deleuze and Guattari (1988: 386) provide the link we need:

> One of the fundamental tasks of the state is to striate the space over which it reigns, or to utilize smooth space as a means of communication in the service of striated space. It is a vital concern of every state not only to vanquish nomadism but to control migrations and, more generally, to establish a zone of rights over an entire 'exterior', over all of the flows traversing the ecumenon. If it can help it, the state does not dissociate itself from a process of capture of flows of all kinds, populations, commodities or commerce, money or capital etc. There is still a need for fixed paths in well-defined directions, which restrict speed, regulate circulation, relativize movement and measure in detail the relative movements of subjects and objects. That is why Paul Virilio's thesis is important, when he shows that

'the political power of the state is polis, police, that is 'management of the public ways' and the 'gates of the city, its levies and duties, are barriers, filters against the fluidity of the masses against the penetration power of migratory packs', people, animals and goods.

We can see the threads here of our argument relating how governance needs to accommodate the metaphors of movement with the fluidity of modern international society to ensure that the maritime sector produces effective policies to restrict or even prevent the death, pollution and inefficiency that still characterises its everyday activity.

References

Agnew, J. (1998). *Geopolitics: Re-visioning world politics*. London: Routledge.

Allen, J. (2003). *Lost geographies of power*. Oxford: Blackwell.

Allmendinger, P. (2001). *Planning in postmodern times*. London: Routledge.

Appadurai, A. (1990). Disjuncture and difference in the global cultural economy. *Public Culture, 2*(2), 1–11, 15–24.

Appadurai, A. (1996). *Modernity at large, cultural dimensions of globalization*. Minneapolis, MN: University of Minnesota Press.

Arrighi, G. (1994). *The long twentieth century*. London: Verso.

Augé, M. (1995). *Non-places: Introduction to an anthology of supermodernity*. London: Verso.

Bale, J., & Sang, J. (1996). *Kenyan running: Movement culture and global change*. Portland, OR: Frank Cass.

Ballve, T. (2011). *Space of flows, territorial masquerades*. Scattered notes and commentary on politics and geography. http://territorialmasquerades.net/space-of-flows.

Barney, D. (2004). *The network society*. Oxford: Polity Press.

Blatter, J. K. (2001). Debordering the world of states: towards a multi-level system in Europe and a multi-polity system in North America? Insights from border regions. *European Journal of International Relations, 7*(2), 175–209.

Brenner, N. (1999). Beyond state-centrism? Space, territoriality, and geographical scale in globalization studies. *Theory and Society, 28*, 39–78.

Carnoy, M., & Castells, M. (2001). Globalization, the knowledge society, and the network state: Poulantzas at the millennium. *Global Networks, 1*(1), 1–18.

Cashore, B., & Howlett, M. (2007). Punctuating which equilibrium? Understanding thermostatic policy dynamics in Pacific Northwest Forestry. *American Journal of Political Science, 51*(3), 532–551.

Castells, M. (1989). *The informational city*. Oxford: Blackwell.

Castells, M. (1993). European cities, the informational society and the global economy. *Tijdschrift voor Economische en Sociale Geografie, 84*(4), 247–257.

Castells, M. (1996). *The rise of the network society*. Oxford: Blackwell.

Castells, M. (1997). *The power of identity*. Oxford: Blackwell.

Castells, M. (1998). *End of the millennium*. Oxford: Blackwell.

Castells, M. (1999). Grassrooting the space of flows. *Urban Geography, 20*(4), 294–302.

Castells, M. (2000). Materials for an exploratory theory of the network society. *British Journal of Sociology, 51*(1), 5–24.

Castells, M. (2002). Local and global: Cities in the network society. *Tijdschrift voor Economische en Sociale Geografie, 93*(5), 548–558.

Chambers, I. (1994). *Migrancy, culture, identity*. London: Routledge.

Christopher, M. (1992). Logistics and supply chain management. London: Pitman Publishing.

Clifford, J. (1997). *Routes, travel and translation in the late twentieth century*. Cambridge, MA: Harvard University Press.

Cresswell, T. (1999). Embodiment, power and the politics of mobility: The case of female tramps and hobos. *Transactions of the Institute of British Geographers NS, 24*, 175–192.

Cresswell, T. (2006). *On the move. Mobility in the modern western world*. London: Routledge.

DeCourville-Nicol, V. (1997). Monstrous ovedlowing: A gothic counter-production of modernity. *Space and Culture, 1*, 67–84.

Delaney, D. (2003). *Law and nature*. Cambridge: Cambridge University Press.

Deleuze, G., & Guattari, F. (1988). *A thousand plateaus. Capitalism and schizophrenia*. London: The Athlone Press.

Dicken, P., Kelly, P. F., Olds, K., & Yeung, H. W. (2001). Chains and networks, territories and scales: Towards a relational framework for analysing the global economy. *Global Networks, 1*(2), 89–112.

Dickens, P., & Ormrod, J. S. (2007). *Cosmic society. Towards a sociology of the universe*. London: Routledge.

Dykrton, J. (1997). The culture of the excluded middle, 1914–1989 a preliminary investigation. *Space and Culture, 1*, 35–48.

Elias, N. (1978). *What is sociology?*. New York: Columbia University Press.

Friedmann, J. (2000). Reading Castells: *Zeitdiagnose* and social theory. *Environment and Planning D, 18*, 111–120.

Graham, S., & Marvin, S. (2001). *Splintering urbanism. Networked infrastructures, technological mobilities and the urban condition*. London: Routledge.

Gritsenko, D. (2013). The Russian dimension of Baltic maritime governance. *Journal of Baltic Studies*, iFirst, 1–25.

Gritsenko, D., & Yliskyla-Peuralahti, J. (2013). Governing shipping externalities: Baltic ports in the process of SO_X emission reduction. *Maritime Studies, 12*, 10.

Grossberg, L. (1993). Cultural studies and/in New Worlds. *Critical Studies in Mass Communication, 10*, 1–22.

Gupta, A., & Ferguson, J. (1992). Beyond culture: Space, identity and the politics of difference. *Cultural Anthropology, 7*(1), 6–22.

Harvey, D. (1989). *The condition of postmodernity: An enquiry into the origins of cultural change*. Oxford: Blackwell.

Harvey, D. (2000). *Spaces of hope*. Edinburgh: Edinburgh University Press.

Hassan, R. (2009). *Crisis time: Networks, acceleration and politics within late capitalism*. www.ctheory.net/printer.aspx?id=618.

Henderson, J., & Castells, M. (1987). Introduction. In J. Henderson & M. Castells (Eds.), *Global restructuring and territorial development*. London: Sage.

Hier, S. P., & Greenberg, J. (2007). Mobility, privacy, ethics and resistance. In S. P. Hier & J. Greenberg (Eds.), *The surveillance studies reader*. Milton Keynes: Open University.

Ironstone-Catterall, P. (1997). Instructions for use unmaking narrative/normative spaces of intelligibility. *Space and Culture, 1*, 31–33.

Jameson, F. (1991). *Postmodernism or the cultural logic of late capitalism*. Durham, NC: Duke University.

Jensen, O. B., & Richardson, T. (2004). Framing mobility and identity: Constructing transnational spatial policy discourses. In J. O. Baerenholdt & K. Simonsen (Eds.), *Space odysseys: Spatiality and social relations in the 21st century* (pp. 83–100). Aldershot: Ashgate.

Knox, P. L. (1995). World cities and the organization of global space. In R. J. Johnston, P. J. Taylor, & M. Watts (Eds.), *Geographies of global change* (pp. 232–247). Oxford: Blackwell.

Lash, S., & Urry, J. (1994). *Economies of signs and space*. London: Sage Publications.

Leamer, E., & Storper, M. (2001). *The economic geography of the internet age, WP8450*. Cambridge, MA: National Bureau of Economic Research.

Lefebvre, H. (1991). *The production of space*. Oxford: Blackwell.

Leinbach, T., & Brunn, S. (2001). *Worlds of e-commerce: Economic, geographical and social dimensions*. Chichester: Wiley.

Leyshon, A. (2001). Time-space (and digital) compression; software formats, musical networks, and the reorganisation of the music industry. *Environment and Planning A, 33*, 49–77.

Lo, V., & Grote, M. (2002). Localisation, urbanization, virtualization and what next? Stock trading in the information age. In H. Kujath & C. Heinrich (Eds.), *Die Bedeutung von externen-Effekten und Kollektivgutern fur die regionale Entwicklung* (pp. 89–106), REGIO Transfer, Erkner.

Lodge, D. (1983). *Small world*. London: Penguin.

Lorraine, T. E. (1999). *Irigaray and deleuze: Experiments in visceral philosophy*. Ithaca, NY: Cornell University Press.

Luke, T. W. (1991). The discipline of security studies and the codes of containment: Learning from Kuwait. *Alternatives, 16*, 315–344.

Luke, T. W. (1992). *New world order or neo-world orders: Power politics and ideology in the informationalizing global order*. Theory, Culture and Society 10th Anniversary Conference, Champion, PA.

MacDonald, F. (2007). AntiAstropolitik: Outer space and the orbit of geography. *Progress in Human Geography, 31*(5), 592–615.

Marston, S. A., Jones, J. P, I. I. I., & Woodward, K. (2005). Human geography without scale. *Transactions of the Institute of British Geographers NS, 30*, 416–432.

Martin, E. (1990). Science and women's bodies: Forms of anthropological knowledge. In M. Jacobus, E. F. Keller, & S. Shuttleworth (Eds.), *Body/Politics: Women and the discourses of science* (pp. 69–82). London: Routledge.

McCarthy, C. (1997). Constructions of a culinary abject. *Space and Culture, 1*, 9–23.

McNay, L. (1994). *Foucault. A critical introduction*. Cambridge: Polity Press.

McLennan, G. (2003). Sociology's complexity. *Sociology, 37*(3), 547–564.

Miller, C. L. (1993). The postidentitarian predicament in the footnotes of *A Thousand Plateaus*: Nomadology, anthropology and authority. *Diacritics, 23*(3), 6–35.

Mol, A. P. J. (2007). Boundless biofuels? Between environmental sustainability and vulnerability. *European Society for Rural Sociology, 47*(4), 297–315.

Moore, A. (2008). Rethinking scale as a geographical category: From analysis to practice. *Progress in Human Geography, 32*(2), 203–225.

Neo, B. S., & Chen, G. (2007). *Dynamic governance: Embedding culture, capabilities and change in Singapore*. Singapore: World Scientific.

O'Brien, R. (1992). *Global financial integration. The end of geography*. London: Pinter.

O'Connor, D. (1997). Lines of (f)light: The visual apparatus in Foucault and Deleuze. *Space and Culture, 1*, 49–66.

Ohmae, K. (1995). *The end of the nation state*. New York: The Free Press.

Paasi, A. (1998). Boundaries as social processes: Territoriality in the world of flows. *Geopolitics, 3*(1), 69–88.

Parks, L. (2005). *Cultures in orbit: Satellites and the televisual*. Durham, NC: Duke University Press.

Pred, A. (1984). Place as historically contingent process: Structuration and the time-geography of becoming places. *Annals of the Association of American Geographers, 74*(2), 279–297.

Prendergast, C. (1992). *Paris in the nineteenth century*. Oxford: Blackwell.

Riddle, D. L. (1986). *Service-led growth. The role of the service sector in world development*. New York: Praeger Publishers.

Roberts, S. (1994). Fictitious capital, fictitious spaces: The geography of offshore financial flows. In S. Corbridge, N. Thrift, & R. Martin (Eds.), *Money, power and space*. Oxford: Blackwell.

Rodrigue, J.-P., Comtois, C., & Slack, B. (2009). *The geography of transport systems*. http://people.hofstra.edu/geotrans/eng/ch5en/conc5en/table_flowsglobalization.html.

Roe, M. S. (2013). *Maritime governance and policy-making*. London: Springer.

Rosow, S. J. (2005). Beyond democratic idealism: Borders, speed and cosmopolitan ethos. International Studies Association Annual Meeting, Honolulu, HI.

Ruggie, J. (1993). Territoriality and beyond: Problematizing modernity in international relations. *International Organization, 47*(1), 139–174.

Sack, R. D. (1986). *Human territoriality*. Cambridge: Cambridge University Press.

Santos, M. (1995). Contemporary acceleration: World-time and world-space. In G. B. Benko & U. Strohmayer (Eds.), *Geography, history and social sciences* (pp. 171–176). Dordrecht: Kluwer.
Sassen, S. (1991). *The global city: New York*. Princeton, NJ: Princeton University Press.
Sassen, S. (2006). *Territory-authority rights. From medieval to global assemblages*. Princeton, NJ: Princeton University Press.
Schatten, G., & Schatten, H. (1996). *Gender. Language and science*. Templeton lecture, University of Sydney, Sydney, Australia.
Serres, M. (1993). *Angels. A modern myth*. Paris: Flammarion.
Shields, R. (1997). Flow as a new paradigm. *Space and Culture, 1*, 1–4.
Sidaway, J. D. (1995). Political geography in the time of cyberspaces: New agendas? *Geoforum, 25*(4), 487–503.
Smith, R. G. (2003). World city topologies. *Progress in Human Geography, 27*(5), 561–582.
Soja, E. (2000). *Postmetropolis: Critical studies of cities and regions*. Oxford: Blackwell.
Spaargaren, G., & Mol, A. P. J. (2008). Greening global consumption: Redefining politics and authority. *Global Environmental Change, 18*, 350–359.
Stalder, F. (2002a). *The status of objects in the space of flows*. McLuhan program in culture and technology, University of Toronto, Canada.
Stalder, F. (2002b). *Spaces of flows: Characteristics and strategies*. Doors of Perception Conference, Amsterdam.
Stalder, F. (2006). *Manuel Castells, The theory of the network society*. Cambridge: Polity Press.
Storper, M. (1997). *The regional world. Territorial development in a global economy*. New York: The Guilford Press.
Sum, N.-L. (2000). Beyond techno-globalism and techno-nationalism. In R. D. Germain (Ed.), Globalization and its critics, perspectives from political economy (pp. 223–244), Political Economy Research Centre, University of Sheffield.
Taylor, P. J. (1996). Territorial absolutism and its evasions. *Geography Research Forum, 16*, 1–12.
Taylor, P. J. (2000a). World cities and territorial states under conditions of contemporary globalisation. *Political Geography, 19*, 5–32.
Taylor, P. J. (2000b). World cities and territorial states under conditions of contemporary globalisation II: Looking forward, looking ahead. *GeoJournal, 52*, 157–162.
Taylor, P. J. (2000c). Embedded statism and the social sciences 2: Geographies (and metageographies) in globalization. *Environment and Planning A, 32*, 1105–1114.
Taylor, P. J. (2005). New political geographies: Global civil society and global governance through world city networks. *Political Geography, 24*, 703–730.
Thrift, N. (1995). A hyperactive world. In R. J. Johnston, P. J. Taylor, & M. Watts (Eds.), *Geographies of global change: Remapping the world* (pp. 18–35). Oxford: Blackwell.
Torpey, J. (2007). Coming and going: on the state monopolization of the legitimate 'means of movement'. In S. P. Hier & J. Greenberg (Eds.), *Mobility, privacy, ethics and resistance, the surveillance studies reader* (pp. 52–66). Milton Keynes: Open University.
Urry, J. (2000). Mobile sociology. *British Journal of Sociology, 51*(1), 185–203.
Urry, J. (2002). The global complexities of September 11. *Theory, Culture and Society, 19*, 57–69.
Urry, J. (2010). Mobile sociology. *The British Journal of Sociology, 61*, 347–366.
Virilio, P. (1999). *Polar inertia*. Thousand Oaks, CA: Sage.
Warf, B. (2007). Geopolitics of the satellite industry. *Tijdschrift voor Economische en Sociale Geografie, 98*(3), 385–397.
Waterman, P. (1999). The brave new world of Manuel Castells: What on Earth (or in the Ether) is going on? *Environment and Change, 30*, 357–380.
Watts, M. J. (1991). Mapping meaning, denoting difference, imagining identity: Dialectical images and postmodern geographies. *Geografiska Annaler B, 73*(1), 7–16.
Webster, F. (2002). *Theories of the information society*. London: Routledge.

Wheeler, J., Aoyama, Y., & Warf, B. (Eds.). (2000). *Cities in the telecommunications age*. New York: Routledge.

Woolf, V. (1928). Orlando: A biography. London: Hogarth.

Yeung, H. W. (1998). The social-spatial constitution of business organisations: A geographical perspective. *Organization, 5*, 101–128.

Yeung, H. W. (2000). Embedding foreign affiliates in transnational business networks: The case of Hong Kong firms in Southeast Asia. *Environment and Planning A, 32*, 201–222.

Young, I. M. (1990). *Throwing like a girl and other essays in feminist philosophy and social theory*. Bloomington, IN: Indiana University Press.

Zook, M. A. (2003). Underground globalization: Mapping the space of flows of the internet adult industry. *Environment and Planning A, 35*, 1261–1286.

Chapter 7
Speed

Abstract And finally to speed, seen by some as important to politics and policy-making as wealth and necessitating a political economy of speed. We have come a long way from the static perception of maritime governance, passing through various concepts which have become increasingly mobile—change, process, flow—and now arrived at our ultimate destination. Cruising at speed in a way that reflects how the world now works and how the maritime sector needs to be viewed. When we have completed our tour of the dynamism of governance, we can move on to the last chapter where we have a number of issues to consider, not least some models of change in governance, the contradictions that shipping exhibits and their impact upon the governance and policy-making process, and to place it all in context. This chapter has to focus on the work of Paul Virilio but also looks at speed and space, power and policy-making in some detail as considered by others, concluding with a discussion of equilibrium in governance.

> Subscribers were not so much buying daily news as they were buying instantaneity, ubiquity – in other words, their own participation in universal contemporaneity, in the movement of the future Planet City. Virilio (1995a: 49).

> Speed has conditioned us to daily scandal. Haynes Johnson, *The Washington Post*.

> There's a new idea in Europe: happiness.
> *Louis Antoine de Saint-Just* in a speech to the National Convention of France, March 3, 1794.

> For the flaneur, it was traffic that did him in. In the relatively tranquil shelter of the arcades, his original habitat, he practiced his trade of not trading, viewing as he loitered the various selection of luxury goods and luxury people displayed before him. 'Around 1840 it was elegant to take turtles for a walk in the arcades (This gives a conception of the tempo of flaneurie)' (Benjamin 1972: 532). By Benjamin's time, taking turtles for urban strolls had become enormously dangerous for turtles, and only somewhat less so for flaneurs. The speed-up principles of mass production had spilled over into the street, waging 'war on flaneurie' (Benjamin 1972: 547). The 'flow of humanity… has lost its gentleness and tranquillity', *Le Temps* reported in 1936. 'Now it is a torrent, where you are tossed, jostled, thrown back, carried to right and left' (Benjamin 1972: 564). With motor transportation still at an elementary stage of evolution, one already risked being lost at sea. Buck-Morris (1986: 102).

M. Roe, *Maritime Governance*, DOI 10.1007/978-3-319-21747-5_7

Introduction

And finally to speed, seen by Virilio as important to politics and policy-making as wealth necessitating a political economy of speed [see Armitage (1999a: 4), citing Virilio and Lotringer (1997: 67)]. We have come a long way from the static perception of maritime governance, passing through various concepts which have become increasingly mobile—change, process, flow—and now arrived at our ultimate destination. Cruising at speed in a way that reflects how the world now works and how the maritime sector needs to be viewed. When we have competed our tour of the dynamism of governance, we can move on to the last chapter where we have a number of issues to consider, not least some models of change in governance, the contradictions that shipping exhibits and their impact upon the governance and policy-making process, and to place it all in context.

Meanwhile, let us take Marx (1973: 349 and 524; originally written 1857–8):

> *Circulation proceeds in space and time*... It is... an essential process of capital... The *constant continuity* of the process, the unobstructed and fluid transition of value from one form into the other, or from one phase of the process into the next, appears as a fundamental condition for production based on capital to a much greater degree than for all earlier forms of production. (original emphasis).

It is notable just how much Marx emphasises the issues with which we have been so concerned—notably process and fluids—and all the issues in dynamism they suggest reflecting that the need for dynamic governance had been noted some 150 years ago. However, not everyone was enamoured with the growth in movement and the inevitable emphasis on speed. Take Le writing (1978):

> In the early evening twilight on the Champs-Elysees it was as though the world had suddenly gone mad... Day after day the fury of traffic grew. To leave your house meant that once you had crossed the threshold you were a possible sacrifice to death in the shape of innumerable motors. I think back 20 years, when I was a student, the road belonged to us then.

Elden (2005: 8) notes Virilio's (1986, 1999) consideration of the significance of speed over space and that the speed of movement and the process of deterritorialisation had changed the whole context for governance. Obando-Rojas et al. (2004: 295) emphasise the importance of the relationship between flow and speed and so provide a convenient link between the last chapter and this one and in the good academic company of Deleuze and Guattari (1988: 386) who refer to Virilio (1986: 12–13) in that:

> the gates of the city, its levies and duties, are barriers, filters against the fluidity of the masses, against the penetration power of migratory packs.

Armitage (1999a: 4) brings much of the significance of speed to philosophical debate together. He emphasises Virilio's scepticism of the 'political economy of wealth' and its substitution by the importance of speed to a dromocratic society in terms of history, politics and revolution. Virilio is heavily influenced by Tzu's (1993) theory on war and the positive (Fascist) and negative (anti-Fascist) political and technical aspects of Marinetti and Futurism (Tisdall and Bozzolla 1977; Virilio and Lotringer 1997: 45). Both Virilio and Sun Tzu see cultural development and socio-political institutions such as the military demonstrating the need for war and the importance of speed compared with wealth as a societal foundation. Speed and institutions are fundamentally interrelated and the need for institutional reform outlined in Chap. 1 makes a discussion of speed that much more important. This is reaffirmed by Armitage (1999a: 7) who sees major technological developments, mostly focussed on increasing speed of communication and fundamental to maritime governance, also challenging the nature of existing institutions and their structures and nowhere is this more apparent than in the problems of maritime safety and the environment, exacerbated by the renewed and ever-increasing ability to use speed-focussed communication systems by an enormously expanded number of stakeholders.

Speed: Preliminaries

> Are we having today, another, a different experience of speed? Is our relation to time and to motion qualitatively different? Or must we speak prudently of an extraordinary – although qualitatively homogenous – acceleration of the same experience? Derrida (1984: 20) quoted in Crogan (1999a: 164).

> The revolutionary contingent attains its ideal form not in the place of production, but in the street, where for a moment it stops being a cog in the technical machine and itself becomes a motor machine of attack, in other words a producer of speed. Virilio (1977: 3).

> Time and space died yesterday. We already live in the absolute because we have created eternal, omnipresent speed. J.G. Ballard, *Crash*.

Derrida (1984: 23) suggests that social and political life have been dominated since 1945 by a 'speed race in search of speed' creating an 'aporia of speed'. He believed that all techno-scientific developments have to be considered in the light of the speed race. 'No single instant, no atom of our life (our relation to the world and to being) is not marked today, directly or indirectly by that speed race' (Derrida 1984: 20).

We do not need to spend too much time on defining speed as this has been covered earlier both directly and indirectly. Commentators on the importance of speed to the transport sector and its governance include Nielsen and Oldrup (2001: 9–11), Armitage and Graham (2001: 121) who emphasise the need for contemporary societies to recognise the centrality (but not uniquely) of speed, and Brunn (1998a: 6) who sees speed as a 'critical component of human organisation, exchange, interaction and governance'. He describes its spatial relevance in terms of how speed dictates the geographical relationships between phenomena. Historically, it is central to how communications and power influenced the world. Speed is central to all transport whatever mode and this is related to an understanding of space through such concepts as time and cost space convergence, time-space compression and human extensibility (Abler et al. 1975). He then relates this to the development of the Internet all of which can be seen closely linked to the developments in shipping that have taken place and had a significant effect upon governance. Its significance has even been expressed in the rather extreme terms of adoration. Santos (1995: 171), for example, comments on the underlying velocity:

> In fact, speed was precisely what struck those who first witnessed the appearance of railways, of steam boats, and who thus, at the end of the XIX century and the beginning of the XX, were reached by the diffusion of automobile, airplane, of wireless telegraph and submarine cable, of telephone and radio.

He also suggests that speed can be identified in many ways, and taking the concept of acceleration and its relationship to speed suggests that:

> it has also added new ingredients to history: a new evolution of potencies and productivity, the use of materials and new forms of energy, the mastering of the electro-magnetic spectrum, demographic expansion… urban explosion, consumption explosion, exponential multiplication of the number of objects and words.

However, one interesting debate centres upon the ideas of the most vocal promoter of the notion of speed and its philosophical importance, Paul Virilio. Despite his contribution to the issue, Deleuze and Guattari (1988: 137–138) in particular are critical of Virilio's definitions of speed and Crogan (1999b: 142) emphasises this issue. Virilio identified three types of speed: that of the nomadic or revolutionary in which he included speed exemplified by riots and guerrilla warfare; speeds that are 'regulated, converted, appropriated by the state' and include as an example the management of public ways (roads, docks, railways, etc.); and thirdly speeds that are instituted by worldwide organisations responsible for 'total war or planetary over-armament'. These latter include naval and air fleets, nuclear strategies and rockets.

Deleuze and Guattari's suspicions about Virilio's interpretation of speed for his own philosophical convenience are supported by Crogan (1999b: 142) and together they see it as his own analysis that makes an over-emphasis of the distinctiveness of speed possible and consider his definition of the character of speed as universally 'Fascist'.

However (or perhaps even because), this type of debate does not detract from the importance that speed has for an understanding of dynamism in governance and policy-making. This is even the case within the shipping industry—not renowned for its focus upon fast speed but nevertheless heavily affected by and affecting the speed of distribution and the choices to be made about acceleration and the slowness or quickness of movement, decision-making, transfer of money, data, etc. Take slow steaming, laying-up, fast-ferries, self-loading in ports, electronic document transfer, Internet platforms and many more for evidence of its significance. And at risk of opening up the debate again let us take Virilio:

> In our situations of televisual experience, we are living in nothing less than the sphere of Einstein's relativity, which wasn't at all the case at the time that he wrote it since that was the world of trolley cars, and at most, the rocket. But today we live in a space of relativity and non-separability. Our image of time is an image of instantaneity and ubiquity.

> And there's a stunning general lack of understanding of speed, a lack of awareness of the essence of speed… And this passage from an extensive to an intensive time will have considerable impact upon all the various conditions of our society: it leads to a radical reorganization both of our social mores and of our image of the world. This is the source of the feeling that we're faced with an epoch in many ways comparable to the Renaissance: it's an epoch in which the real world and our image of the real world no longer coincide… (in Sans 1991: 139–140, re-quoted in Der Derian 1999: 220).

Dickens and Ormrod (2010: 537) stress the importance of speed in the context of war, an approach familiar to those who have read much of Virilio. Considering the use of missiles guided by satellites, they suggest that these:

> are supposed to not only annihilate an enemy but reduce the surviving population into stunned submission. Military proponents of 'shock and awe' tactics make clear that the purpose of speed on a global and now galactic scale is not necessarily to kill large numbers of people, but to shock wider populations into mental defeat through unrelenting attacks on simultaneous targets. This type of 'shock and awe' war has been simultaneously made part of the 'society of the spectacle', one in which massive public events are commodified in the form of live television broadcasts via satellite to a global audience (Debord 1967; Retort 2005; Baudrillard 1995).

Urry (2004: 20) considers that power is closely related to speed in almost all circumstances and although he also relates power to lightness, distance and weightlessness it is speed that dominates. Taking the central role that speed forms in shipping markets and how either increasing or decreasing speed (e.g. fast-ferries and slow steaming) have such an important impact on the market. This is exacerbated by the increasing significance of speed in the everyday world of communication and monitoring, much of which also impacts on shipping. Power is also always characterised by mobility which in turn frequently features speed as a central trend.

Power in business is related to speed as much as anything else, and shipping is in many ways one most central to the core of business. Change, process, flow and then speed suggest a logical sequence of characteristics that governance needs to accommodate in shipping as the progress of globalisation continues to take effect and the anachronistic features of existing maritime governance becomes evermore apparent.

And speed is not going away. In fact if anything, it is commonly increasing. ['Events move too quickly'—Angelides and Caiden (1994: 223)]. Take the whole character of supply chains of which shipping is a major part and the continuous struggle to minimise stock. Just in time logistics manifests itself in the desire to increase speed. Meanwhile, Santos (1995: 171) sees it all as something much bigger:

> Accelerations are culminating moments in history which seem to concentrate powers that explode to create something new. The run of time (as mentioned by Michelet in the preface to his history of the XIXth century) is marked by these great commotions which at times seem meaningless. Thus, each time that we believe to have reached a definite stage, the unusual, the strange strikes us. The problem is to understand and find a new set of concepts able to express the new order created by this *acceleration*.

Hesse (1999: 33) (with apologies for the appalling English that characterises the original) identifies three processes that require societal speed to continue increasing:

- Individualisation and flexibilisation characterised by more and more complex activities, trip chains and action spaces.
- Spatial fragmentation exemplified by disaggregated land uses (e.g. the suburbanisation of retailing and the location of housing in dock areas), including the creation of the urban functional region (*Zwischenstadt*).
- Globalisation. Need we say more?

Increasing speed is an 'outcome of and a requirement for these processes' as the inevitable stretching of space needing faster communication with modern society requiring acceleration in overcoming time and distance. Armitage (1999b) sees it as unrelenting acceleration, suggesting that all societies are pyramidal with the higher reaches reflecting higher speeds and vice versa. 'Ever-increasing speed lies at the heart of the organisation and transformation of the contemporary world', and along with Urry (2000: 128), he goes on to emphasise Virilio's (1989: 9–28) concerns about the relationship between productive interruptions, suggestions, jumps and creative dynamics that characterise changes in levels (Armitage 1999a: 1, 6).

This continuous acceleration has attracted a variety of attention. This ranges from Heidegger, disturbed by the 'seemingly uncontrolled acceleration of temporal processes' (Harvey 1990: 209) occurring in Germany in the 1930s, to Wark (1988: 82) and his consideration of Virilio's vectors, technology and acceleration, and to Brugger (1999: 17). Sidaway (1994: 491–492) quotes Walker (1993: 5):

> The experience of temporality, of speed, velocity and acceleration, is more and more bewildering. Despite the bewilderment, this experience is now richly inscribed in the contemporary imagination. Discourses of military strategy express worries about contracting response times and instantaneous decisions... Discourses of political economy speak about the enhanced mobility of capital compared with the territorial constraints experienced by governments and labour.

Brugger in particular suggests that any specific movement has to take place within or upon a specific matter; this gives the characteristic of space to its extension and time its duration. As acceleration increases then:

- The extent of space will get smaller until eventually it becomes a point.
- The time duration will eventually become immediacy.
- Matter becomes immaterial.
- Finally, the object that is accelerated will become available in more and more places, achieving ubiquity.

In turn, this leads to movement becoming inertia, as things that are everywhere do not move; when space becomes a point, it is effectively no longer space; when time is immediacy, then there is no time; and immaterial matter becomes non-matter. The globalising world shows all the signs of moving in this direction with acceleration all around us and affecting every aspect of the maritime supply chain. There is some way to go before non-materiality is with us (and therefore no need for shipping), but the progressive disappearance of space and time is clearly apparent.

Virilio

> The minimum speed on this road was greater than I had previously driven… Instructions screamed at me from the road once: 'Do not stop! No stopping. Maintain speed! Trucks as long as freighters went roaring by, delivering wind like the blow of a fist. These great roads are wonderful for moving goods but not for inspection of a countryside. You are bound to the wheel and your eyes to the car ahead and to the rear-view mirror for the car behind and the side mirror for the car or truck about to pass, and at the same time you must read all the signs for fear you may miss some instructions or orders. No roadside stands selling squash juice, no antiques stores, no farm products or factory outlets. When we get these thruways across the whole country, as we will and must, it will be possible to drive from New York to California without seeing a single thing. John Steinbeck (1962) *Travels with Charley in Search of America*, pp. 70–73; quoted in Virilio (2007: 75–76).

Without wishing to focus excessively or exclusively upon Paul Virilio and his contribution to understanding speed and society, like our dealings with Manuel Castells and flows, there is no way he can (or should) be avoided. The aim is not to provide a comprehensive analysis of speed through Virilio's eyes but merely to point the reader in the right direction if more is wanted and to indicate the character, size and importance of the contribution made. John Armitage provides much of what is needed and as an introduction to what Virilio has written and others about him then there is little better than to see his bibliographic review (Armitage 1999c) although there are many other commentators not least Paasi (1998) and Wark (1988).

Der Derian (1999: 215) indicates just how significant to the debate on speed that Virilio is seen by some:

> His take on a deterritorialized, accelerated, hyper-mediated world redefines outlandish. Nonetheless, when shit happens – events that defy conventional language, fit no familiar pattern, follow no notion of causality – I reach for Virilio's conceptual cosmology.

Der Derian continues to show how Virilio uses speed as a variable, chronopolitics as a concept and dromology as a method to produce new understandings of an ever-accelerating global politics (Der Derian 1999: 218). Virilio links his ideas on speed and its importance to globalisation, the decline of the state and postmodernism. Whilst issues relating speed and the nation-state are also discussed by Brunn (1998b: 112) who recognises that states once disparate and largely unconnected can now be fully integrated as distances come to mean nothing, the role of postmodernism is taken up by Porter (1968: 6) in a very early paper considering changes in society and their relationship to speed (McLuhan 1964), and Der Derian (1990: 297) and his discussion of the three forces that characterise the postmodern, one of which centres on speed. Each issues that have direct relevance to maritime governance. Thus:

> In effect, the geopolitics of nations that yesterday still presupposed the hierarchical privilege of the center over its peripheries, of the summit over the base, the 'radiocentrism' of exchanges and horizontal communication, loses its value in the same way as does the extreme vertical densification to the benefit of an inapparent morphological configuration. The NODAL succeeds the CENTRAL in a preponderantly electronic environment, 'telelocalization' favouring the deployment of a generalized eccentricity, endless periphery, forerunner of the overtaking of the industrial urban form, but especially of the decline of the sedentary character of the metropolis to the advantage of an obligatory *interactive confinement*, a sort of inertia of human populations for which the name *teleconcentrism* may be proposed, while waiting for that of 'homeland' to replace that of the large suburb. The secular opposition city/country is being lost while the geomorphological uniqueness of the state is dissipating. Virilio (1984: 156).

Virilio has contributed extensively to the debate (see, e.g. Virilio 1986, 1995a, b, 1999, 2000 and 2007 amongst many others), focussing repeatedly upon the relationship between speed, politics and war which in turn has clear relevance to any debate on governance reform. These connections particularly with war are re-emphasised by Crogan (1999b: 142), Kellner (1999) and Dickens and Ormrod (2007: 89–90). The last cited emphasises the significance of technology and war (reminiscent of Burroughs (1991: 95): 'This is a war universe. War all the time. That is its nature. There may be other universes based on all sorts of principles, but ours seems to be based on war') and its inevitable focus on speed including the development of aircraft, rockets, missiles, satellites and the Internet as a response to the Cold War. Armitage (1999a: 3) continues the theme relating the importance of the military and their desire for speed for all types of policy across many contexts. Shipping is clearly one that could be included as the development and current characteristics of many commercial shipping activities stem directly and indirectly from military needs and desires not the least of which is the need to control the speed of both a domestic and opposing fleet. Take the USA Jones Act, the strategic interpretation of commercial fleets in the EU, and port policies as diverse as those in Poland, China and France where the strategic implications of privatisation have reduced the extent of proposals. The relationship between war, power and speed (and consequently policy-making and governance) has also been pointed out by Pokrovsky (1959), Arrighi et al. (1999) and Ropp (2000), whilst Armitage and Graham (2001: 112) also indicate how international trade and war have come together in ambition through the vehicle of speed, both desirous of its control (and commonly its increase).

> It has long been recognised that, while trade is dependent on the overproduction of speed, capitalism is also based on systemic economic excess. Indeed, the systematic and conscious production of massive excess which, according to Virilio and Marx, is founded firstly on 'the increasing speed of information transmission', and secondly on production for export for the external market. Armitage and Graham (2001: 115).

And no industry is possibly more bound up with exports than shipping. Meanwhile, others provide extensive reviews of Virilio's perception of speed as central to political ambitions and policy design (and consequently the design of governance). They include Cubitt (1999), Der Derian (1999: 216), Gane (1999), Kellner (1999), Conley (1999), Leach (1999), Zurbrugg (1999), Armitage (2000),

Van Ham (2001: 8), Dickens and Ormrod (2007: 61) and Bogard (2007: 95), the latter focussing on contemporary power and its concentration on observation and how this is a dromological function—the race to see 'first'. Meanwhile, clear connections between development policies including those for the maritime sector and the role of speed can be identified. Virilio takes the 'terrestrial city' which he sees as a 'mere concentration of passengers in transit' with ports, railway stations, bus termini and airports acting as the 'archaeology' of future society, concentrated in the 'vector of transportation'. 'The new capital is a city at the intersection of practicabilities of time, in other words of speed' (Virilio and Lotringer 1997: 67).

Brugger (1999: 18) however, provides some critical comments. He feels that Virilio is over one-dimensional focussing on a single phenomenon—speed and acceleration—which he claims controls the organisation of the world and the progress of civilisation. Starting from the assumption that speed is central to everything he then proceeds to prove its totality, blind to alternative influences and impacts. 'To put it in a paradoxical way: a theory that makes speed and its acceleration the dominant pivot can explain almost everything; and what it cannot explain, it simply does not explain' (Brugger 1999: 18).

Despite these criticisms, Virilio's interpretation of modern society remains both popular and pertinent. Crogan (1999a: 161) summarises it neatly. For Virilio, speed:

> ...is a crucial if neglected factor in all historical and political developments. He suggests that the ever increasing speeds of contemporary technologies of transportation, communication, destruction and representation threaten to overwhelm the capacity for conventional historical modes of interpretation to make sense of them. These speeds challenge the ability of historic-critical discourse to account adequately for their impacts on society, politics, culture and even the perception of reality. In exploring this challenge, Virilio has questioned the mainstream discourses of a 'future perfect' that accompany, promote and even generate techno-scientific advance. At a more profound level, Virilio's work addresses not only these technophilic discourses but also the humanist conception of history that allows both positivist and critical discourse to be articulated – including necessarily his own.

Speed and Space

The relationship between speed and space is clearly a close one and the recent emphasis that has been placed upon speed has changed the nature of space in many ways, not least effectively reducing the space between locations through time-space compression. We shall return to this in a short while meanwhile noting that the relationship is two way. As resources and locations for production become more dispersed, so the speed of distribution becomes that bit more difficult to accommodate satisfactorily. Thus, space and location and speed become inseparable. Armitage (1999b: 35) makes this clear outlining what he sees as the pyramidal nature of society with higher speeds occupying the territory and space of the higher reaches and the slower the bottom. This is taken up further by Brugger (1999: 13) who discusses the relationships that exist between movement, time, space and matter:

> Any movement takes place in or on a specific matter, which gives space its extension and time its duration. How movement, space, time and matter are configured is an effect of a property of the thing that is moved, namely that it is moved with a certain speed.

We return to Virilio briefly who sees space 'suspended' by speed (Virilio 1995b: 2). Agnew (1994: 72) notes the increasing permeability of boundaries as globalisation continues apace particularly those for nation-states (Nye 1988; Stopford and Strange 1991) and Virilio picked up on this suggesting that space would disappear and that 'territory has lost its significance in favour of the projectile. In fact, the strategic value of the non-place of speed has definitely supplanted that of place' (Virilio 1986: 133). Der Derian (1990) continues the theme replacing geopolitics with chronopolitics and the 'ground' with a 'cathode ray'.

Forsberg (1996: 369–370), along with Rosow (undated), considers that speed and mobility has undermined the role of place (and hence space) with particular reference to the political sector. He suggests that the whole process of deterritorialisation that has been a central cause and effect of globalisation (and thus central to maritime policy) is related to the move towards an unstable nation-state system which in turn is a consequence of the substitution of the central place of space by one of speed and movement. He focuses on the importance of information flow, fragmentation and pace, replacing homogeneity, community and place. However, he also places some question marks against this postmodern perception of change suggesting that since time and space are intrinsically interlinked, there is never going to be a substitution of one with another (a dichotomy he calls 'false') but rather a change in balance between them.

The significance of space as well as speed and time is also stressed by Jameson (quoted in Keith and Pile 1993: 2):

> I think it is at least empirically arguable that our daily life, our psychic experience, our cultural languages, are today dominated by categories of space rather than by categories of time, as in the preceding of high modernism.

Morris (1988: 7) provides a link between speed, space and our earlier discussion on the role of metaphors. He considers the use of a motel as a metaphor for a number of things including the use of community, sports and leisure facilities as mechanisms for understanding change in life. At the motel, there is a 'constant intermingling of the host family's domestic life, the social activities of the town's residents, and the passing diversion of tourists. The motel's solidity as *place* is founded by its flexibility as *frame* for varying practices of space, time and speed'. This theme is continued, comparing it with the work of Venturi et al. (1977: 34–35) who took the cityscape of Las Vegas and used a metaphorical approach to understand the differences between the auto-orientated fronts and service-orientated backs of many of the buildings.

Der Derian (1990: 297) is emphatic about the importance of speed over space, identifying two major forces (in his case in international relations but these are equally applicable in international policy-making and governance). The *chronopolitical* is significant in that chronology has been elevated over space in its political context. In this, he refers extensively to Virilio (1986). In addition, they are *technostrategic* in that they use technology extensively (particularly for war but taking Virilio's interpretation of war as central to all other policy ambitions, this can be applied across any area including the maritime). He refers to Clausewitz (1976: 128 and 177), Der Derian (1987) and Klein (1989).

He continues, quoting the Futurists and in particular their leader Filippo Marinetti:

> The Futurist writers will make use of free verse, an orchestration of images and sounds in motion to express our contemporary life, intensified by the speeds made possible by steam and electricity, on land, on the seas and in the air. (see Lista 1986: 12–14).

The Futurists placed speed as central to their manifesto in the 1920s, expressing themselves through paintings, poetry and literature before allying themselves to Mussolini's Fascism and ultimately falling to other art movements but their message of the significance of speed over all other forces, its close relationship with change, process and dynamism and its growth in importance with globalisation has clear resonance with maritime governance. This emphasis on speed, however, was not to be resurrected until Virilio in the 1980s, reflecting our earlier consideration by Santos (1995: 171):

> Up until the nineteenth century, society was founded on the brake. Means of gathering speed were very scant. You had ships, but sailing ships evolved little between Antiquity and Napoleon's time. The only machine to pursue speed with any sophistication was the optical telegraph and then the electric telegraph. In general up to the nineteenth century there was no production of speed. They could produce brakes by means of ramparts, the law, rules, interdictions, etc.... Then suddenly there's the great revolution that others have called the Industrial Revolution or the Transportation Revolution. I call it a dromocratic revolution because what was invented was... a means of fabricating speed with the steam engine, then the combustion engine. And so they can pass from the age of brakes to the age of the accelerator. In other words, power will be invested in acceleration itself. Virilio (1983: 44–45).

Brunn (1998b: 108) considers the relationship between space and speed, noting the increasing role of technology in reducing the time it takes to cross political space. Taking this as referring to policies as much as anything else, the clear relevance to shipping policy-making is apparent and since then the process itself has speeded up. In particular, they have affected the way that states transact policy business with its citizens and between government and organisations translating distant places into those much more accessible. He goes on to discuss the relationship between distance and proximity both of which he related to the technology of speed and the impact of a 'highly networked and integrated information and communication system' (Brunn 1998b: 117–118). In fact in many ways, the speed of communication has made much of distance irrelevant as communication across the world is as fast (effectively) as to a colleague across a room. This is clearly significant as not only quicker decisions can be made about and as a consequence of policies but also it means that the same information is available to all stakeholders at the same time generating greater competition and restricting the value which can be gained through time lags. Meanwhile, Nielsen and Jespersen (2001: 67) continue the theme of space and speed referring in turn to Gudmundsson (2000) and Urry (2000).

Finally in respect of space, discussion of the relationship between space and speed has already led us into the issue of time and speed. McQuire (1999: 145 and 154) sees it as 'displacement in space by the technological control of time' occurring in an 'era of time-light media (which) instantiates a society governed by an inordinate privilege granted to the present moment'. Wark (1988: 7) agrees quoting Venturi et al. (1977: 151): 'time moves very fast these days' but also emphasising that this is not simple: 'even on the bullet train of post-modernity we're in for a bumpy ride'.

We are now in danger of returning to issues of time-space compression which although highly significant in considering the design of modern maritime governance, have been much dwelt upon elsewhere not least by the author (e.g. Roe 2013). Here, we shall just mention an additional contribution by Dickens and Ormrod (2006: 11) who outline the contribution of Marx and the 'annihilation of space through time'. Overcoming spatial barriers is commonly achieved by the application of new technology which itself in turn focuses on speed and new fixes are then achievable of destinations further from markets. Thus, the penalty of time is reduced, the friction of space is ameliorated, and speed and time are intrinsically related. As Virilio (1998) suggests the technology of speed is central to the spread of empire, not least in the world of shipping and the goods it carries.

Speed, Policy, Power and Postmodernism

Sloot and Verschuren (1990) place the issue of speed and policy-making directly into a real-life context by examining in some depth the speed of generation of polices in the (then) European Community. Thomas Sloot is a reminder to us all of the speed of time passing and how this must be placed in perspective. His life was short as he died unexpectedly on 10 March 1990 at the age of 26. In our consideration of the significance of governance, shipping and the maritime sector, it is a timely reminder of everyone's fragility and transience.

Angelides and Caiden (1994: 223–224) remind us of the changes that are taking place in society and the need for policy-making to accommodate the increase in speed of almost everything around us. A static maritime governance cannot be appropriate for this new world:

> Everywhere people complain that their leaders do not seem to be up to the task, and their governments seem to be drifting without clear objectives and consistent paths of action, that public policies are inadequate to meet contemporary issues and problems; in short nobody seems to be in charge or knows how to tackle major challenges. Events move too quickly. Too much is happening all at once, giving no chance to catch one's breath. An unprecedented rate of change outdates traditional modes of thinking and overtakes

assumptions, orientation and practices. Governance as a societal steering process, acting as interpreter of the past, shaper of the present and decider for the future, lags behind events. Increasing uncertainty, turbulence and adversity leads to hazy, confused and haphazard policy decisions. Traditional policy thinking is often inadequate and probably counter-productive. New orientations, skills and knowledge are needed to deal more effectively with the discontinuities and non-linearity that characterize the emerging world order, in which misjudgement, miscalculation and misperception could well result in disastrous outcomes for everybody What is badly needed are fundamental changes in the way policy-makers and people think about policy.

Quite so and this was 1994, and so little has been done. The difference here is that the emphasis is on speed of change as well as change itself. At present, maritime governance and those responsible for policy have yet even to recognise the need for change let alone the rate at which it is taking place and its characteristic acceleration.

Armitage and Graham (2001: 111–112) use Virilio's political economy of speed to focus on cultural and economic policy-making in advanced countries and suggest that the focus on speed in both war and trade make dromoeconomics an essential component of policy-making in any sector since 'the whole development of wealth rests in the creation of disposable time' (Marx 1973: 398). Whilst economic growth and contemporary capitalism are reliant upon 'faster processes of production', then speed must be central to economic policy-making, not least for the maritime sector (Armitage and Graham 2001: 116).

Kim and Prescott (2005: 414) note the relationship of speed to the adaptation of governance measures and approaches, albeit in the context of economic deregulation (hardly an area irrelevant to modern shipping interests following the mass privatisations of ports and national shipping companies from the 1980s) (Reger et al. 1992). Urry (2004: 20) follows this up by examining the relationship of speed to policy and power. Effective governance is all about power and its application, distribution and effectiveness. Urry suggests that speed is central to achieving effective power promoting its lightness and weightlessness, the distance across which it can be applied and to whom. He sees power and governance as mobile, and this mobility is focussed upon its accommodation of increasing speed.

As Virilio suggests, 'the speed of light does not merely transform the World. It becomes the World. Globalisation is the speed of light'. Murder at twice the speed of sound, beyond the horizon of murderers, is juxtaposed to and complemented by the global integration of the telecommunications media through which speeded-of-light speculation in financial abstractions forms by far the largest and most 'productive' sector of the global economy. It would seem humanity has reached the apotheosis of an almost universal system of irrational rationality, the logic of the hypermodern manageralism. Armitage and Graham (2001: 119).

Equilibrium

Much of what we have considered across the chapters so far has been directed towards issues of change and a central feature has been how maritime governance is an excellent example of how as society has changed there have been inadequacies in the policy-making framework which in turn have generated substantial issues of maritime failure. Without change to the nature and structure of maritime governance these issues cannot be addressed adequately and the consequence is a series of patched-up solutions which lend themselves to abuse by those who benefit from the inability of the industry to govern itself or be governed by others. Globalisation has intensified these trends, and shipping, central as it is to globalisation, has suffered disproportionately compared with other sectors.

In drawing these themes together, we shall focus on a number of issues in the next chapter but for the moment it is time to introduce a temporary element of stability into the situation and a consideration of equilibrium. Note, not stasis or stagnation but stability and control of change for it is not change in itself which is needed, nor dynamism for its own sake but the ability of maritime governance to accommodate change within a controlled framework.

Ahnert (1994: 126) is very straightforward:

> Literally 'equilibrium' means 'equal weight'. In physical terms, weight is a force. The term equilibrium identifies therefore, in its primary meaning an equality of forces that, like equal weights on a pair of scales, compensate each other. In a secondary sense, the term can denote also the relationship between the processes which these forces bring about, if such compensation exists between them.

Equilibrium is a well-documented concept with a particular history in earth sciences and the study of processes in geology and geomorphology. Bracken and Wainwright (2006: 167–169) provide a detailed introduction suggesting it has been a central theme for many years with origins in the nineteenth century (Gilbert 1877, 1909, 1914). They complain of the variety of terms that have been used citing dynamic equilibrium, quasi equilibrium, steady state and time independence as just some and the confusion this has caused taking examples from geomorphology (Small 1970; Gregory and Walling 1973; Twidale 1976; Schumm 1977, 1991; Ahnert 1996). This debate has taken a central place in the discussion on equilibrium rather than its value in understanding processes (see, e.g. Phillips 1992; Kennedy 1992; Renwick 1992; Ahnert 1994; Thorn and Welford 1994: 666; Vale 2003).

They also stress that its use in other disciplines to their own of geomorphology is both valuable and established and even note its role as a metaphor in simplification of real life, something which we have considered in some depth in an earlier chapter (Bracken and Wainwright 2006: 168, 176). For example, they refer to Klamor and Leonard's (1994) consideration of metaphors across three levels: a simple metaphor which helps to clarify an argument; heuristic metaphors, used to develop thinking in a new direction and thus open to new ideas through further analogy; and constitutive metaphors used to interpret a 'world that is unknowable or at least unknown' (1994: 29). All three types of metaphors are embedded in language and rarely noticed in common usage, and equilibrium also fits into this pattern with each of the three types represented in the processes of the maritime sector to illustrate more clearly what is known, to develop new concepts and to represent what is unknown (at least for the moment) until proper understanding can be achieved.

Renwick (1992: 266) is largely in agreement with Bracken and Wainwright. Taking equilibrium, he suggests that each is dependent on the timescale of analysis and that numerous forms have been identified (Schumm and Lichty 1965; Chorley and Kennedy 1971). He defines equilibrium as 'a constant relation between input and output or form, within some range of those parameters', citing Howard's 'consensual degree of accuracy' with respect to the form retaining relatively stable characteristics, returning to those characteristics following 'minor perturbation' (1982, 1988). Going on he considers disequilibrium, whereby the tendency is towards equilibrium but has yet to be reached (perhaps the marine environment if not severely impacted by shipping) and here the disequilibrium condition can be seen as part of a grander equilibrium process and the reversion to a stable state may be predictable (Graf 1977; Montgomery 1989). Non-equilibrium is seen as the situation where moves towards equilibrium are no longer occurring (e.g. where the marine climate has over-warmed) and negative or positive feedback may prevent the return to equilibrium (see, e.g. Bull 1975; Culling 1987; Huggett 1988; Turcotte 1990; Malanson et al. 1990, 1992; Renwick 1992: 268–269).

Chorley and Kennedy (1971: 201) provide examples of the types of equilibrium that can be identified:

1. Static equilibrium. 'A balance of tendencies brings about a static condition of certain system properties'. We could suggest the role of market forces in stabilising freight rates, port charges or ship construction costs (Giancoli 1985; Halliday and Resnick 1981; Renwick 1992: 267).
2. Stable equilibrium. The 'tendency for a system to move back towards a previous equilibrium condition after being disturbed by limited external forces'. As (1) but following an economic shock such as a sizeable oil price rise.
3. Unstable equilibrium. Where a 'small displacement leads to a greater displacement usually terminated by the achievement of a new stable equilibrium'. The impact of new technology in shipping and supply chain management.
4. Meta-stable equilibrium. 'When stable equilibrium obtains only in the absence of a suitable trigger, catalyst or minimum force, which carries the system state over some threshold into a new equilibrium regime' (e.g. globalisation and its profound impact). It is in this context that new governance mechanisms and structures become essential.

Meanwhile, dynamic equilibrium can be found anywhere that processes operate that tend to minimise the effect of change—and in the maritime sector, there is no better example than the maritime environment (e.g. oceans, rivers, weather systems, marine life) which within bounds is self-regulating and has the ability to return to a stable (if changing) position (Giancoli 1985; Halliday and Resnick 1981; Thorn and Welford 1994: 666).

In many ways, dynamic equilibrium is the more appropriate term in that an equilibrium is commonly far from static but is a state of changing conditions (and hence is particularly relevant to maritime governance and policy-making). This is because an equilibrium is commonly maintained by achieving a balance of the forces acting upon it—and it is here that its relevance to maritime governance is clear. The latter is a process that is constantly buffeted by changing forces that need to be controlled and therefore to be kept in equilibrium—not stopped or stultified, but understood, maintained, adjusted but most essentially balanced to prevent getting out of control. And the situation at present is out of control (note the prevalence of pollution, monopolistic markets and seafarer death as just some of the evidence) and the presence of a new understanding and appreciation of equilibrium within governance could contribute to finding a solution. As Ahnert implies (1994: 126):

> In an open system, the input of energy/and or matter from outside (eksystemic input) keeps the process components within the system (ensystemic response) active. The process rates adjust themselves to that input and, by negative feedbacks, to each other with a tendency towards establishing a dynamic equilibrium between them. Correspondingly, the form and material components interact with the processes and tend towards a steady state that is, towards constancy of process rates and of the parameters of form and materials.

Otherwise, we have a problem. In formal terms, Wilson (1981: 1) sees it as one or the other—catastrophe or equilibrium. At present, maritime governance is firmly orientated towards catastrophe with the speed of the forces characterising the sector accelerating into an evermore catastrophic and dynamic equilibrium with unstoppable and unpredictable results. That is unless something is done. And soon.

> The violence of speed has become both the location and the law, the world's destiny and its destination. Virilio (1977: 151).

References

Abler, R., Janelle, D., Philbrick, A., & Sommer, J. (Eds.). (1975). *Human geography in a shrinking world.* North Scituate, MA: Duxbury Press.

Agnew, J. (1994). The territorial trap; the geographical assumptions of international relations theory. *Review of International Political Economy, 1*(1), 53–80.

Ahnert, F. (1994). Equilibrium, scale and inheritance in geomorphology. *Geomorphology, 11*, 125–140.

Ahnert, F. (1996). *Introduction to geomorphology*. London: Arnold.

Angelides, C., & Caiden, G. (1994). Adjusting policy-thinking to global pragmatics and future problematics. *Public Administration and Development, 14*, 223–239.

Armitage, J. (1999a). Paul Virilio. An introduction. *Theory, Culture and Society, 16*(5–6), 1–23.

Armitage, J. (1999b). From modernism to hypermodernism and beyond. An interview with Paul Virilio. *Theory, Culture and Society, 16*(5–6), 25–55.

Armitage, J. (1999c). Paul Virilio. A select bibliography. *Theory, Culture and Society, 16*(5–6), 229–240.

Armitage, J. (2000). *Beyond Postmodernism? Paul Virilio's Hypermodern cultural theory.* www.ctheory.net.

Armitage, J., & Graham, P. (2001). Dromoeconomics: Towards a political economy of speed. *Parallax, 7*(1), 111–123.

Arrighi, G., Hui, P., Ray, K., & Reifer, T. (1999). Geopolitics and high finance. In G. Arrrighi & B. Silver (Eds.), *Chaos and governance in the modern world system.* Minneapolis, MN: University of Minnesota.

Baudrillard, J. (1995). *The gulf war did not take place*. Bloomington, IN: Indiana University Press.
Benjamin, W. (1972). Passagen-Werk. In R. Tiedermann & H. Schweppenhauser (Eds.), *Gesammelte Schriften* (Vol. V). Franfurt am Main: Suhrkamp Verlag.
Bogard, W. (2007). Surveillance, its simulation, and hypercontrol in virtual systems. In P. Hier & J. Greenberg (Eds.), *The surveillance studies reader* (pp. 95–103). Milton Keynes: Open University.
Bracken, L. J., & Wainwright, J. (2006). Geomorphological equilibrium: Myth and metaphor? *Transactions of the Institute of British Geographers NS, 31*, 167–178.
Brugger, N. (1999). Critical introduction to the work of Paul Virilio. In J. Beckmann (Ed.), *Speed—A workshop on time, space and mobility, note 99-05* (pp. 11–21). Copenhagen: Danish Transport Council.
Brunn, S. D. (1998a). The internet as the 'new world' of and for geography: Speed, structures, volumes, humility and civility. *GeoJournal, 45*(1–2), 5–15.
Brunn, S. D. (1998b). A treaty of Silicon for the treaty of Westphalia? New territorial dimensions of modern statehood. *Geopolitics, 3*(1), 106–131.
Buck-Morris, S. (1986). The flaneur, the sandwichman and the whore; the politics of loitering. *New German Critique, 39*, 99–140.
Bull, W. B. (1975). Landforms that do not tend toward a steady state. In W. N. Melhorn & R. C. Flemal (Eds.), *Theories of landform development* (pp. 111–128). Binghamton: Binghamton University Publications in Geomorphology.
Burroughs, W. (1991). The war universe (Interviews with Raymond Foye). *Grand Street, 57*, 92–108.
Chorley, R. J., & Kennedy, B. A. (1971). *Physical geography. A systems approach*. London: Prentice-Hall.
Clausewitz, C. (1976). *On war*. Princeton, NJ: Princeton University Press.
Conley, V. A. (1999). The passenger: Paul Virilio and feminism. *Theory, Culture and Society, 16*(5–6), 201–214.
Crogan, P. (1999a). The tendency, the accident and the untimely. Paul Virilio's engagement with the future. *Theory, Culture and Society, 16*(5–6), 161–176.
Crogan, P. (1999b). Theory of state. Deleuze, Guatarri and Virilio on the state, technology and speed. *Angelaki: Journal of the Theoretical Humanities, 4*(2), 137–148.
Cubitt, S. (1999). Virilio and new media. *Theory, Culture and Society, 16*(5–6), 127–142.
Culling, W. E. H. (1987). Equifinality: Modern approaches to dynamical systems and their potential for geographical thought. *Transactions of the Institute of British Geographers NS, 12*, 57–72.
Debord, G. (1967). *The society of the spectacle*. New York: Rebel Press.
Deleuze, G., & Guattari, F. (1988). *A Thousand Plateaus. Capitalism and Schizophrena*. London: The Athlone Press.
Der Derian, J. (1987). *On diplomacy: A genealogy of western estrangement*. Oxford: Basil Blackwell.
Der Derian, J. (1990). The (s)pace of international relations: Simulation, surveillance and speed. *International Studies Quarterly, 34*, 295–310.
Der Derian, J. (1999). The conceptual cosmology of Paul Virilio. *Theory, Culture and Society, 16*(5–6), 215–227.
Derrida, J. (1984). No apocalypse. Not now (full speed ahead, seven missiles, seven missives). *Diacritics, 14*, 20–31 (Summer).
Dickens, P., & Ormrod, J. S. (2006). *The outer spatial fix*. Paper presented to the 'Critical Approaches to Outer Space' Panel, BISA Annual Conference, Cork, Ireland.
Dickens, P., & Ormrod, J. (2007). *Cosmic society: Towards a sociology of the universe*. London: Routledge.
Dickens, P., & Ormrod, J. (2010). Globalization of space. From the global to the galactic. In B. S. Turner (Ed.), *The Routledge international handbook of globalization studies* (pp. 531–553). London: Routledge.

Elden, S. (2005). Missing the point: Globalization, deterritorialization and the space of the world. *Transactions of the Institute of British Geographers NS, 30*, 8–19.

Forsberg, T. (1996). Beyond sovereignty, within territoriality. Mapping the space of late-modern (geo)politics. *Cooperation and Conflict, 31*(4), 355–386.

Gane, M. (1999). Paul Virilio's bunker theorizing. *Theory, Culture and Society, 16*(5–6), 85–102.

Giancoli, D. C. (1985). *Physics, principles with applications*. Englewood Cliffs, NJ: Prentice-Hall.

Gilbert, G. K. (1877). *Report on the Geology of the Henry Mountains, United States*. Geographical and Geological Survey of the Rocky Mountains Region, US Printing Office, Washington, DC.

Gilbert, G. K. (1909). The convexity of hilltops. *Journal of Geology, 17*, 344–350.

Gilbert, G. K. (1914). *The Transportation of Debris by Running Water*. United States Geological Survey, Professional Paper 86.

Graf, W. L. (1977). The rate law in fluvial geomorphology. *American Journal of Science, 277*, 178–191.

Gregory, K. J., & Walling, D. E. (1973). *Drainage basin form and process*. London: Arnold.

Gudmundsson, H. (2000). *Mobilitet og baeredygtighed – srtategier, mal og institutioner I reguleringen af persontransport*. Copenhagen: Samfundslitteratur.

Halliday, D., & Resnick, R. (1981). *Fundamentals of physics*. New York: Wiley.

Harvey, D. (1990). *The condition of postmodernity*. Oxford: Blackwell.

Hesse, M. (1999). Globalocal. On the significance of space for time, speed and mobility. In J. Beckmann (Ed.), *Speed—A workshop on time, space and mobility, note 99-05* (pp. 31–39). Copenhagen: Danish Transport Council.

Howard, A. D. (1982). Equilibrium and time scales in geomorphology: Application to sand-bed alluvial streams. *Earth Surface Process Landforms, 7*, 303–325.

Howard, A. D. (1988). Equilibrium models in geomorphology. In M. G. Anderson (Ed.), *Modelling geomorphological systems* (pp. 49–72). New York: Wiley.

Huggett, R. J. (1988). Dissipative systems. *Implications for geomorphology, earth surface process landforms, 13*, 45–49.

Keith, M., & Pile, S. (1993). The politics of place. In S. Pile & M. Keith (Eds.), *Place and the politics of identity* (pp. 1–21). London: Routledge.

Kellner, D. (1999). Virilio, war and technology. Some critical reflections. *Theory, Culture and Society, 16*(5–6), 103–125.

Kennedy, B. A. (1992). Hutton to Horton—Views of sequence, progression and equilibrium in geomorphology. *Geomorphology, 5*, 231–250.

Kim, B., & Prescott, J. E. (2005). Deregulatory forms, variations in the speed of governance adaptation and firm performance. *Academy of Management Review, 30*(2), 414–425.

Klamor, A., & Leonard, T. C. (1994). So what's an economic metaphor? In P. Mirowski (Ed.), *Natural images in economic thought: Markets read in tooth and claw* (pp. 20–51). Cambridge: Cambridge University Press.

Klein, B. S. (1989). The textual strategies of military strategy: Or have you read any good defense manuals lately? In J. Der Derian & M. J. Shapiro (Eds.), *International/intertextual relations: Postmodern readings of world politics* (pp. 97–112). Lexington, MA: Lexington Books.

Leach, N. (1999). Virilio and architecture. *Theory, Culture and Society, 16*(5–6), 71–84.

Le Corbusier (1978). *The city of tomorrow*. London: The Architectural Press.

Lista, G. (1986). *Futurism*. New York: Universe Books.

Malanson, G. P., Butler, D. R., & Georgakakos, K. P. (1990). Chaos theory in physical geography. *Physical Geography, 11*, 293–304.

Malanson, G. P., Butler, D. R., & Georgakakos, K. P. (1992). Nonequilibrium geomorphic processes and deterministic chaos. *Geomorphology, 5*, 311–322.

Marx, K. (1973). *Grundisse*. London: Penguin.

McLuhan, M. (1964). *Understanding media: The extensions of man*. New York: McGraw-Hill.

McQuire, S. (1999). Blinded by the (speed of) light. *Theory, Culture and Society, 16*(5–6), 143–159.

Montgomery, K. (1989). Concepts of equilibrium and evolution in geomorphology: The model of branch systems. *Progress in Physical Geography, 13*, 47–66.

Morris, M. (1988). At Henry Parkes Motel. *Cultural Studies, 2*, 1–47.

Nielsen, L. D., & Jespersen, P. H. (2001). Time and space in freight transport. In L. D. Nielsen & H. H. Oldrup (Eds.), *Mobility and transport* (pp. 63–72). Denmark: Transportradet, Aalborg University, Aalborg.

Nielsen, L. D., & Oldrup, H. H. (Eds.). (2001). *Mobility and transport*. Denmark: Transportradet, Aalborg University, Aalborg.

Nye, J. (1988). Neolrealism and neoliberalism. *World Politics, 40*, 235–251.

Obando-Rojas, B., Welsh, I., Bloor, M., Lane, T., Bedigannavar, V., & Maguire, M. (2004). The political economy of fraud in a globalised industry: The case of seafarers' certifications. *The Sociological Review, 52*(3), 295–313.

Paasi, A. (1998). Boundaries as social processes: Territoriality in the world of flows. *Geopolitics, 3*(1), 69–88.

Phillips, J. D. (1992). Nonlinear dynamical systems in geomorphology: Revolution or evolution? *Geomorphology, 5*, 219–229.

Pokrovsky, G. (1959). *Science and technology in contemporary war*. London: Atlantic.

Porter, J. (1968). The future of upward mobility. *American Sociological Review, 33*(1), 5–19.

Reger, R. K., Duhaime, I. M., & Stimpert, J. L. (1992). Deregulation, strategic choice, risk and financial performance. *Strategic Management Journal, 13*, 189–204.

Renwick, W. H. (1992). Equilibrium, disequilibrium, and nonequilibrium landforms in the landscape. *Geomorphology, 5*, 265–276.

Retort, I. (2005). *Afflicted powers: Capital and spectacle in a new age of war*. London: Verso.

Roe, M. S. (2013). *Maritime governance and policy-making*. London: Springer.

Ropp, T. (2000). *War in the modern world*. Baltimore, MD: John Hopkins University Press.

Rosow, S. J. (undated). *Deterritorialization and Democracy*. Unpublished Lecture Notes, State University of Oswego, New York.

Sans, J. (1991). Paul Virilio. In H. Martin (Ed.), *AA and philosophy* (pp. 139–150). Milan: Giancarlo Politi Editore.

Santos, M. (1995). Contemporary acceleration: World time and world space. In G. B. Benko & U. Strohmayer (Eds.), *Geography, history and social sciences* (pp. 171–176). Dordrecht: Kluwer.

Schumm, S. A. (1977). *The fluvial system*. New York: Wiley.

Schumm, S. A. (1991). *To Interpret the earth: Ten ways to be wrong*. Cambridge: Cambridge University Press.

Schumm, S. A., & Lichty, R. W. (1965). *Channel widening and flood plain construction along Cimarron River in Southwestern Kansas*. US Geological Survey Professional Paper, 352D.

Sidaway, J. D. (1994). Political geography in the time of cyberspaces: New agendas? *Geoforum, 25*(4), 487–503.

Sloot, T., & Verschuren, P. (1990). Decision-making speed in the European Community. *Journal of Common Market Studies, XXIX*(1), 75–85.

Small, R. J. (1970). *The study of landforms*. Cambridge: Cambridge University Press.

Steinbeck, J. (1962). *Travels with Charley: in search of America*. London: Viking.

Stopford, J. M., & Strange, S. (1991). *Rival states, rival firms: Competition for world market shares*. Cambridge: Cambridge University Press.

Thorn, C. E., & Welford, M. R. (1994). The equilibrium concept in geomorphology. *Annals of the Association of American Geographers, 84*(4), 666–696.

Tisdall, C., & Bozzolla, A. (1977). *Futurism*. London: Thames and Hudson.

Turcotte, D. L. (1990). Implications of chaos, scale-invariance and fractal statistics in geology. *Global Planetary Change, 89*, 301–308.

Twidale, C. R. (1976). *Analysis of landforms*. Sydney: Wiley.

Tzu, Sun. (1993). *The art of war*. Hereford: Wordsworth Editions.
Urry, J. (2000). *Sociology beyond societies*. London: Routledge.
Urry, J. (2004). The complex spaces of scandal. In K. Simonsen & J. O. Baerenholdt (Eds.), *Space odysseys: Spatiality and social relations in the 21st century* (pp. 15–25). Aldershot: Ashgate.
Vale, T. R. (2003). Scales and explanations, balances and histories: Musings of a physical geography teacher. *Physical Geography, 24*, 248–270.
Van Ham, P. (2001). *European integration and the postmodern condition*. London: Routledge.
Venturi, R., Scott Brown, D., & Izenour, S. (1977). *Learning from Las Vegas: The forgotten symbolism of architectural form*. Cambridge, MA: MIT Press.
Virilio, P. (1977). *Speed and politics, an essay on dromology*. New York: Semiotext(e).
Virilio, P. (1983). *Pure war*. New York: Semiotext(e).
Virilio, P. (1984). *L'Espace critique*. Paris: Christian Bourgois.
Virilio, P. (1986). *Speed and politics*. New York: Semiotext(e).
Virilio, P. (1989). *War and cinema: The logistics of perception*. London: Verso.
Virilio, P. (1995a). *The art of the motor*. Minneapolis MI: University of Minnesota Press.
Virilio, P. (1995b). *Speed and information: Cyberspace alarm*! www.ctheory.net.
Virilio, P. (1998). Military space. In J. Der Derian (Ed.), *The Virilio reader*. Oxford: Blackwell.
Virilio, P. (1999). Indirect light. *Theory, Culture and Society, 16*(5–6), 57–70.
Virilio, P. (2000). *Polar Inertia*. London: Sage.
Virilio, P. (2007). *The University of disaster*. Cambridge: Polity Press.
Virilio, P., & Lotringer, S. (1997). *Pure war*. New York: Semiotext(e).
Walker, R. B. J. (1993). *Inside/outside international relations as political theory*. Cambridge: Cambridge University Press.
Wark, M. (1988). On technological time: Cruising Virilio's over-exposed city. *Arena, 83*, 82–100.
Wilson, A. G. (1981). *Catastrophe theory and bifurcation. Applications to urban and regional systems*. London: Croom Helm.
Zurbrugg, N. (1999). Virilio, Stelare and 'terminal' technoculture. *Theory, Culture and Society, 16*(5–6), 177–199.

Chapter 8
So?

Abstract For those of you brave enough to have survived to this stage, perhaps it is time to emphasise where we are trying to go. The discussion has circled around the issues that define governance and in fact could have been applied to many other sectors than the maritime. Far from an accident, this is a deliberate strategy in attempting to illustrate that governance should be considered in a much wider context than is conventional in the maritime sector, where it has become focussed almost entirely upon ownership and the issues that stem from this. A result of an obsession with privatisation that characterised the maritime sector as much as any other in the 1970s and 1980s, it has taken central place in the discussion of governance at the expense of issues which are more substantial and thus important to any and every sector; issues discussed both in this book and its predecessor *Maritime Governance and Policy-Making*. Whilst the issue of private or public ownership is significant in determining the effectiveness of the maritime sector, it is more important to understand the underlying issues that have been going on in terms of globalisation, the Postmodern revolution, change, fluidity and process and the relationship that exists between them and the stagnant and immobile characteristics of policy-making in the maritime industry. This stasis and its contrast to the changes taking place within the shipping industry and more widely has generated the problems identified earlier in safety, security, the environment and efficiency, and unless addressed, these will not go away, regardless of the efforts of policy-makers and the quantity or quality of policies produced. This final chapter begins with a trawl through what others have said and how this can help understand the changes needed and the environment within which governance has to operate. It goes on to look closely at one approach that might have validity—time geography—borrowed from the Swedish (Lund) School of Geography before placing all of the discussion in a framework derived from David Harvey (Seventeen contradictions and the end of capitalism. Profile Books, London 2014) contradictions for capitalism. Finally, we look ahead.

M. Roe, *Maritime Governance*, DOI 10.1007/978-3-319-21747-5_8

> It is necessary to reassess the role and effectiveness of central political authority embodied in the primacy given to national governments in organizing society in general and ocean governance in particular. Wirth (2012: 240).

> Some people change lovers like they change their sheets. But I won't change you honey, you're for keeps.
>
> Sophie Ellis-Bextor (2003), *I Won't Change You.*

For those of you brave enough to have survived to this stage, perhaps it is time to emphasise where we are trying to go. The discussion has circled around the issues that define governance and in fact could have been applied to many other sectors than the maritime. Far from an accident, this is a deliberate strategy in attempting to illustrate that governance should be considered in a much wider context than is conventional in the maritime sector, where it has become focussed almost entirely upon ownership and the issues that stem from this. A result of an obsession with privatisation that characterised the maritime sector as much as any other in the 1970s and 1980s, it has taken central place in the discussion of governance at the expense of issues which are more substantial and thus important to any and every sector; issues discussed both in this book and its predecessor *Maritime Governance and Policy-Making*. Whilst the issue of private or public ownership is significant in determining the effectiveness of the maritime sector, it is more important to understand the underlying issues that have been going on in terms of globalisation, the Postmodern revolution, change, fluidity and process and the relationship that exists between them and the stagnant and immobile characteristics of policy-making in the maritime industry. This stasis and its contrast to the changes taking place within the shipping industry and more widely has generated the problems identified earlier in safety, security, the environment and efficiency, and unless addressed, these will not go away, regardless of the efforts of policy-makers and the quantity or quality of policies produced.

Focussing on one industry—maritime—is also deliberate. By so doing, not only does this sector provide an excellent example of the problems that contradictions within governance can produce, but also it helps to illustrate the specific practical problems in governance and the application of a suitable framework. The theoretical issues considered are in fact highly relevant to the design of modern and appropriate governance mechanisms, and unless debate at this level is introduced, nothing will change. There are too many vested interests for this to occur spontaneously—too many stakeholders gaining substantially from the inadequacies of governance. Consequently, meaningful governance change requires a changed approach.

And how little evidence there is of this at the moment. It is revealing to look at contributions through research institutions and universities expressed through the ubiquitous refereed journal and conference papers. Repeatedly, well-designed, carefully conducted research studies of ports and shipping, consisting of masses of assembled data, modelled and manipulated to produce more data which can then be discussed. Endless reams of numbers and equations without any consideration of the philosophical underpinning for the industry and the structures which support its operation and which are clearly inadequate for the task. Data assembly, manipulation and interpretation are useful training but will do little to address the inefficiency, filth and death that characterises shipping largely because it is ungoverned.

Finally, in this introduction, a warning! This is only volume two of three dealing with maritime governance. The three volumes reflect the three dimensions of maritime governance that need to be addressed if it is to have the potential to provide an adequate framework for the industry. The previous volume (*Maritime Governance and Policy-Making*) reflected the need to move on from the existing, unidimensional situation which characterises shipping today (2015) and to recognise the impact of globalisation and the need to accommodate the implications it has had for nation-states and global governance of which shipping is a significant part. This volume (*Maritime Governance. Form, Flow, Speed, Process*) assumes that globalised position and moves the governance of shipping into its second dimension—a fluid, changing, process-orientated framework that allows policy change and adaptation to the globalised world with its continuous flux. The third volume (conceived but as yet unborn and thus unnamed) will address the third dimension and move the discussion onto the relationship between policies and a governance framework to accommodate policy juxtaposition, plasticity in policy-making and the continuously changing intercourse of governance, polices and stakeholders. Now, there is something to look forward to.

This final chapter begins with a trawl through what others have said and how this can help understand the changes needed and the environment within which governance has to operate. It goes on to look closely at one approach that might have validity—time geography—borrowed from the Swedish (Lund) School of Geography before placing all of the discussion in a framework of Harvey's (2014) contradictions for capitalism. Finally, we look ahead.

Others

> With his hands clasped in his lap he let his eyes swim in the wideness of the sea, his gaze lose focus, blur, and grow vague in the misty immensity of space. His love of the ocean had profound sources: the hard-worked artist's longing for rest, his yearning to seek refuge from the thronging manifold shape of his fancy in the bosom of the simple and vast; and another yearning, opposed to his art and perhaps for the reason a lure, for the unorganised, the immeasurable, the eternal – in short, for nothingness. He whose preoccupation is with excellence longs fervently to find rest in perfection: and is not nothingness a form of perfection? As he sat there dreaming this, deep, deep, into the void, suddenly the margin line of the shore was cut by a human form.
>
> Thomas Mann, *Death in Venice.*

This section is aimed at placing the whole of governance in the maritime sector in some kind of broad perspective. This is a deliberate attempt to move away from the excessively narrowly focussed tradition of maritime governance which at best considers the issues of safety, security, the environment and efficiency and at its least helpful appears to consider that the only issue of concern is that of ownership and particularly the relationship between private and public involvement. Governance not only means so much more, but also needs to be viewed far more catholically if it is to be structured at its best. Rather incautiously, we will venture here into morals and ethics as well as much else. Maritime governance requires a philosophical perspective as well as an operational, political, financial, technical, and whatever else dimension. Without it, it is doomed:

> Earth is such a pretty blue and pink and white pearl in the pictures NASA sent me. It looks so clean. You can't see all the hungry, angry earthlings down there – and the smoke and the sewage and the trash and sophisticated weaponry.
>
> Vonnegut (1990), Quoted in Klinkowitz (1998: p. 71).

Shipping has suffered seemingly forever with a split personality—on the one hand revered for its romanticism and contribution to global prosperity and on the other always seen as something discreditable. Dangerous, dirty, corrupt, conservative and disreputable:

> To gain an advantage from better knowledge of facilities of communication or transport is sometimes regarded as almost dishonest, although it is quite as important that society make use of the best opportunities in this respect as in using the latest scientific discoveries.
>
> Hayek (1945: 522).

This has never been helped by the industry's obsession with confidentiality and unwillingness to share information for the common good. This position is exacerbated by its international nature which curiously has both made communication and publicity difficult and provided the opportunity for shipping to hide behind the mask of globalisation—flag-hopping, internationalised ownership, opportunities for money laundering, etc. Accurate information (knowledge) of the sector has become less rather than more easy with modern communication, and its reputation has declined alongside. This in turn provides a substantial incentive for governance reform for policy-makers and the opposite for the industry itself.

> Knowledge is no longer based in eternal 'truths' but is simply 'an archipelago of islands of epistemic stability in a sea of disorder, fluctuations, noise, randomness and chaos' (Thrift 1998).
>
> O'Tuathail et al. (1998: 4).

Russell (1918) provides commentary that is wholly applicable to the governance of the maritime sector—something that would seem at first sight to be eminently straightforward—cleaner seas, safer ships, a more secure environment and an efficient industry—all for everyone's benefit. The trouble is that of the prisoner's dilemma or alternatively the issue of collective good. The decision that an individual makes may well make them better off at the expense of the community and as such provides no incentive to approve good governance and what this involves. Maritime governance starts with principles as simple as this but has become so increasingly complex that almost no one believes in what it now represents:

> …the point of philosophy is to start with something simple as not to seem to be worth stating, and to end with something so paradoxical that no one will believe it.
>
> Russell (1918: 514).

Globalisation has not helped of course, and the mess that maritime governance represents is not of its own design—at least not originally. The increased pace of globalisation witnessed over the past 50 or more years has provided endless opportunities for shipping to become inherently corrupt, and whilst not alone in this, it does present a fine example. The solution lies not in more laws, agreements, rules; not even in more governance, but in a new attitude and approach to the fundamentals of life; not bin Laden's (although his extreme view of the potential of globalisation was thought-provoking) necessarily, but something that does not build on what we have, but on what we have not; what we need.

> If Russia can be destroyed, the United States can also be beheaded. They are like little mice.
>
> Osama bin Laden (quoted in Reeve and Foden 2001).

And it is here that Ustinov's comments on Polish–Soviet relations at the time of the Cold War have some indirect relevance:

> A desire for safety against powers whose enmity and double-dealing has been, in the Russian view, amply proved, explains perhaps that harshness with which Russian influence has been imposed on the peoples of her client states, in Poland, in Czechoslovakia and in Hungary. A story from over there illustrates the point. A Russian and a Polish labourer repairing a derelict house, chance on a horde of gold. The Russian says eagerly, 'We'll share it like brothers.' 'No,' replies the Pole, 'fifty-fifty'.
>
> Ustinov (1987: 37).

Globalisation has had a marked effect upon the power and nature of the nation-state, and the anchoring of maritime governance within a state-dominated structure has been fundamental to its ineffectiveness. Nation-states have changed, and perhaps most surprising (even disappointing) is that with the substantial changes stimulated by globalisation, they have not disappeared. As an anachronism, they are difficult to beat and to leave the design and operation of maritime governance to their ineffective structure and political inadequacies is nothing short of criminal. And these inadequacies reflect those of maritime governance themselves. The same politicians and institutions find it convenient to allow the inadequate frameworks to remain unchanged, thus perpetuating their own power at the expense of life, health, prosperity and security:

> When [General de Gaulle] took power there were a million television sets in France… When he left there were 10 million… The state is always a show-biz affair. But yesterday's theatre-state was a very different matter from the TV-state that exists today.

Debray's comments may now be relatively dated but remain relevant. TV sets themselves may no longer suggest globalisation directly, although their output and manufacture certainly do, but Debray's message remains pertinent. Shipping was central to the spread of cheap television and remains so today, central to the manufacture, distribution and consumption of all manner of pads, pods, tablets and the rest reflected in the world described by Roy:

> She presided over the World in her drawing room on satellite TV… It happened overnight. Blondes, wars, famines, football, sex, music, coups d'état – they all arrived on the same train. They unpacked together. They stayed at the same hotel. And in Ayemenem, where once the loudest sound had been a musical bus horn, now whole wars, famines, picturesque massacres and Bill Clinton could be summoned up like servants.
>
> Roy (1997: 27).

Globalisation in some ways is best expressed through the social changes that have taken place, and shipping has done its bit in spreading the message of different cultures around the world. Without it, the range of available cultures would be much thinner (shipping has always been central to migration from slavery to modern-day movements across the EU); the range of culturally diverse products would be much less (take bananas as a really good example of the maritime contribution); and it remains a form of transport central to globalisation despite its relative slowness and long history. Carmen Maura is perhaps an extreme example, but without shipping, his/her story would be impossible:

> In the film, Carmen Maura plays a man who's had a transsexual operation, and, due to an unhappy love-affair with his/her father, has given up men to have a lesbian (I guess) relationship with a woman, who is played by a famous Madrid transvestite.
>
> Film Review in the *Village Voice*, Berman (1987: 572).

The importance of the social context is central to maritime governance if it is to offer more to the industry and in achieving the aims of maritime policy-making. Inteles is clear about the failure of policy development in focussing too often on the facts rather than their explanation and on the models rather than the policies:

> Answering in 1964 the question 'What is Sociology?', Alex Inteles offered a similar contrast: 'The historian prides himself on the explicitness, the concreteness of detail which characterises his discipline. The sociologist is more likely to abstract from concrete reality, to categorize and generalize, to be interested in what is true not only of a particular people's history but of the histories of many different peoples'. For some reason sociologists did not recognize the condescension in that distinction between those who gather the facts and those who explain them, those who describe and those who analyze, those who grab and those whom pluck, those who scrub and those who polish.
>
> Tilly (1988: 705).

We have spent some energy on time in earlier chapters and have noted its significance in the governance of shipping. An appreciation of the need for speed (or otherwise), the effect of differing times on the governance process and the relationship of time to value, price, opportunity and coordination cannot be ignored in designing an effective maritime governance framework. The movement we have noted that is lacking in the existing governance structures and which is clearly a central feature of effective policy-making and the processes which characterise it is clearly time-related. And change is a central feature of this:

> Without change there is no history; without regularity there is no time. Time and history are related as rule and variation; time is the regular setting for the vagaries of history.
>
> Kubler (1962: 72).

Kubler's emphasis is wholly appropriate noting that although governance may well have to be centred around change if it is to be meaningful, it also has to recognise the historical background, something emphasised in our discussion of path dependency and lock-in in Chap. 2. Process is movement; movement takes time; and time is an essential feature of history which in turn directs change.

The emphasis on change and the relationships between the elements that can be seen to exhibit change in the shipping industry—the technical (vessel design, new fuels), political (the rise and fall of nation-states and their maritime influence), legal (new rules and regulations emanating from the IMO, OECD, UN, national governments, etc.), managerial (e.g. changes in vessel and port ownership, new methods of ship management), spatial (the rise of new ports, maritime nations, commodities, economic centres, clusters), operational (approaches to navigation, new regulations for ship operation), environmental (more new regulations for managing the maritime environment and also for the production of energy and commodities), financial (the rise and fall of the €, the US$ and the availability of credit), economic (the impact of changes in the world economy and stock markets), social (what is acceptable for the shipping industry to do, own, be owned by, support and encourage?) and organisational (who owns what, the role of professional organisations)—is an area that we turn to in the next volume of this series, but for now, the interconnectedness in the changes that characterise the sector should be noted. Huggett and Perkins have it about right in emphasising that nothing remains still (although it may appear to for a while) and that fluidity and juxtaposed change is a central feature:

> As cultural landscapes, the political, legal, social, economic, and cultural context is also always fluid. Interconnections occur at all levels and change is continuous.
>
> Huggett and Perkins (2004: 230).

Governance does not operate in a vacuum, and here, we do not simply mean that it is affected by all the issues just listed and of course by the overarching maritime context. It also has to work within a world framework of governance that overlaps considerably with other sectors—economic, industrial, agricultural, health, social and so on. Metagovernance provides a way of understanding this broader context taking on board and within its remit all manner of governance activities.

There remains some debate about how metagovernance actually operates and what it does, and we shall return to this again in the next volume with respect to the maritime sector. However, Whitehead provides some indication of a central theme:

> Metagovernance differs substantially from the concept of governance. The fundamental difference between governance and metagovernance is that the former draws attention to the processes that dislocate political organisation from government and the state, the latter

> focuses explicitly on practices and procedures that secure governmental influence, command and control within governance regimes.
>
> Whitehead (2003: 8).

And all this is possible only through relationships which form a major strand of how governance does (or does not) work. Ineffective relationships (and these include friendly and unfriendly, discipline and reward based, and long term and short term) will ensure that maritime governance cannot work. Take the current situation of the development of IMO regulations, EU directives and national policies for international and global issues. None reflect good and effective relationships between partners. Nadel suggests that this also requires relationships over a longer timescale with elements or repetition to reinforce trust. Shipping in many ways suggests much the opposite:

> We can only define social positions in terms of behaviour sequences which consume time and happen on a time scale: relationships can only be abstracted from successive, repetitive actions.
>
> Nadel (1957: 127).

Resistance to change in governance is to be expected. The only reason maritime governance looks like it does is because that is how it is commonly viewed to be as it should look. And that is because that is how it does look. Thus, an existing situation (effective or otherwise) is commonly self-reinforcing—think back to the discussion on path dependence and lock-in in Chap. 2. Or to put it another way, is there anybody out there with an ounce of brain who if they had a blank sheet of paper would design maritime governance anything like what it has become, characterised by ineffective and stagnant institutions, minority stakeholder involvement and a corrupt constituency provided with an ideal platform to exploit?

New things will always be viewed with suspicion even though they may be based on processes and objects that have been around forever. Maritime governance based upon process and change involving new stakeholders and recognising the subversive role of the industry in its operation and application may well appear 'queer and disquieting':

> Tradition branded the force of gravity in particular as occult. It seemed that it could not be observed, that it acted at a distance across empty space, and that it was propagated at infinite speed. Magnetism, too, seemed queer and disquieting.
>
> Bigelow et al. (1988: 614).

We have seen as well how important the role of institutions can be in the operation of effective governance, and central to any form of institution is its organisation. The IMO provides an ideal example and in principle admirable. In aims and objectives, it is unquestionably heading in a widely acceptable direction. But in terms of the way, it operates—its organisation—inept; it is structured around anachronistic nation-states squabbling over principles and measures, whilst the shipping industry sits back and fiddles. One can only assume that reflected in its ineffectuality, the institutions of maritime governance will achieve good over evil but only when they sort out their organisation.

> There is no reason why good cannot triumph as often as evil. The triumph of anything is a matter of organization. If there are such things as angels, I hope that they are organized along the lines of the Mafia.
>
> Kurt Vonnegut (in Griffiths 1980: 107).

Change requires change in mentality and an acceptance that things will not be the same and that other views of the world may be different but valuable. Whilst accepting that shipping's history is glorious at times and there is much to be admired and learnt from what has gone before, there is also the need to recognise that other approaches to its governance may have much to offer. Shipping's deep roots in tradition are commonly obstructive, providing not a reason but an excuse to prevaricate, delay, avoid and ignore. Overcoming this legacy, this deadweight of the past is proving difficult and it is not a new feature:

> '(A)nimals are divided into: (a) those that belong to the emperor; (b) embalmed ones; (c) those that are trained; (d) suckling pigs; (e) mermaids; (f) fabulous ones; (g) stray dogs; (h) those that are included in this classification; (i) those that tremble as though they were mad; (j) innumerable ones; (k) those drawn with a very fine camel's-hair brush; (l) *et cetera*; (m) those that have just broken the flower vase; (n) those that at a distance resemble flies'. (Taken from the *Heavenly Emporium of Benevolent Knowledge; Chinese Encyclopaedia*, undated).
>
> When he read this, Foucault said that he roared with laughter, a laughter that seemed to shatter all the familiar landmarks of European thought, breaking all the ordered surfaces and all the planes with which we are accustomed to tame the wild profusion of existing things. In his wonderment at this strange taxonomy, Foucault claimed to recognise the limitation of his own, 'our' own – system of thought: 'the stark impossibility of thinking *that*'. Foucault, *Order of Things* (1966).

What is significant here is not that the Chinese were odd, wrong or just plain stupid but that an alternative vision existed from an advanced (at the time) society, suggesting not only that there are different, defensible views of the world but also that they are no more odd or stupid than our own.

Meanwhile, since shipping is undoubtedly heavily involved in economics, Veblen's question is perhaps unnecessary. Economics does not need to be evolutionary as at least one of its main customers (shipping) certainly does not appear to be so:

> Why is economics not an evolutionary science?
>
> Veblen (1898).

Shipping has shown considerable resilience in generating new markets and serving those established by others. Part of the process of capitalism (more appropriately termed globalisation), the process of market growth and expansion is an inevitable and permanent requirement within which virtually all of us operate. Once new markets cannot be generated, there can be no more expansion to use up the excess capital produced through the global economic system and the consequence of which is economic depression and decline. The perpetual growth of GDP is a requirement of all modern and developed nation-states and without which the capitalist system must fail.

Shipping's role in this growth has been as substantial as any and more than most. It has a deep-seated need to be part of it and to maintain its upward cycle. However impossible, it is to sustain this in perpetuity (and we return to the problems of compound growth demanded by the existing global economic system in a later section of this chapter), and in the short term, the shipping industry shows it as a central aim.

However, in time, alternative markets will be needed as the Earth exhausts its capacity to grow forever. Hudgins has a solution:

> In the past patriots fought to establish political and economic conditions of free exchange and private property rights. These conditions opened commercial frontiers on earth and allowed us to create material wealth and technical capacities never dreamed of. By establishing these conditions throughout the solar system, we will open boundless new commercial frontiers.
>
> Hudgins (2002: xxv).

Although Zubrin, some years earlier, had hit on the same sort of theme:

> This hope is even more apparent in real-estate developer Chuck Lauer's plans to build 'space business parks' in orbit with modules to rent out to scientists, businesses and tourists.
>
> Zubrin (1999: 65).

The potential for shipping to expand its horizons into outer space has been noted in Roe (2013: 422–423) considering the possibilities of endless territorial opportunities across the cosmos. Clearly, there are benefits if some dangers—limitless space, almost limitless raw materials, a fragile or even non-existent legal regime (which sounds familiar and something to which international shipping could clearly adapt with ease) and an inadequate jurisdiction. OK, the vessels themselves may be sea-locked and thus can take no real part in the market, but the market for shipping services—finance, registration, regulation and eventually raw materials—is full of possibilities. And the maritime industry could always diversify into spaceships…

The difficulties of stimulating change in maritime governance should not be underestimated. As in most sectors, there is considerable resistance to overcome and whilst one might expect there to be considerable advice out there to prompt and guide in the right direction, in fact the inherent conservatism of the industry means that it takes something special, most probably from political sources to make something happen rather from within the sector itself. The industry is happy and comfortable in the familiar. Kissinger said it all:

> Most foreign policies that history has marked highly, in whatever country, have been originated by leaders who were opposed by experts. It is, after all, the responsibility of the expert to operate the familiar and that of the leader to transcend it.
>
> Kissinger (1972), *Years of Upheaval*.

Marx summarised it rather well in identifying that the capitalist economy actually accommodated a wide range of players whose individual needs and demands are subsumed within 'universal values' that generalise and thus smooth over the difficulties that a catholic range of desires and needs might present:

> Capitalism survives largely because capitalist states' practices are couched in terms of general, universal values rather than in terms of the factional interests and struggles of which a capitalist economy is composed. Different groups within the imperializing powers engage in a range of universalizing practices.
>
> Marx, *Das Kapital* (1970).

Or summarised neatly:

> Most of the time what we do is what we do most of the time.
>
> Townsend and Bever (2001: 2).

The inability to be able to change or even see the need for change in maritime governance is not surprising. Familiar, well-trodden paths are so much easier to follow than new ones which, although they hold the potential to solve some or all of accepted problems in the sector, also hold the possibility of danger and of introducing new systems and structures that might well lead to the destruction of what is traditional and often in the short term and locally beneficial. Damner et al. imply that this type of inertia and that of path dependency discussed much earlier, is more to do with laziness than fear; with finding it easier to not confront but to hide within known habits:

> (Once formed in those circumstances of both high frequency and stability, habits then become a strong predictor of behaviour) "over and above intentions suggesting that such behaviour is initiated without much deliberation and thought".
>
> Danner et al. (2008: 246) in Maréchal (2010: 1104).

Maritime governance could always wait of course. For with time, it is possible that new ideas would be accepted and opposition to change would disappear as those who resist retire and die. Well, that is how Max Planck saw it:

> A new scientific truth does not triumph by convincing its opponents and making them see the light, but rather because its opponents eventually die, and a new generation grows up that is familiar with it.
>
> Planck (1948).

However, the story so far does little to suggest that the shipping industry ever catches up with the problems that it faces. Change occurs and progress over the years has been seen in safety, security, environment and efficiency but not without much kicking and screaming to the extent that by the time change has occurred, it is long past its immediate need. Thus, undoubtedly, the shipping industry is considerably safer and cleaner than it was 100 years ago, but the existing governance system reflects the needs of the early twentieth century and not the twenty-first, and its ability to govern and the remedies it produces to the problems of today reflect this distant time as well:

> The tendency is apparently involuntary and immediate to protect oneself against the shock of change by continuing in the presence of altered situations the familiar habits, however incongruous, of the past.
>
> Morrison (1966).

Shipping remains characterised by specific spatial contexts, and despite the efforts of globalisation, ship registration, flags, the location of ports and commodities and the origins of seafarers are still central to the industry. Globalisation, if anything, has intensified this relationship whilst at the same time facilitating its abuse. It is this contradiction perhaps more than any other that sums up maritime governance today and reflects the same contradiction that we can see moving through the heart of capitalist society and to which we turn later in this chapter. Rich hits it on the head:

> A place on the map is also a place in history.
>
> Rich (1986: 212).

These spatial issues have been around a long time and their relevance to the maritime sector similarly so. Take Virilio's interpretation of space and war:

> An English cartoon from the nineteenth century shows Bonaparte and Pitt cutting chunks out of an enormous globe-shaped pudding with their sabres, the Frenchman taking the continents while the Englishman claims the sea (see Fig. 8.1). This is another way of parcelling out the universe rather than confronting each other on the same terrain, within the limits of the battlefield, the adversaries choose to create a fundamental physical struggle between two types of humanity, one populating the land, the other the oceans. They invent nations that are no longer terrestrial, homelands in which no-one could ever set foot;

Fig. 8.1 William Pitt, the UK Prime Minister and Napoleon Divide the World (1805). *Source* http://www.historyhome.co.uk/c-eight/france/coalit3.htm

> homelands that are no longer countries. The sea is open, the joining of the demos and the element of freedom (of movement). The 'right to the sea', it seems, is a particularly Western creation, just as, later, the 'right to air space' will be the element in which Air Force Marshall Goering dreams of installing *die fliegende nation*, the Nazi demos. 'Every German must learn to fly... Wings hang dormant under men's skins. 'Watching the launch of the first rockets, Hitler, who feels military defeat coming, tells Dornberger, 'If I had believed in your work, there would have been no need for war...- *or at least there would have been no need for combat!*'
>
> Virilio (1986: 37–38).

Merleau-Ponty continues the spatial theme emphasising the importance of space to individuals and by implication to the wider world:

> Our body is not a space like things: it inhabits or haunts space. It applies itself to space like a hand to an instrument, and when we wish to move about we do not move the body, as we move an object. We transport it without instruments... since it is us and because, through it, we have access to space.
>
> Merleau-Ponty (1962: 5).

Whilst Grosz, alluding to Merleau-Ponty, takes up the same spatial theme:

> For example, it is not by means of access to a Cartesian abstract or geometrical space that one knows where to scratch in order to satisfy an itch on one's back. This is true, even if I use an instrument like a stick. From this point, Merleau-Ponty claims, the stick is no longer an object for me but has been absorbed or incorporated into my perceptual faculties or body parts.
>
> Grosz (1994: 91).

Shipping and movement and its relationship to space are thus fundamental to maritime governance. It is not a matter of providing structure and support to a fixed and stable activity that will lie back and conform; shipping is resistant and slippery, difficult to pin down and difficult even to find on occasions as flags, vessel names and owners change incessantly and uncontrollably almost to the tune of Baudelaire's consideration of the human soul:

> It seems to me that I would always be better off where I am not, and this question of moving is one of those I discuss incessantly with my soul.
>
> Baudelaire (1970: 99–100).

Space, time and process are clearly interlinked and central to the design of effective governance in the maritime sector. The idea that any spatial arrangement has boundaries and margins presents a series of issues that needs to be accommodated especially as globalisation has encouraged the spread of useful space (markets). These marginal areas are where shipping has most to offer in opening up new markets and providing improved accessibility to areas previously unreachable, but it is also the areas where governance may break down. Take the difficulties of monitoring maritime activities in the mid- and more extreme ocean areas (e.g. the Arctic, Antarctic, southern Atlantic and southern Indian Oceans) but also the political margins of the EU. The importance of spatial issues is not just the steady increase of useable economic space and the problems faced by continuous expansion but also the difficulties inherent in making governance effective in a global industry, nationally policed across a variety of ill-defined jurisdictions. As Douglas suggests:

> All margins are dangerous.. [A]ny structure of ideas is vulnerable at its margins.
>
> Douglas (1966: 121).

Much of what is wrong with maritime governance has stemmed from an over-reliance and overemphasis upon the economic value of the seas rather than the social contribution that they make to human existence. The seas are seen as an extended possession of those who use them—largely shipping in our case—rather than a possession of all and something that those who use them should be looking after for everyone's benefit. This seems to hint of Proudhon's (1840) 'property is theft', the revolutionary idea that nothing actually belongs to anyone but everything is temporarily borrowed and should be conserved for the benefits of those who are to use it in the future. The seas cannot be owned by those who use them or the nation-states in closest proximity any more than the clouds, sky, sunshine or air that we breathe. Any form of permanent possession must be theft from others around us.

The idea that the seas are the property of all rather than those who are using them or have jurisdictional rights over them may well be an anathema to many. However, it brings us back to the issue of care and the role of governance in the maritime sector. The difficulty of bringing together the concepts of safety, security, the environment and efficiency then becomes clear. Whilst each of the first three is clear human values that most would accept as possessing virtue even though in individual lives they are frequently abused, the last is clearly less virtuous although it may have indirect effect upon the other three in that an efficient industry may well be able to afford to better look after safety, security and the environment because it is efficient. The virtuous characteristics of shipping are commonly relegated to second best because they are each similarly unmarketable and as such their value is difficult to calculate and consequently commonly devalued or ignored. Efficiency meanwhile remains top dog—measurable, calculable, priceable and therefore inclusive. The safety, security and the environment of the oceans, however, are each intrinsically valuable if impossible to value, but this makes the role of good governance much more significant—to ensure that these values are recognised. Thus, maritime governance must accommodate the intangible evidence of these humane characteristics and of the need to care as well as to price. Perhaps Gandhi reflects this sort of ideal:

> I cannot imagine anything nobler or more rational than that, say one hour in the day, we should do all the labour that the poor man must do and thus identify ourselves with them and, through them, with all mankind.
>
> Mahatma Gandhi.

These ideas have been supported by others. The fundamental link between economics (e.g. the efficiency of shipping) and the human spirit (e.g. concern for the long term, for the unmarketable and for others) was noted by the Italian economist Giarini:

> The final link between economics and the spiritual aspects is that people have the right to live under the best possible minimal economic conditions to respect their physical and consequently also their spiritual identity. It is here that economics becomes ethical because it deals with the respect of people and their right to develop their physical well-being, which is inevitably linked to emotions, dreams and spirituality.
>
> Orio Giarini.

And perhaps the overall, underlying objective of maritime governance and the policies that are derived should be to reflect the caretaker role of the shipping industry and its participants:

> People do not in fact own things, for the real owner is their Creator: they only enjoy the usufruct of things, subject to the Divine law.
>
> Abu al-Faraj, quoted in Khalid and O'Brien (1992: 7).

This is re-emphasised by the futility and common failure of current maritime governance to recognise the wider perspective of its role as a means of caretaking resources for other and future uses:

> See they not how many
> Of those before them
> We did destroy?
> Generations We had established
> On the earth, in strength
> Such as We have not given
> To you – for whom
> We poured out rain
> From the skies in abundance,
> And gave [fertile] streams
> Flowing beneath their [feet]:
> Yet for their sins
> We destroyed them,
> And raised in their wake
> Fresh generations
> [To succeed them].
>
> *The Koran* (6:6)

There is a history to what is being governed, and as such, there is a responsibility to those before and also to those who come in the future. This presents problems to those who see it in efficiency terms primarily as the here and now (and possibly the immediate tomorrow) is all then that matters. The issues which take longer have a longer history, and greater ramifications can be relegated too easily to the appendices of real life, and yet:

> Children, only animals live entirely in the Here and Now. Only nature knows neither meaning nor history. But man [*sic*] – let me offer you a definition – is a story-telling animal. Wherever he goes he wants to leave behind him not a chaotic wake and an empty space, but the comforting marker-buoys and trail-signs of stories. He has to go on telling stories. As long as there's a story, it's all right. Even in his last moments, it's said, in the split second of a fatal fall – or when he's about to drown – he sees, passing quickly before him, the story of his whole life.
>
> Swift (1983: 53–54).

Unfortunately, even if the inadequacy of maritime governance to accommodate or understand the need for a wide and long-term vision is recognised by those with the responsibility for its design and operation, a major concern is that it is those with the greatest short-term desires (the shipping industry itself) who seemingly inevitably end up taking the decisions. Hayek recognised the significance of familiarity in governance and the difficulty of separating interests from knowledge so that a degree of responsibility and care can be guaranteed:

> If we can agree that the economic problem of society is mainly one of rapid adaptation to changes in the particular circumstances of time and place, it would seem to follow that the ultimate decisions must be left to the people who are familiar with these circumstances, who know directly of the relevant changes and of the resources immediately available to meet them.
>
> Hayek (1945: 524).

However, maritime governance cannot rest entirely on its moral credentials even if these were to turn out impeccably. Presently, they are far from this, but a well-designed approach to governance that turns away from the immediate obsession with profit could generate the appropriate circumstances to accommodate a much wider range of issues. Obviously, some are out of order:

> There are five trades that a lay follower should not ply. What five? They are: trading in weapons, in breathing things, meat, liquor and poisons.
>
> Anguttara Nikaya, note 13, V.177.

Meanwhile, to achieve a more embracing attitude, those responsible will need to take a more pragmatic approach to policy-making design and operation and to the design of maritime governance and its presentation to those who will be affected. For example:

> Successful demonstrations are not necessarily those which mobilize the greatest number of people, but those which attract the greatest interest among journalists. Exaggerating only slightly, one might say that fifty clever folk who can make a successful 'happening' get five minutes on TV, can have as a much political effect as half a million demonstrators.
>
> Bourdieu (1994).

A new vision of what is achievable will be needed by whoever is responsible at whatever level of the jurisdictional hierarchy. Too often at present the response to suggesting a redesign of the governance institutions or the framework in which they work is one of incredulity. Stoppard sums up the need for a new vision:

> Meeting a friend in a corridor, Wittgenstein said: 'Tell me, why do people always say it was *natural* for men to assume that the sun went around the earth, rather than the earth was rotating?' His friend said, 'Well, obviously, because it just *looks* as if the sun is going round the earth.' To which the philosopher replied: 'Well, what would it have looked like if it had looked as if the earth was rotating?'
>
> Stoppard, *Jumpers* (1972).

The process of globalisation continues its (currently) unstoppable path almost inevitably bringing what were distant places closer together. However, the design for maritime governance needs to recognise that there may well be alternatives but that to realise them, some action will need to be taken:

> 'Alas,' said the mouse, 'the world is growing smaller every day. At the beginning it was so big that I was afraid, I kept running and running, and I was glad when at last I saw walls far away to the right and left, but these long walls have narrowed so quickly that I am in the last chamber and there in the corner stands the last trap that I must run into'. 'You only need to change your direction', said the cat, and ate it up.
>
> Kafka, *A Little Fable* (1946).

And globalisation affects everything not just the shipping industry. Any attempts to move maritime governance forward to reflect the changes taking place globally need to accommodate the political changes that have occurred as well. Isolationism and nationalism are inadequate stances to take when attempting to reconfigure a globalised industry working in a globalised political environment. This should of course come naturally. If the inadequacy of modern maritime governance is its failure to recognise the globalisation of the industry it is attempting to govern, then it would seem obvious that the new governance framework should itself recognise the existence of global influences in how it can work:

> Politics everywhere, it would seem are related to politics everywhere else. Where the functioning of any political unit was once sustained by structures within its boundaries, now the roots of its political life can be traced to remote corners of the globe.
>
> Li and Thompson (1975: 63–64).

Meanwhile, Kant brings us back to the moral requirements of maritime governance and the relationship between them. We have seen already that there is a moral dimension to the shipping industry that should be reflected in care for its use of the oceans and in response to the needs of those who will take responsibility for them in the future. Kant suggests that these moral imperatives are with us all the time:

> A true system of politics cannot therefore take a single step without first paying tribute to morality.
>
> Kant (1970).

The continued failure of maritime governance to make significant progress towards achieving a safer, cleaner and more efficient industry—at least to the extent that might be possible by now and the evidence of an industry that appears to be committed to obstructing progress on all three fronts wherever it is financially convenient leads to cynicism when suggestions for improvement are made. Apart from hiding behind the cloak of tradition and establishment, there is also a tendency by those committed to change to lose heart or even turn to ridicule. Globalisation has reinforced these trends devaluing the significant and diluting the important so that a universal low is sustained, even welcomed:

> It is the Houdini aspects of the space program which reward most earthlings – the dumb ones, the drop-outs, the elevator operators and stenographers and so-on. They are too dense even to care about the causes of craters on the moon. Tell them about the radio signals coming from Jupiter, and they forget right away. What they like are shows where people get killed.
>
> Killed.
>
> And they get them, too.
>
> Vonnegut (1990).

Parker continues the cynical line:

> In A.S. Byatt's *Angels and Insects*, one of the characters suggests that angels would need a breastbone protruding by several feet to counter-balance the wings, and another remembers her brother commenting that 'angels are only a clumsy form of poultry' (1993: 202). My colleague Peter Armstrong calculated that the average angel would have a wing loading of about 35 lb per square foot, about the same as a WW2 Spitfire. This would require a take-off speed of about 110 mph, which means that they would have to be able to run quite quickly.
>
> Parker (2009: 1297).

But cynicism need not prevent at least an attempt to do something about the problems that besiege shipping and its governance at a time of revolutionary change in globalisation. Much is right with the sector, and there have been achievements in addressing some of the problems it faces. However, it would be complacent to ignore the failings that remain and in particular the tendency for issues to be tackled much later than they need be because of prevarication, obstruction, ignorance or disbelief. The industry remains characterised by death, injury, filth, inefficiency and insecurity, and little is being done to address the fundamental reasons why these problems remain to such a degree. The time has come for something to happen:

> But there come to us moments in life when about some things we need no proof from without. A little voice within us tells us, 'You are on the right track, turn neither to your left nor right, but keep to the straight and narrow way'. With such help we march forward, slowly indeed, but surely and steadily. That is my position.
>
> Mohandas K. Gandhi, 22nd December, 1916.

And in this process to incorporate principles that can be defended on the grandest of stages because it is there that shipping operates—worldwide, across all cultures, races and nationalities, serving everyone of every creed, colour, religion and sex. It is the universal industry and as such should be directed and guided by the finest of ambitions and motives:

> Do not judge unfairly,
> God abhors partiality,
> Regard one you know like one you don't know,
> One near you like one far from you.
> *Instructions to the Vizier Rekhmire, Official of Pharaoh Thutmose III, c. 1500 BC*, quoted in Soloman (1995: 287).

That leads us to the notion of care and its relationship to governance:

> Since everyone is vulnerable both to oppression and to abandonment, two moral visions – one of justice and one of care – recur in human experience. The moral injunctions, not to act unfairly towards others, and not to turn away from someone in need, capture these different concerns.
>
> Gilligan (1987: 20).

We have touched on this already, perhaps not directly but certainly by implication. Governance is a framework for delivering policies which emanate from a wide range of stakeholders and which ultimately have impact, particularly in the maritime sector, on a vast range of individuals, institutions and organisations. These policies have only one basic aim—to care for society whether it be through direct means—health, shelter, cleanliness, safety and so on—or indirectly through a more efficient industry delivering greater wealth to more people who can translate this into better living standards for themselves and others. Governance is not a mechanism for delivering greater financial profit for private interests unless in turn this delivers greater social wealth for both all and our loved ones. But as the former chief economist to the World Bank Joseph Stiglitz suggested in 2015 at a lecture at the London School of Economics, there is no evidence to show that feeding the rich and assuming trickle down benefits to others have ever occurred. Solomon confirms:

> It is ultimately caring that counts, and it is not reason (as opposed to emotion) that allows us to extend our reach to the universal but rather the expansive scope of the emotions themselves.
>
> Solomon (1995: 263).

The issue of care is a fraught one but must remain central to the whole process of governance. However, we cannot care effectively for everyone, certainly not personally. Effective maritime governance can ensure that we have the constructs that will maximise the care of most, but individually, it is those closest for whom we must primarily care. This is not favouritism but logical sense as care for those closest and most precious could overall ensure the best for all especially if undertaken within a broad framework of appropriately designed governance.

> We have limited resources for caring. We cannot care for everyone or do everything that a caring approach suggests. We need moral guidelines for ordering priorities, Though the hunger of our own children comes before the hunger of children we do not know, the hunger of children in Africa ought to come before some of the expensive amusements we may feel like providing for our own children. These are moral problems calling to some extent for principled answers.
>
> Held (1993).

And perhaps better put by Sidgwick (whilst noting the date of original publication and its cultural significance):

> We should all agree that each of us is bound to show kindness to his parents and spouse and children, and to other kinsmen in a less degree: and to those who have rendered services to him, and any others whom he may have admitted to his intimacy and called friends: and to neighbours and fellow countrymen more than others: and perhaps we may say to those of our own race more than to black or yellow men, and generally to human beings in proportion to their affinity to ourselves.
>
> Henry Sidgwick, *The Methods of Ethics* (1874) (quoted in Belsey 1992: 38).

The maritime governance framework at present fails to achieve this and effectively places efficiency and profit above all else. The maritime sector is not alone in this but that does not excuse the approach taken. We need to focus upon those closest and consider them above others to maximise the effectiveness of governance to all:

> Hardly any moral philosopher, these days would deny that we are each entitled to favour our loved ones. Some would say, even more strongly, that we ought to favour them, that it is not simply a moral option… intimacy and close relationships require partiality.
>
> Friedman (1991: 818).

And the universality of some of the issues that care centres upon is clear. Corbridge's selection could also include safety, security and cleanliness:

> I am not willing to deconstruct further certain minimally universalist claims, of the type that involuntary death from hunger, or involuntary malnutrition, or involuntary homelessness, or slavery, or torture are bad things which must be struggled against… The postmodern dilemma is avoided as and when we accept that certain human needs and rights, at least can be taken to be 'universal', and when we learn that in attending to these needs and rights we are not so much dictating to others as dictating to ourselves.
>
> Corbridge (1993: 466, 469).

The issue of proximity is central to maritime governance as much of what is going wrong in the sector goes on at a distance either from the nation-state of those involved or from anywhere which might be considered civilised. Consequently, in a curious way, these issues become less important as they do not impact upon the relevant nation-state nor the majority of people. Few are concerned about distant territories—and certainly not so much as locally—and even less about activities in mid-ocean, which are unseen, unrecognised, unbelievable and unimportant. Pollution from shipping across the oceans, death and injury of seafarers from distant lands and occurring far way, and the unseen world of maritime security with materials locked away in containers are far from the majority of peoples' minds. Thus, 'proximity' is central to governance and the need to develop a framework that relates to people and their locality, to individuals whom they know and can see and meet, is one facet that needs incorporating in the development of a mobile, dynamic maritime governance:

> This *universal Benevolence* towards all Men, we may compare to that Principle of Gravitation, which perhaps extends to all Bodys in the *Universe*, but, the *Love of Benevolence*, increases as the Distance is diminish'd, and is *strongest* when Bodys come to *touch* each other.
>
> Hutcheson (1726).

MacIntyre reaffirms the central issue of proximity, familiarity and society that characterise all activities and individuals and which need to be recognised in the design of effective maritime governance even in times (or perhaps because) of intense and accelerating globalisation.

> In many pre-modern, traditional societies it is through his or her membership in a variety of social groups, that the individual identifies himself or herself and is identified by others. I am both cousin and grandson, member of this household, that village, this tribe. These are not characteristics that belong to human beings accidentally, to be stripped away in order to discover 'the real me'. They are part of my substance, defining partially at least and sometimes wholly my obligations and my duties. Individuals inherit a particular space within an interlocking set of social relationships; lacking that space, they are nobody, or at best a stranger or outcast.
>
> MacIntyre (1981: 33–34).

The lack of knowledge or experience of groups, individuals and issues happening out of sight is a well-known phenomenon, immediately reducing the impact of unwanted events such as pollution, death, injury and insecurity. We are all well aware of the diminishing effect of mass communication which on the one hand brings these events to our attention and on the other lessens their impact. Shipping is no different except that whilst these issues are commonly remote from us all (in mid-ocean), their impact can be intensely local affecting families, weather systems and the seas, the latter spreading across the globe. Thus, well-reported adverse impacts of shipping have little effect on our views and opinions but can still have major local impacts. This contradiction is a serious characteristic of globalisation. Understandably, our role as individuals is primarily to look after those closest and this is commonly defined by location either immediately or in wider terms whereby those temporarily faraway can remain central. Maritime governance has a role to play here in effectively bridging the gaps between people (seafarers and families for example) whilst also attempting to illustrate the significance of the global impacts of shipping which may seem less important. However, O'Neill was dubious that we should look any further than those closest:

> The two groups lived in unconnected worlds, ignorant of one another's very existence; their activities assumed no connection to those in the other group, whom they accorded no ethical standing... Members of each group could legitimately limit their ethical considerations to exclude their contemporaries of whom we now know, but who lived beyond their horizons.
>
> O'Neill (1996: 105) considering the inhabitants of Anglo-Saxon England and their contemporaries in T'ang China.

Hume agrees:

> We find in common life that men are principally concerned about those objects, which are not much remov'd either in space or time, enjoying the present and leaving what is far off to the care of chance and fortune... The breaking of a mirror gives us more concern when at home, than the burning of a house, when abroad, and some hundred leagues distant.
>
> David Hume, *A Treatise of Human Nature* (1738) (quoted in Ginzburg 1994: 116–117).

There are those who see the focus upon those closest as a lost cause with the development of globalised communications and technology:

> People in communities use words on screen to exchange pleasantries and argue, engage in intellectual discourse, conduct commerce, exchange knowledge, share emotional support, make plans, brainstorm, gossip, feud, fall in love, find friends and lose them, play games, flirt, create a little high art and a lot of idle talk.
>
> Rheingold (1993: 3).

Others meanwhile reject the proximity debate and suggest that it is important to recognise the needs of all others, regardless of location and also of differences in style, culture and background—as well as colour, race, religion, disability and sex:

> Feelings of solidarity are necessarily a matter of which similarities and dissimilarities strike as salient.... There is such a thing as moral progress (which is) in the direction of greater human solidarity. But that solidarity is not thought of as recognition of a core self, the human essence, in all human beings. Rather it is thought of as the ability to see more and more traditional differences (of tribe, religion, race, customs, and the like) as unimportant when compared with similarities with respect to pain and humiliation- the ability to think of people wildly different from ourselves as included in the range of 'us'.
>
> Rorty (1989: 192).

Meanwhile, Eagleton suggests that our consideration of other individuals and cultures should be wider than just measured by proximity, and as long as there are mutual communication skills and abilities, then there can be no suggestion of not treating everyone on an equal basis. This raises questions about maritime policies that favour the rich and successful above others (e.g. the difficulties faced by developing countries in dealing with multinational shipping corporations and negotiations with developed countries over shipping regulation and the freedoms of trade):

> If another creature is able in principle to speak to us, to engage in material labour alongside us, sexually interact with us, produce something which looks vaguely like art in the sense that it appears fairly pointless, suffer, joke and die, then we can deduce from these biological facts a huge number of moral and even political consequences… because of the form of their bodies, we would know more or less what attitudes to these animals it was appropriate to take up, such as respect, compassion, not cutting off their feet for the fun of it and the like.
>
> Eagleton (1996: 47).

Singer says much the same. Seafarers are equally as important wherever they are and wherever they come from. The problem is that naturally we associate more easily with those closest, whilst globalisation both brings people closer together whilst also forcing them further apart physically, societally and mentally. It is a central task of maritime governance to reconcile these contradictions:

> The fact that a person is physically near to us, so that we have personal contact with him, may make it more likely that we *shall* assist him, but this does not show that we *ought* to help him rather than another who happens to be further away. If we accept any principle of impartiality, universalizability, equality, or whatever, we cannot discriminate against someone merely because he is far away from us.
>
> Singer (1972: 24).

Clement also spreads the discussion beyond care to that of looking at those who deserve to be cared for, and the issue of vulnerability raises its head. This synchronises well with the issue of proximity as it is those with whom we are physically and metaphorically nearest that are most vulnerable to us and our decisions. Thus, in the world of globalised shipping, a problem emerges in that the nature of the industry is one that is continuously making the proximate less so and as a consequence making them less worthy of consideration as vulnerable. The seafarer thus becomes disposable and the foreign seas of less vital importance:

> Our obligations to care for family and friends are based on the fact that our family and friends are particularly *vulnerable* to our actions and choices… But many people beyond our family and friends are *also* vulnerable to our actions and choices, and thus the ethic of care has implications beyond our sphere of personal relationships. Those closest to us will tend to be more vulnerable to our actions and choices than those distant from us, and thus we are not obliged to weigh everyone's interest exactly equally. Yet insofar as those from us are particularly vulnerable to our actions and choices, we have special obligations to care for them. And to that extent, the conventional boundaries of the ethic of care break down.
>
> Clement (1996: 73), following Goodin (1985).

And to place in the context of the debate on nation-state, the process whereby the state is becoming less well defined and less influential has impact upon society and the care of individuals. Thus, a modern flag (of convenience or otherwise) can do little for its seafarers—and perhaps has no incentive to do so. Globalised shipping is what it says it is; non-allied to any state in any real meaningful way and thus also unattached to any societal framework:

> To be a morally good person requires, among other things, that a person strives to meet the demands of caring that present themselves in his or her life. For a society to be judged as a morally admirable society, it must, among other things, adequately provide for care of its members and its territory.
>
> Tronto (1987: 126).

And maritime governance needs to show some accommodation not only of care but justice as well:

> While care is essential to a morally adequate life and society, so too is justice.
>
> Clement (1996: 116).

Maritime governance has also been victim of postmodern globalised society characterised as it is by little incentive to do anything novel or risky. Thus:

> The human donkey requires either a carrot in front or a stick behind to goad it into activity... The whole drift of... society for two generations has been to whittle away both at the carrot and the stick until now very little of either is left. It is the passion for equality – excellent in itself – that has removed the carrot. The regard of success has not merely been shrivelled, they have been poisoned, since commercial success has been turned, in the eyes of wide circles of society, into a positive disgrace. There is a conspiracy of labor, capital, and the state to deny enterprise its reward.
>
> The process of both removing the carrot and the stick has culminated in the extraordinary circumstances of today. Shrunken as were the incentives and sanctions of pre-war days, they have now for the time being, vanished completely. Nobody gains anything from activity or suffers anything from inactivity. There is hardly the flavour of a carrot or a shadow of a stick. And yet we wonder why the donkey does not break into a trot.
>
> *The Economist*, quoted in Terborgh (1950: 120).

Despite Spretnak's enthusiasm for postmodernism, maritime governance has been important in displaying both strong postmodern characteristics and a failure to place the shipping industry within its 'cosmic whole'. The modern features of maritime governance and its excessive cautious reliance on traditional form have done nothing to associate the role of the industry not only in its contribution to economic development but at the same time its place in destroying the environment and lives, and dividing communities and promoting the unequal distribution of wealth. Maritime governance has a task here in reflecting and promoting the need for cohesion and understanding:

> Modernity situates human kind in a glass box *on top of* nature, insisting on a radical discontinuity between humans and the rest of the natural world. It traces the human story apart from the larger unfolding story of the Earth community. To be truly postmodern is to reject that discontinuity by opening the box to connect anew with our larger context: the Earth, the cosmos, the sacred whole.
>
> Spretnak (1997: 66).

And maritime governance reflects Williams rather gloomy interpretation of the selfish individual, gleefully ignoring its wider implications and influence:

> At every moment each of us finds himself the apparent centre of the world, enjoying a little bit of foreground of the here and now, while around him there looms, thing beyond thing, event beyond event, the plethora of the universe.
>
> Donald Carey Williams, in Hagerstrand (1989a: 1).

Although stretching it a little, Martin's interpretation of maleness reflects that of traditional shipping with its features of strength, durability, prowess and confidence. In fact, the opposite may well be more appropriate as the industry hides behind a false virility rather than recognising its vulnerability and inadequacies:

> One depiction of sperm as weak and timid instead of strong and powerful – the only such representation in western civilization, so far as I know – occurs in Woody Allen's 1972 movie *Everything You Always Wanted to Know About Sex* (*But Were Afraid To Ask)*. Allen, playing the part of an apprehensive sperm inside a man's testicles, is scared of the man's approaching orgasm. He is reluctant to launch himself into the darkness, afraid of contraceptive devices, afraid of winding up on the ceiling if the man masturbates. Martin (1991: 491).

But perhaps it is time to move on.

Time Geography

> On the continuum between biography and aggregate statistics there is a twilight zone to be explored, an area where the fundamental notion is that people retain their identity over time, where the life of an individual is his foremost project, and where aggregated behaviour cannot escape these facts. Hagerstrand (1970a: 7).

And so to an approach to the reform of maritime governance that has notable potential and despite its age, retains relevance through its holistic approach and understanding of the role of society in policy-making.

The origins of time geography can be traced back many decades, but its true value was only recognised following the work of Torsten Hagerstrand, long-standing professor of cultural geography at the Lund School of Geography at the University of Lund in Sweden (Martensson 1981; Neutens et al. 2011). Hagerstrand was introduced to the contributions of von Thunen and Christaller to mathematical modelling in geography whilst working with Edgar Kant at the University of Lund in Sweden in the 1940s (Unwin 1992: 121). Hagerstrand died in 2004 after a long career focussing from the 1960s on the relationship between time, space and activities and the constraints that each placed upon each other and the consequent patterns of activity. Hannah (1997: 349) suggests that this embryonic time geography was a way of representing and helping to understand peoples' daily movements through time and space. Liepitz (1997: 270) takes a similar view—time geography can be viewed as a:

> micro-sociology, based on routinized *choreographies* of individual trajectories by agents subject to the constraints of their spatial temporal materiality. These trajectories combine to form bundles at the *stations* where they interact. The projection of these trajectories into planar-space produces the structuration of space.

Although certainly not directly related to the issue of maritime governance (or even governance more generally), much of what was analysed and developed can be adapted and many of the issues involved reflect similar problems in design and application if not in example. The geographical approach to understanding broader policy problems and the potential for their reinvigoration has been noted by Harvey (1974), Peck (1999: 131, 2000: 255–7) and his extensive discussion of 'grey geography' and the significance of the discipline to public policy, Berry (1994), Pacione (1999: v and viii) who saw geography as having a major goal of influencing forces and agents that condition the well-being of people, Buttimer (1999) and her consideration of humanism and geography, Pollard et al. (2000: 243–5), and Banks and MacKiant (2000: 250) who returned to the issue of grey geography and stressed the inherent fluidity of policy and its need for a medium like geography. Consequently, time geography has something to offer in the redesign of maritime governance.

Consequently, we consider it here although the associated issues of structuration (Carlstein 1981a; Pred 1982: 158–160, 163, 1984: 280–281, 1985a; Hallin 1991: 199; Rose 1993: 20) and constraints (Hagerstrand 1970a: 12–18, 1973: 79; Parkes and Thrift 1980: 244, 248–9; Rose 1993: 21–22; Miller 2005a: 19; Neutens et al. 2011: 27; Yin et al. 2011: 423; Ellegard and Svedin 2012: 23; McQuoid and Djist 2012: 27) which form significant parts of time geography are left for others to consider. Hagerstrand's substantial contribution in theory and practice to the development of time geography models has been widely recognised [see for example Parkes and Thrift (1980); Johnston (2005) and Oberg (2005) amongst many others]. His publications on the concept are extensive and include the seminal '*What About People in Regional Science*' (Hagerstrand 1970a) as well as a multitude of others (e.g. 1967, 1970b, 1973, 1974a, b, 1975, 1978a, b, 1982, 1986, 1989a, b, 1991, 1992, 2004). His overall contribution is summarised in Taylor (2003: 153–5), Lenntorp et al. (2004), Lenntorp (2004) and Persson and Ellegard (2012), but again there have been many contributions related to time geography from Swedish scholars in particular (e.g. Carlstein 1978, 1981a, Carlstein et al. 1978, Carlstein and Thrift 1978, Cederlund 1977, Lenntorp 1974, 1978, 1981, Martensson 1974a, 1977, 1981, Nilsson 1985, Oberg 1981, 2005, Olander and Carlstein 1978). Here, we can consider only a relatively small part of the work that was developed but hope that by so doing can point towards its potential contribution in understanding some of the issues we have been discussing.

Buttimer (1976: 287) suggests that Hagerstrand recognised the conceptual weakness of an exclusive focus on space and the need to incorporate considerations of 'time, people and finitude', issues which resonate with our consideration of the inadequacies of maritime governance, its fixation on space (nation-states) and its neglect of temporal issues (time) and wider stakeholders' involvement (people). Ellegard and Svedin (2012: 21) reaffirmed the value of Hagerstrand's work emphasising its consistency over a long period of time creating a 'coherent philosophical world view in which new concepts… are also introduced' centring on holistic issues that are undivided by the artificial constructs of national borders, existing institutions and preselected actors.

Hallin (1991) suggests that Hagerstrand (1983) noted the work of Arthur Edington in the early 1900s as an early inspiration, whilst Rose (1977: 43) points to Reichenbach (1958) who recognised that space and time should be treated as a single item rather than separate entities emerging from extensive work on space-time budgets [e.g. Strauss 1961, Nystuen 1963, Chapin 1965, Converse 1968, Anderson 1971, Lynch 1972, Holly 1976, and Thrift (1977b)] although this in turn had stemmed from much earlier work on understanding the relationships between human activity and time exemplified by Christaller (Openshaw and Veneris 2003) and von Thunen (Sasaki and Box 2003). Even earlier, time budgets had derived their rationale from the work of Marx (1970) and Engels (1969) on the characteristics of the working day. Thrift (1977c: 413) identifies the earliest, what he terms 'real' time budget studies as those carried out by Strumlin on workers in Moscow in 1924, whilst at the same time, studies of housewives' time in the USA were being undertaken. Time budgets focussed explicitly on the finite resource of time (commonly split between work and leisure), how this can be allocated by individuals and how their environment affects and can be affected by them.

Lundberg et al. (1934) pursued leisure time budget studies, Sorokin and Berger (1939) looked at the motivation for activities and social contacts, whilst the Mass Observation Unit (1939) in Thrift's eyes attempted to accommodate everything. However, these studies paled into insignificance with the activity in Russia and the USSR where time budgeting was viewed as a basic form of social accounting generating a standard set of codes and a sizeable range of literature to support the research. Despite this, there is no review available for the work in the USSR although reviews of Western studies do exist—for example MacMurray (1971), Ottensman (1972) and Szalai (1972).

Time geography continued to be of interest throughout the 1970s with, for example, Pred (1977a: 211) agreeing with Hagerstrand's (1976) claim that it could be used to uncover 'structural patterns and outcomes of processes which can seldom be derived from the laws of science as they are formulated today'. Hoppe and Langton (1986: 116) suggest that in the 1980s, it was considered the key for human geographers to enter social theory, providing access to active debates in the social sciences (Pred 1981a: 5, 1984, 1985a; Gregory 1978). Meentemeyer (1989: 165) is one example of its application within the behavioural sciences.

Meanwhile, time geography has seen a resurgence of interest in recent years focusing on the capabilities of space-time paths and prisms (Miller 1991, 1999; Forer 1998; Kwan and Hong 1998). Miller (2005a: 17) sees the potential only growing particularly in association with location-based services and location-aware technologies and the development of geographical information systems (GIS) (Shekhar and Chawla 2002). Miller and Bridwell (2009: 50) emphasise its continued relevance, suggesting that it retains a 'powerful framework' which has been progressively applied to contemporary problems. These include applications to transportation networks (Miller 1991, 1999; Kwan 1998; Kwan and Hong 1998; Wu and Miller 2001; Weber and Kwan 2003), to planar space (Forer 1998; Forer and Huisman 2000; O'Sullivan et al. 2000), visualisation of space-time paths (Kwan 2000), women and transport (Law 1999: 569) and software design (Frihida et al. 2004; Yu 2006).

So what is time geography and how does it relate to governance and the problems of maritime governance in particular? There are countless definitions and explanations reflecting in many ways the significance of the model and its adaptability. Carlstein et al. (1978: 117–118) suggest that rather than a technique it is an approach, 'an origin and a place to start anew', regardless of whether there is any idea of the ultimate destination or objective. Taking a basic model of reality, time geography addresses elements which have been long neglected in social studies. These surprisingly simple but fundamental tenets (Davies 2003: 133) were outlined by Hagerstrand (1975):

1. The indivisibility of the human being;
2. The limited length of each human life;
3. The limited ability of the human being to take part in more than one task at a time;
4. Every task has a duration;
5. Every movement between points in space consumes time;
6. The limited packing capacity of space;
7. The limited outer size of terrestrial space;
8. Every situation is inevitably rooted in past situations.

Clearly, the degree of significance of each of these factors varies with different situations and also over time—Hagerstrand was writing in the 1970s before the emergence of the Internet with all its associated impacts upon time, multitasking, space and so on, but the principles basically remain the same and with adaptation are still very relevant.

Hagerstrand went on:

> I believe that the interaction between these fundamental conditions could be and ought to be the object of precise theoretical research. I feel that this research is the starting point from which more practical considerations should develop concerning better or worse institutions (capitalism versus socialism, bureaucracy versus participation), concerning better or worse technologies (private transport versus public, videophones versus letters), and concerning better or worse cities (circular, multi-nuclear, band-like or not cities at all)… this conclusion is a clear-cut value consideration. I believe that the criteria for a good socio-technical organization are not to be found along the spatial cross-section but along the time axis and in the particular experience of events which makes up the life of each individual human being. It is the biographies of people that should count.

To arrive at these interactions between humans and their environment, a common language was needed based on a space-time ecology which would be designed around deductive powers to analyse system states and transformations something recognised by Giddens (1985: 265) who describes it as a convergence of geography and the wider social sciences drawing on the established traditions of social theory.

Gregory (1978: 140) provides a concise summary of time geography as envisaged by Hagerstrand:

> The failure of modern science, he argues, is a result of its neglect of synchronization and synchronization; the inescapable necessity for space-time 'packing' in the conduct of practical life. Hagerstrand ties the structures of the social and the natural worlds into the

> organization of space-time through the identification of paths, which define, via coupling constraints, when and for how long, the individual has to join other individuals, tools and materials in order to produce, consume and transact', and of domains, which define, via authority constraints, systems of regulations which govern the actions of particular individuals or groups. (Hagerstrand in Pred 1977b: 39–42).

These ideas are visualised in Hagerstrand's famous diagrammatic explanations, a derivation from which can be found in Fig. 8.2 and described by Crang (2003: 193) as 'something of a ballet of lines of motion'.

Now, whilst this may seem far away from maritime governance, there is actually much in common, especially the recognition of the importance of space-time relationships, the need to focus on the individual and the relevance of many of the

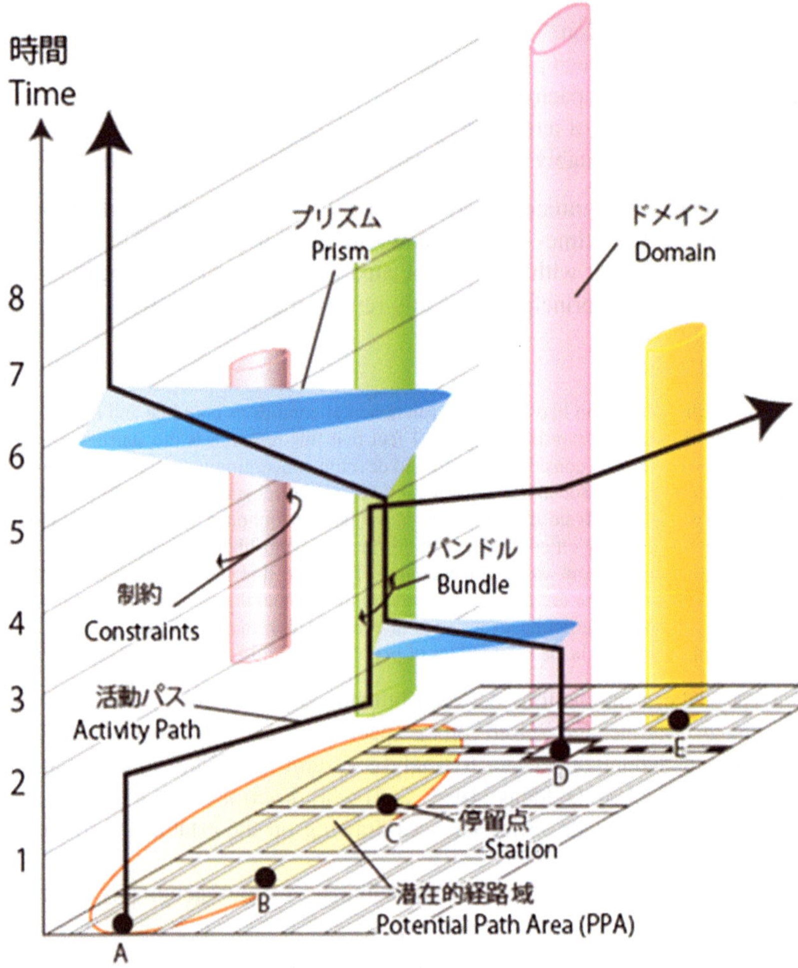

Fig. 8.2 Time geography. Miyuki Meinaka, 2013

principles noted above in designing workable structures for governance in any sector. The importance of institutions and technology was also stressed, and although designed to apply to urban and rural planning situations, its adaptability to maritime planning as well is clear. The focus could then be the relationship between time and space in the maritime sector—clearly significant as we have seen in the impact of globalisation on maritime activities and the connections between the shipping industry, markets for international goods and commodities, international labour and administration amongst many others.

In particular, a time geographical model might provide the framework for recognising the relationships between policies over space and time within a single governance framework—something that is not recognised at all at present. Policies emerge from the current hierarchical framework commonly as independent entities, and regardless of their individual merit, there is a tendency to fail to see the spatial and temporal relationship to other polices. Failure to do this can result in policies which are weak or even work against each other, and time geographical models might provide a way of understanding the significance of policy juxtaposition.

Hagerstrand went on to develop his model, further suggesting that not only human beings but also 'domestic organisms, tools, vehicles, equipment and indeed buildings have paths in time-space which interact and form bundles over time all constituting a web or texture of paths having a complicated structure' (Carlstein et al. 1978: 119). This could be applied to ships, ports, all manner of maritime institutions, seafarers, maritime interest groups and so on and used to understand the relationships that exist between them and the policies to which they have to adapt, providing the basis for a governance framework specifically related to globalisation and its space-time dimensions. As Davies (2003: 134) suggests, this model allows us to understand how daily lives are shaped whilst focussing on the constraints and enabling factors that influence choice and action. Policies can be analysed similarly and potentially a governance framework designed to accommodate this.

Ellegard and Svedin (2012: 25) emphasise the relevance that time geography might have for maritime governance, noting its multidimensionality and its focus on processes within time-space, both essential features of the maritime community. They see this as an essential part of transport research. Thus:

> different individuals (grains in Hagerstrand's terms) belonging to different types of populations, are in touch with each other at a geographical place over time, thus filling the landscape in all its variety. They encounter each other and stay together at a place for a while, and they leave each other and the place. Non-living articles are moved by living individuals, by artefacts, or by natural forces. The landscape is, from this perspective, occupied by many populations - not only human. Hagerstrand hereby also recognizes an implicit power relation: 'The room-occupation and duration of one grain establish boundaries over which other grains cannot move without causing some sort of transformation to occur' (Hagerstrand 1985: 186). Ellegard and Svedin (2012: 20).

Hagerstrand's great contribution to what was the development of a relatively simple model was to recognise the social context of spatial and temporal studies (including both the opportunities and constraints that this implies), and maritime governance currently has limited recognition of the societal relationship that exists between space-time and effective policy-making. The maritime sector is characterised by spatial and temporal impacts within societies. Hagerstrand (1985: 6) hinted at this wider application of his model and its significance throughout planning all sectors:

> The landscape (*in our case the maritime landscape of ports, vessels and associated activities but also including for example the environment*), as we see it, or the region, which we only get to know more indirectly, is the zone of action as it appears at a particular moment. It is not enough to add time to the spatial dimension. Not only the visible participates but everything that is present. To a great extent human actors (*in all their maritime manifestations: seafarers, officers, brokers, owners, bankers, lawyers, consumers* etc.), their knowledge and intentions, belong here naturally, which not even natural geographers can ignore any longer. (*italics added*).

There have been many interpretations of Hagerstrand's time geographical work over the years since it was first introduced. These include Thrift with a number of detailed considerations (e.g. 1977a: 7–33, 1977c: 428–431, 1996: 9), Gren (2003: 209) who questions the very nature of the concept, Tornqvist (2004: 236) who concentrates upon time geography's simultaneous whilst separate view of time and space, and Miller (2005b: 382), Yin et al. (2011: 423) and Harvey (1990: 211–212) who, whilst recognising its value and originality, was also critical of its lack of interpretation of causality of patterns of activity in society. Others who have considered time geography include Rose (1977: 43); Landau et al. (1982); Hesse (1999); Miller (2004, 2005a); Raubal et al. (2004: 247–249); Arentze and Timmermans (2005); Scott (2010), Shaw (2012); and Scott and He (2012: 62).

Hoppe and Langton (1986: 115–116) similarly recognise its contribution and that it could be applied not only to individuals but also to projects (and by inference to policies, thus providing a model for governance). However, apart from some generalised statements [e.g. from Hagerstrand (1973, 1975) and Pred (1977a, 1981a)], consideration of both people and resources in a time geographical framework had been largely ignored. McQuoid and Dijst (2012: 26 and 27) reflect on its consideration of time and space (see also Dijst 2009) and in particular the significance of constraints in determining the range of opportunities for individuals, an issue that has particular relevance for the possibilities of effective governance. Meanwhile, Miller (2005a: 19) and Neutens et al. (2011: 27) also focus heavily upon the constraints that time geography places central to the debate of societal organisation and Parkes and Thrift (1979: 354) suggest that time geography and its consideration of constraints has clear roots in Huntingdon (1926, 1931, 1945), Whittesley (1945) and Hawley's (1950) interest in periodicities as a central point of geographical study.

Pred (1977a: 209) considers that time geography is so flexible that it can accommodate virtually all forms of interaction between individuals, groups and objects and as such is so malleable it could play a useful role in the design and organisation of governance—including that of the maritime sector needing an approach that is both catholic and accommodating. By being able to 'treat both the individual and society (or large segments thereof) as a whole', it is both unusual and flexible. Pred (1977a: 210, b: 36–50) continues by emphasising its ability to consider space and time jointly rather than in a combined form and looking at the relationship between the two rather than what he terms the 'conventional' way which focuses upon bundling them together into a single force.

Time geography has many proponents reflecting the advantages that they see it can bring to analysis of time and space and their interrelationship. Hallin (1991: 19, 20) suggests that it met with success because it broke with 'the spatial science that prevailed in the 1960s and introduced possibilities to analyse the individual'. However, also as important was that it retained the advantages of traditional scientific objectivism and physicalism (Schwanen 2007: 9). Hallin goes on to identify four main contributions:

- It reveals that all social science incorporates temporal and spatial conditions.
- It contains analytical instruments to reflect the importance of constraints on individual life.
- It reveals how time and space are scarce resources and that this is reflected in the ways people shape their activities.
- It has developed a unique and useful descriptive graphical method.

Each of these attributes has potential to be adapted to model the activities and operation of governance and provide a framework for its development, increasing the understanding, largely lacking at the moment, of the relationships between people, institutions, time and space. By so doing, it might contribute to overcoming the inadequacies of the current maritime governance system. Constraints in the possibilities of alternative policies might be revealed, the graphical models might reveal the relationships between actors that much better, and the importance of the delicate balance between time, space, individuals and institutions might be better understood.

Pred (1977a: 209) is equally as enthusiastic as time geography 'can specify the necessary… conditions for virtually all forms of interaction—social and otherwise—involving human beings' with implications for all the social and behavioural sciences. He goes on seeing it as presenting:

> a humanistic concern with the 'quality of life' and everyday freedom of action implications for individuals of both existing and alternative technologies, institutions, organizations, and urban forms. (Pred 1977a: 210).

In addition, it offers an opportunity to overcome the fragmented nature of much policy-making and governance in the maritime sector, facilitating the accommodation of a wider range of individual, group and institutional stakeholders within a process that allows for their integration and that of policies that affect them across space and time. As such, it offers much for providing a structure for policy-making in any sector.

Pred (1977a: 218) provides more comment: time geography simultaneously concentrates upon 'the dynamics of population-system, activity-system matching i.e. on the collective good of those who are subject to the institutions and organisations which control jobs and technology', and we might add much more in the maritime sector. Sui (2012: 6) has no doubts:

> Time geography is widely considered a major breakthrough in the history of geographic thought because it establishes human geography as a vibrant field for investigating localized structures and individual agency. Path, project, pocket, prism bundle, constraints and station are core concepts of time geography that have been used to weave a web of interconnected and shared social experiences... to elucidate the key sociospatial bundles and bound interactions in space-time, which produce the material and symbolic landscapes (Wong and Shaw 2011).

There are others who are less convinced. Schwanen (2007: 9) senses that some feel that time geography only concentrates upon 'constraints, is based on Cartesian space and Newtonian time and (is) characterized by underdeveloped notions of power and agency' (Giddens 1984; Harvey 1990; Rose 1993).

Thrift and Pred (1981) outline and then refute many criticisms of time geography, particularly those put forward by Baker (1977, 1978, 1979, 1981). Giddens (1985: 270–271) is particularly critical noting that it is a naïve approach in how it perceives the individual, detached from their social setting. The main features of the approach—stations and domains—are taken as given and the 'outcome of uninterrupted processes of institutional formation and change' and the individual is left to work around them. Little attention is paid to the transformational capabilities of the individual. Constraints are never seen as opportunities but merely controlling factors, and Hagerstrand's theory of power is weakly developed.

Johnston et al. (2000: 832) refer to Hoppe and Langton (1986: 116–117, 1994) who note two main criticisms which stem from the negativity of the approach: firstly the focus on constraints in terms of how they affect the individual from the aspects of capability, coupling and authority and secondly constraints that emanate from the finiteness of time and space for any individual which in the short term can have dramatic effects inhibiting and preventing actions and decisions which in the longer term might be profitable. The range of alternative configurations, plans and ambitions is consequently restricted by the requisites of the approach. As a consequence, the 'holistic or human ecological intention of Hagerstrand (1974b) cannot, in fact, be realised empirically' (Hoppe and Langton 1986: 117).

Meanwhile, Hallin (1991: 199) notes five main criticisms:

- It is more related to physical than social sciences (Jensen-Butler 1981: 47; Gregory 1985: 309f).
- Individual and social conditions are not given sufficient consideration. Individuals become 'social atoms' (Giddens 1984: 116; Gregory 1985: 324).
- The individual is not viewed as a truly acting subject (Giddens 1984: 117; Gregory 1985: 320).
- The transformation of society remains unexplained (Giddens 1984: 117; Gregory 1985: 321; Van Paassen 1976: 339).
- It has 'an undeveloped perception of opposition, conflict and power and an unproblematic perception of time' (Gregory 1985: 323; Giddens 1984: 117; Rose 1977: 46).

Time budgets can be seen as dehumanising lacking the interplay of life and focussing upon sequences of acts (Thrift 1977a: 5). Meanwhile, Sui (2012: 6) is particularly critical noting the comments of both Gregory (1994) and Gren (2003). Sui suggests that the criticisms can be divided into three types. Humanists argue that it is reductionist and suggests physicalism reducing individuals to actors following travelling paths between stations, determined by various constraints. It is thus no more than a graphical exercise. It is data intensive and requires substantial resources to map adequately the time paths of individuals (or by extension, institutions). Finally, it is embedded within the larger structure–agency debate that dominated geography in the 1970s and 1980s (Gregory 2000; Gatrell 2006). Critics accuse it of being too dependent on either structure or agency depending on their perspective. Meanwhile, feminists accuse it of over-masculinity in its depiction of social reality (Dyck 1990; Rose 1993).

Gren (2003: 209) questions what time geography really is:

> Is time geography a philosophy, an ideology, an ontology, or an epistemology? A research programme, a discourse of practice and power-relations, a method or simply time-spatial distractions of matter (like texts, diagrams, physical bodies coordinated in certain ways in certain places)? Something else or in-between?

However, this inherent doubt about what time geography really is also hints at how useful it might be in accommodating the wide range of issues, demands, expectations and cultures which characterise the maritime sector and its governance. Maritime governance needs an approach into which research, physical bodies, time-space distributions, power relations and much more can be situated. Maybe this is a way ahead, especially as it also focuses highly on the coordination of these things across space and time, an issue severely neglected and which will be a central focus of the next volume.

Despite these criticisms, time geography remains an important and influential theory which has much to offer to the interpretation of time, space and individuals and institutions and in so doing to the development of maritime governance.

The relationship between time geography and policy has been considered on a number of occasions reflecting its potential to offer a novel approach to the problems of governance. Carlstein and Thrift (1978: 228) cite Hagerstrand (1974c) who suggested that:

> organisms and artefacts can be visualized as paths in time-space which forms a web, how each path is subject to individual or environmental time-space constraints, how entities must be synchronized and synchorized, and how all these activities must take place in pockets of operations and domains in time-space which ensure the completion of projects.

This hints closely at the relationship between time, space, individuals and artefacts with policy-making which is the mechanism which can be used to draw these things together to achieve aims and objectives. Governance provides the overarching structure within which these policies must work and determines their effectiveness in the process. A political dimension also needs to be introduced as decisions are made in situational contexts which possess domains that characterise the policy-making process and its effectiveness. Thus, time geography can provide a structural model for the relationship between each of these constituent parts and then also a mechanism to understand the relationships between them. Policy-making and thus governance can be placed within this time geographical framework, and the intricate relationships between the constituent parts can be better understood. This policy and political dimension of time geography has some history dating back to Lee (1968) and his study of administrations (including policy-makers) and Schwartz (1974, 1975) who interprets the relationships between time and social systems.

Pred (1977a: 211) along with Thrift (1977a: 6) emphasises the relationship between time geography and the planning process (incidentally stressing the importance of movement and change), emerging from Swedish government agencies and their interests in regional development policy and the provision of social and cultural services in particular (Oberg 1969; Hagerstrand 1970c; Martensson 1974a, b).

Pred (1981b: 9) continues to be enthusiastic about the application of time geographical approaches to policy-making, and it is possible to see how the central concepts of constraints, paths and projects can be applied to policy interpretation and design. Individuals and institutions:

> react to political and economic events occurring outside the institution, and from their absorption or rejection of prevailing values and norms, all of which, in turn, cannot be divorced from their own uniquely accumulated path history, or intersections with the workings of society. Especially within both the family and today's larger scale economic and governmental organizations, the segments of a power-wielder's earlier path falling within the institution in which he or she holds sway are apt to exercise a complex and strong... effect on his or her consciousness and way of thinking, and thereby on the project defining goals that person (*or institution*) reaches singly or in collaboration. (*italics* added).

Pred (1981b: 17) continues, and although this quotation is along one, it is better to retain the accuracy and pertinence of his comments than to risk misinterpretation:

> Since social reproduction and the dialectics of practice and structure can only be expressed through an unending sequence of temporally and spatially specific intersections between individual paths and institutional projects, social transformation and altered structural relations can only occur through the introduction, disappearance or modification of institutional projects. Or socially change and altered structural relations can only appear through the addition, elimination or recasting of the temporally and spatially specific path couplings demanded by institutions from day to day. As a corollary, the most far reaching social transformations come about through the inauguration, discarding or significant adjustment of the dominant projects. This is true whether dominant projects are initiated or otherwise affected owing: to the unintended contradictions and crises resulting from earlier project implementation; to the intergroup conflicts intrinsic to previous project enactment; to capital accumulation and other power-yielding motivations; to imposition or radical disruption from outside a society; to natural catastrophe; to information exposure and the diffusion of technological or other innovations; or to some more or less complex combination of such circumstances. In any event, whether or not the inception, abandonment or modification of a dominant project has simple, complex or over-determined origins, the inventiveness, goals and rules brought to bear by institutional decision-makers are indissolubly connected with their own external-internal and life path - daily path dialectics, and hence with the dialectics of practice and structure of which they are part. This... also holds for the inventiveness, goals and rules brought to bear on the introduction, elimination, or

> redefinition of all non-dominant institutional projects, which, on a day-to-day basis incrementally, without drama, and unbeknownst to their perpetrators contribute to the making of history, to social transformation and the mutation of structural relations.

This is not the last we hear of Pred who emphasises the relationship between activities at different times and places, the individuals and institutions involved and the projects and policies being undertaken (Pred 1982: 165). Both projects and policies are made up of a series of tasks which have to be completed within specific and constrained time and space, and time geography can help to interpret the best sequence of events to optimise these processes. Referring to a number of sources (Hagerstrand 1974c; Cederlund 1977; Olander and Carlstein 1978; Pred 1978, 1981a, b; ERU 1980; Carlstein 1981b), he stresses the structure that a time geographical framework can offer to the policy-maker and project manager, clarifying the relationships that exist between time, space, individuals and institutions, and it is these relationships and their manipulation that maritime governance needs to address effectively. Thus, effective maritime governance needs to accommodate the complex relationships between (for example) seafarers, shipowners, cargo promoters, port operators, national governments, environmental pressure groups, maritime training institutions, the EU, financiers, lawyers and security services within the constricts of time and the need for an appropriate speed of implementation and space in the form of the vessel, port, nation-state, protected waters, etc.

The interest in time geography's application to policy-making continues to this day. Neutens et al. (2011: 27), for example, note that in particular, Hagerstrand's concept of constraint (what controls and directs what is actually possible in space-time terms) is really the only type of issue that policy-makers can affect to any degree. Thus, issues such as physical capacity of the Suez Canal, the speed of container ship turnaround, the environmental quality of ships' bunkers and the capabilities of seafarers are all susceptible to policy change and all are constraints in a time-space context. The latter of course changes over time and between location, but in principle, it is against these constraints on individual and institutional activity that policies and thus governance are aimed.

Sui (2012: 8) indicates the value that time geography might have for policy-makers and for the derivation of a new and more effective maritime governance by outlining some thoughts of Hagerstrand and the need for further research into the potential of the technique. Stressing that it was only by moving from symbolic to embedded time that we could hope to understand the constraints that time and space place upon individuals and institutions, he identifies four aspects of embedded time that need more consideration (Hagerstrand 1988):

- The accelerated speed by which distance is covered;
- The colonisation of the future;
- The relationship between the natural population of organisms (including humans) and the fabricated population of material artefacts;
- The problems of piecemeal engineering of innovations.

The relevance of these issues remains central to effective maritime governance, all the more remarkable as this was written in 1988.

Time geography also has a close relationship to process, as we saw earlier an issue central to the reinvigoration of maritime governance. Pred (1977a: 210–211) offers an opening in considering time geography as a contextual rather than a compositional approach to human activity and experience. Compositional refers to the consideration of phenomena as a hierarchy of component parts and then how these are joined together into a whole thus centring on form and structure, something we noted much earlier in Chap. 2. A contextual approach looks much more closely at the situation in which an individual or object is placed and the relationships between them with the definitive issues of space and time, something that rings very true with the processes that characterise maritime governance (Dewey 1925; James 1940; Pepper 1942: 268–279; Lowenthal 1961; Wright 1966; Schutz 1973; Tuan 1977; Buttimer 1982). Pred notes (1977a: 211) that the 'contextual synthesis centers on structure and process. And in Hagerstrand's version of such synthesis, it is explicit that process ought to extend to general types of event sequences rather than just specific chains of occurrences'.

Meanwhile, Thrift (2005: 339) further emphasises the relationship of time geography to process:

> Paintings that insist that you focus on parts but still see the whole. The air as ground (Ingaray 2002). Time geography in a world of flows. A new way of saying 'hello' to the world, one which, in its own way is just as organic as that valley in the inner woodlands of southern Sweden… (Hagerstrand's) ethical dynamics of the cosmopolitan moment, still trying to touch the feral beauty of social life.

Unfortunately, we are unable to continue quite so flamboyantly. For example, Wolch and DeVerteuil (2003: 161) consider time geography both a 'notation and a heuristic device designed to contextualize collateral processes in time and space', emphasising the central part played by time paths and the trajectory of movements that are followed (Jackson and Smith 1984: 50). Schwanen (2007: 10) is similarly as convinced of time geography's close relationship to process, movement and change—process-oriented, 'concerned with movement and circulation of materiality', a form of communication which would lend itself to the processes of globalisation that characterise the modern maritime sector (Pred 2005). And to bring it all up to date, Ellegard and Svedin (2012: 20) consider time geography to be a 'multidimensional and abstract view of the world in terms of processes in… timespace'.

There are many examples of the application of time geography to a multitude of situations although none confined to the maritime sector. However, the large variety of applications itself is indicative of its flexibility and potential for maritime governance and policy-making. Taking just a selection, they include Pred (1977a: 211) who looked at the application of time geography to urban and domestic planning issues in Sweden, Olander (1977) who applied the methodology to industrial administration, Martensson (1977) and his consideration of childhood interaction and temporal organisation, Ellegard et al. (1977) focusing on daily travel patterns, Aldskogius (1977) who examined the relationship between recreational behaviour and the environment in Sweden, Olander and Carlstein (1978) who applied time geography methods to studying the geomorphology of the quaternary sector, Lenntorp (1978) simulating an individual's activities,

Carlstein (1978) studying the relationship between innovation, time allocation and time space packing in Mexico, Tornqvist (1979: 7) who examined regional applications, Thrift and Pred (1981: 280) who looked at integrating time geographical approaches into business functions, Pred (1981a, b: 9) in examining social reproduction in everyday life, Oberg (1981) and labour migration, Miller (1982) who considered the characteristics of household activity patterns in nineteenth-century suburbs, Gregory (1985) who provided a number of examples, and Pred (1985a, b) who looked at the relationship between human agency and regional spatial structures with particular reference to southern Sweden.

Others include Adams (1995) (personal boundaries), Hannah (1997) (prisoners), Miller and Bridwell (2009) who cited a wide range of examples from transport including Miller (1991, 1999), Forer (1998), Kwan (1998), Kwan and Hong (1998), Forer and Huisman (2000), O'Sullivan et al. (2000), Wu and Miller (2001), Weber and Kwan (2003), Frihida et al. (2004), and Yu (2006), Yin et al. (2011) (ICT and its potential effect upon face-to-face meetings), Tornqvist (2004) (innovation), Andersson et al. (2012) (travel to school), Downs and Horner (2012) (vehicle tracking), McQuoid and Dijst (2012) (poverty), Scott and He (2012) (shopping destination choice) and Fang et al. (2012) (transport and bridge links in China).

Time geography continues to raise strong feelings, and this in itself suggests it has something to offer and not least to governance, maritime or otherwise although Pred's (1977a: 209) suggestion that 'time geography can specify the necessary… conditions for virtually all forms of interaction—social and otherwise' may be a bit over the top. However, its structure reflecting the nature of real life and its flexibility to accommodate the variety and movement that is apparent in the maritime sector gives it characteristics that could form the basis for a new governance model. Many see that it has a significant future. Hallin (1991: 200) suggests that it could be applied to many other areas (including in our case maritime governance) and in particular by allying its approach with other theories of the social sciences. Couclelis (2009), for example, analyses its redevelopment in the light of changes in ICT, whilst Schwanen (2007: 11) suggests that time geography offers much to circumstances where *routinisation* and *stabilisation* are important, 'enhancing the predictability of projects'. As such, it could provide the basis for a governance framework for a market where routine and stability are both needed but are often sadly lacking. Hagerstrand placed much emphasis on material things (which included policies) as they act as mediators for social structures and relations (Hagerstrand 1995a; b) and by so doing can help to make policy-making more effective. Material forms were considered as essential in constraining and enabling social processes.

Time geography has clear links with much we have discussed earlier in providing a structure for an understanding of movement and change in the context of society and the individual and its ability to accommodate the processes of governance that we have identified. Cresswell (2006: 45–46), for example, begins to link all these ideas together by emphasising its relationship to nomadology and the work of Deleuze and Guatarri (1988) that we saw earlier along with the concepts of 'place ballets'—'intricate, repeated and habitual movements', ideas which can be applied as much to the maritime industry as to, for example, an individual undertaking family life or a society comprised of educators and learners. Whilst it may seem divorced from the reality of maritime governance, there are lessons to be learnt in accommodating the future for shipping and its policy-making that may well help to accommodate the demands of the environment, safety, security and the economy that are currently very poorly served.

Contradictions

Shipping is full of contradictions, and this is commonly one of the causes of the problems that make effective maritime governance so difficult to achieve. The argument here is that governance needs to move from a unidimensional, static character to one that is more dynamic and all-embracing if it is to reflect the characteristics of the industry and the problems it faces; much of this book has been about that movement. However, underlying the unidimensionality and static nature of maritime governance is a series of contradictions that, if addressed, might make the difference between failure and success. It is to these contradictions that we now turn.

The foundation of this contradictory framework comes from the work of Harvey (2014) and his consideration of *Seventeen Contradictions and the End of Capitalism*, an overtly Marxist approach to resolving at least some of the issues that trouble modern society. One not so popular nowadays and commonly (and incorrectly) overshadowed by the fall of the Soviet Union in 1991, but in the light of recent difficulties in the financial sector, in many privatised industries, and in the world economy in general, perhaps an approach with some merit. The 17 contradictions will be reduced to 16 in placing them within a maritime context and they are not proposed as a solution to the problems of maritime governance—more as an attempt to understand underlying causes and pressures that need to be addressed if there is to be any progress. We will consider them here only briefly, and the reader is referred to the original for a more detailed discussion of the fundamental principles and outside of a specifically maritime application.

Contradiction 1: Use and Exchange Value. The use value of shipping is its worth as a form of transport, as a provider of storage and carriage which by so doing opens up utility in a commodity which otherwise would be less or even non-existent. Thus, crude oil lying beneath the North Sea has no value as a fuel until it is shipped; iPods sitting in a warehouse in China are worthless until they arrive in markets in South America, Europe, etc.

The exchange value of shipping is another thing altogether. Ships and shipping are used as a marketable commodity similar to land and property and traded on their own merit with their value fluctuating according to a wide range of issues, many of which having little to do with their use value. Thus, the variation in other commodity prices (oil or money for example) can have a dramatic effect on a ship's exchange value—that is the price at which shipping capacity is traded, made more complex and less related to use value by futures trading, essentially gambling on prospective costs some time in the future.

This discrepancy between use and exchange value matters. Use value is what might be termed the 'true' price of shipping, whilst the wildly fluctuating exchange value generates the booms and busts in shipping markets which cause so much disruption and distortion. These cycles of prices of course characterise many other sectors (if not all to a certain extent) largely because they in turn exhibit use and exchange values, and to the detriment of a stable, meaningful and rational market. And the result—a contribution to the instability in the world economy which has derived to a large part through the existence of two values with little relationship to each other. Affected is the price of shipping and commodities and indirectly the location of power and finance in the shipping industry and the jobs that follow from this. Supply and demand of both ships and commodities are fundamentally affected as well with obvious implications for individuals both rich (who tend to benefit from all fluctuations) and poor (who do not) and for the spatial distribution of wealth. Shipping plays a major role in acting both as a provider of transport and as a commodified form of exchange and speculation.

Contradiction 2: Money and the Social Value of Labour. Value is a social concept and has no absolute meaning but simply reflects the value placed on an object or event by an individual. It can change over time, with circumstances and contexts. Money attempts to represent this value and thus can be considered a measure of value. Thus, prices change even when the characteristics of an object or service fundamentally do not. The price of shipping varies even when the ship itself does not. However, despite this, money can be bought and sold like a commodity (a ship) even though it is actually a measure (like dwt) of value. Money has been commodified. The shipping industry (amongst all forms of capitalist activity) reacts to this valuation of money, and its activities, location, propensity to maintain standards, and to reform, renovate and innovate are similarly dictated by the inappropriate use of its measure of value and something which can be valued in itself. Thus, for example, the upward spiralling price of oil provides opportunities for speculators to make money but adds no value to the oil in itself. The money price therefore ceases to represent the value of the commodity.

This separation of money from its true purpose has been made worse by its 'electronification', which allows for limitless money to be generated in virtual form and compares markedly with money's original manifestation in limited quantities of metal (in particular gold). To prevent excessive 'easing' (the generation of unlimited quantities of electronic money), we have central banks which intervene in the 'free' markets. Shipping is no less affected by this contradiction than any other sector, and its international nature in some ways makes it even more susceptible in that investment in new ships at inappropriate times for the commodity markets can be encouraged by the characteristics of the money marketplace rather than the need of the commodity sector. The shipbuilding sector is a classic example where the price of money is an important factor (amongst others) in determining investment levels and future deliveries and ultimately affecting market cycles.

Contradiction 3: Private Property. There is a considerable contradiction between the existence of both private property (e.g. a ship or terminal) and the nation-state, one of the main functions of which is to protect private property rights. This is prevalent in the maritime sector where a free market in shipping has to be extensively regulated (safety, the environment, security, competition, finance, law, etc.) because it is unable to regulate itself. The contradictory problems this presents between the ambitions of a free market and those of regulators are exacerbated by the inability to take effective legal action against corporations compared with those possible against individuals (exemplified by the *Herald of Free Enterprise*, Zeebrugge ferry disaster and that of the *Costa Concordia*), whilst corporation rights remain guaranteed by the state. The privatisation (commodification) of water, energy, telecommunications, etc. has worsened this situation as what might be seen as collectively owned goods have become private possessions, owned by corporations needing effective regulation but which remain outside much of the judicial system. Interestingly, the inability to regulate effectively international waters presents an extreme version of this issue where corporations (shipping) can do what they wish, with little fear of legal redress, in areas of the earth's surface that in effect belong to us all. All this is reflected in the erroneous relationship claimed between capitalism and democracy. Capitalism is the protection of private property; democracy is collective governance. The nation-state is the nominal guardian of democracy but the actual perpetrator of capitalism. Shipping cannot exist without the nation-state.

Contradiction 4: Private Appropriation and Common Wealth. Polanyi (1957) suggested that there are both legitimate and illegitimate sources of social wealth. Land, labour and money are not legitimate as they are not commodities; despite this, they have been commodified and so they are traded between private individuals and corporations. Examples in the maritime sector abound canals, rivers, territorial waters, beaches, port land and seafarers etc.

Polanyi was convinced that this leads to violence, cheating and theft, and although the intensity of his beliefs might be questioned, their fabric cannot. Territorial waters have been enclosed so that they are effectively owned by nation-states; land is a natural phenomenon and should be no more owned than air, rain, or vistas; and labour has been stolen from individuals through a process of compulsion which denies them the value that they have created through their efforts. Harvey sees this as a process of dispossession and one that is central to the capitalist society of which shipping is a central part. The contradiction lies in the legitimation by capitalist society of possession of illegitimately commodified products. This is a fundamental contradiction that has become so unquestionable that the privatisation (commodification) of energy, water, land, transport, and even outer space has become common and accepted. But in so doing, it exacerbates all the problems of commodification, the unpredictability and extremes of cycles of economic prosperity and failure, and the moves towards a more divided society that we continue to see. Shipping plays its part in this.

Contradiction 5: Capital and Labour. This contradiction reaffirms much what we have said already but is so fundamental to the workings of the capitalist shipping industry that it bears repeating. Labour is universally treated as a commodity, to be bought and sold, and used when profitable and discarded when not; although paid its use value (what it is worth in terms of what it contributes), its exchange value only manifests itself as profit for the owner in a way that resembles that of the vessel, bunkers and port facilities each of which has a use value but which also produces an exchange value represented as profit.

Labour retains some freedom which means that its contribution of exchange value is not exactly compulsory: it has the freedom not to work, but it has no freedom to choose to retain its exchange value (unless of course self-employed, something not common amongst seafarers for example). Further contradictions arise because the state controls taxation, minimum wages and employment conditions, but the private sector is far more independent in setting the conditions, opportunities and characteristics of work.

Contradiction 6: Capital as a Process or Thing? This contradiction is particularly relevant to our discussion of the nature of maritime governance and the need to move from a static to a dynamic vision. Capital (or in our case governance) is both. It flows through an industry, institution, individual or activity as a process that manifests itself as product, finance, data, policy, information and opinion. However, none of these things have any material value until they are sold which means that the dynamic flows only translate into value when they are stabilised into objects which can be traded. The result is a contradiction between the dynamism of the flows (which need to be maintained) and the stasis of the items which represent the value.

The ultimate aim is always to minimise the flows in terms of time whilst maximising the quantity of what flows, and the shipping industry is a good example of this contradiction in action as it attempts to maximise the supply chain in the minimum time. There are exceptions to this of course—slow steaming and cruise shipping for example—but these merely emphasise the normal relationship that exists between speed and flow, dynamism and stasis. The contradiction that this represents is made further clear because the flow of materials, money and information is only of any value when its flow ceases.

Contradiction 7: Unity of Production and Realisation. Here, we have a major contradiction experienced by the shipping industry as much as any. Capitalism as a system needs lots of cheap labour (seafarers, dockers, etc.) and an effective way of keeping it well disciplined. But at the same time, it needs a market with lots of money to purchase the products that it produces. This is no less the case for shipping which relies on the market demanding the goods it transports at the same time as minimising the costs of employing a coercive labour force.

The result is a rather neat division of society into compartments, between the rich (who realise the value) and the poor (who produce it). This is exemplified by the shipping industry between the far more wealthy Western shipowners and the poorer Asian seafarers from Myanmar, Vietnam, Bangladesh and North Korea. Between the rich Northern Hemisphere from whence come the large majority of shipowners, and the poorer south who have to rely upon their ships.

Contradiction 8: Technology, Work and Human Disposability. This contradiction resonates closely with Contradiction 7. An increasingly high-tech industry, shipping is adopting technology at least in part to reduce the labour employed, with obvious impact upon markets. Increased income for those better qualified does not compensate sufficiently for the loss of numbers creating a contradiction between the technological trend and the market needs. A similar effect is occurring through an ever slimmed-down more general supply chain requiring a smaller but better qualified labour force.

Contradiction 9: The Division of Labour. Adam Smith's vision of the division of labour, subsequently encapsulated in Taylorism, was a simple idea that was pertinent and meaningful. Today's major corporations are global, and their interpretation of the concept results in a distribution of manufacturing over the globe with production commonly separated across Europe, the Far East and beyond. One result of this is that the supply chain becomes much more vulnerable to failure as no one location is permitted to do everything and the activities of terrorists, the vagaries of the weather or just plain technological failure or labour unrest can interrupt the carefully planned just-in-time logistics with cataclysmic effect. Shipping is a major part of the supply chain (possibly *the* major part). The contradictions between the benefits of specialisation and the fragility of the supply chain.

Coupled with this is the growth of Taylor's 'trained gorillas': the workforce that has the pleasure of facing a specialised but narrow and repetitive task. Essentially mind-numbing, this presents a further contradiction of specialisation with boredom, the latter an identifiable cause of error. The modern seafarer's lot is part of this trend and that of the officer increasingly so and both can see the future characterised by insecurity as machines continue to take over skills bringing immediate and direct economic benefits compensated by psychological drawbacks. Self-loading of vessels, IT remote monitoring of vessel operation and maintenance and the moves towards a human-less ship are clear examples.

Contradiction 10: Monopoly, Competition, Centralisation and Dispersion. In some ways the most classic of all contradictions, the tendency for intense and free competition to result in monopolies or near monopolies is well documented. This is mirrored by the concentration of power and decision-making compared with the process of decentralisation of physical facilities which together presents contradictions of space and location that in turn may have ramifications for the ability to control effectively labour supply, productivity, product quality, efficiency, financial stability and the like. The shipping industry exhibits the practically universal commercial trend towards consolidation through the supply chain—vertically and horizontally—in a quest for control. This in turn is accompanied by the need to manage an increasingly large and complex organisation across many continents and an even greater number of countries and jurisdictions presenting a contradiction to the simplicity and control which was sought through consolidation. Maersk, for example, attempts to manage vessels throughout the world along with a range of trucking companies, supply chain organisations and international ports bringing advantages of economies of scale but at the expense of local knowledge and influence. The same can be said of international port operators and possibly also of international banks, insurers and ship management companies. Curiously, competition is ultimately reduced as a consequence of allowing and encouraging competition, and along with this comes dispersal and concentration of activities all at the same time. Contradictions abound.

One aspect of an intensively competitive market is immediately apparent. It cannot possibly work without a regulator, set up by some sort of state or combination of states, to prevent the market abuse and tendency to monopolise that competition inevitably brings. Examples are common in shipping—a multitude of UN agencies (UNCTAD, IMO, etc.), the OECD, World Bank, WTO and IMF, each with shipping interests. The EU with its range of Directorate Generals and from a national standpoint, the inevitable ministry in each country with a responsibility for the maritime sector. The tendency of a seemingly market led industry such as shipping to need this degree of regulation is a consequence of the monopolisation and lack of competitiveness that ultimately always emerges in free markets. And shipping is not alone—take banking, insurance, the airline industry, water, power, rail transport and many more. Each has been exposed to the marketplace; each has had to be regulated as a consequence.

The theory suggests that all economic activity takes place on 'the head of a pin' and therefore is spatially competitive. However, within the industry, and shipping no less than any other, this is not only never the case, but also the industry continues to aim to generate as much spatial monopoly as possible. This is reflected in the trends towards clusters in international ports (Hong Kong, Rotterdam, London, Singapore, Piraeus, etc.) as well as supply chain integration and shipping company consolidation and cooperation (e.g. in the liner trade).

Contradiction 11: Uneven Geographical Development. Capitalism constantly requires new markets to use up the surplus labour and capital that it creates, and shipping finds itself an intrinsic part of all this in that it is partially responsible for the creation and opening up of new markets in new areas of the world. Take China, for example, where the products of capitalism (e.g. iPods, iPads, iPhones) have been exported to the world by ships and the raw materials needed by China to generate these products (e.g. oil, iron ore, chemicals) are similarly made available.

One inevitable consequence of this is that the traditional places of production are damaged by the exploitation of new markets for production and consumption. Examples abound—Docklands almost everywhere, Detroit in the USA, the industrial north of the UK and France and many more. The contradiction lies in that the development of new markets, and the improvement in income and facilities that (may) follow from this, is always matched by a decline in income and standards elsewhere. Capital (and shipping playing its major part in facilitating this) has no empathy and no time or space for considering the impact of its desperate desire for a new 'spatial fix', and the uneven geographical development that results is neither new nor avoidable and simply inevitable.

Contradiction 12: The Disparities of Income and Wealth. Whilst it might at first sight seem likely that increasing world prosperity (which despite recent economic depressions has been the case across much of the developed world almost every year) would always be a good thing, in fact, for many, it is the opposite as the contradiction between the wealthy and the poor continues to increase as wealth becomes increasingly concentrated with the very few. This in part is a consequence of the fact that increasing wealth and investment is seen as a convenient way to shed rather than create employment. Shipping is no exception to this contributing to the inequitable redistribution of income across the globe through its service to the capitalist market. More efficient shipping always leads to a reduction in employment through mechanisation, improved communications and the like. More workers who are employed are taken from markets where labour costs are lower but without contributing sufficiently to those markets to reduce the discrepancy between the rich, traditional sources of employment (West Europe, USA, etc.), and the poorer, cheaper new sources (Cambodia, Vietnam, North Korea, etc.). Faster ships mean less vessels and further concentration of income to those lucky enough to be employed on them, or more so, to own them.

Contradiction 13: Social Issues. For a capitalist market to work, it needs competent and skilled workers, but this also leads to problems. An educated workforce is more demanding, generates more disputes, is aware of inadequacies, inefficiencies and inequities around them, and demands more pay. An indirect impact is that an educated workforce can also do more for themselves and thus requires less service provision. The maritime sector is not exempt from these contradictions which on the one hand make education and training more important and on the other an obstruction to the exploitation of labour. Multitasking and multilingual seafarers are inevitably more useful, but also more able to question, demand and obstruct when they see the need to.

Contradiction 14: Compound Growth. Capital demands constant growth. Note the wails of horror in the early twenty-first century when growth dared to slow (NB not disappear) and the talk of economic depression and stagnation. Growth is an essential element of capitalism. And also note how growth is measured in percentage terms—in other words, it is compounded. All fine except constant compound growth is impossible to sustain. Shipping is a major player in the inevitably failed attempts to sustain compound growth forever both as a major source of growth generation and as a beneficiary from its occurrence.

Meanwhile, the major contributors to growth are not exhibiting compound growth—population, raw materials, consumption and food—and therefore, we have a contradiction between what is both expected and demanded and what is known to be possible over time. In the short term, maybe; in the long term, no. It is not as if this problem is unknown; solutions are sought periodically including:

- war (Greek shipowners benefitted massively post-war through the availability of cheap vessels and the sizeable demand for raw materials for reconstruction);
- relocating industry (new spatial fixes for production to the Far East);
- privatising public assets (the privatisation of ports in the 1980–2000 period);
- enclosing the commons (commodifying the sea);
- dynamic obsolescence (vessel design);
- the creation of new products and generating demand for them (all manner of hand devices for communication).

However, all have their ultimate limitation, and the continuous demand of the economic community for the compound growth of wealth presents a contradiction that will inevitably fail. As Harvey (2014: 245) suggests:

> The World Bank is fond of reassuring us that a rising tide of economic development is bound to lift all boats. Maybe a truer metaphor would be that exponentially rising sea levels and intensifying storms are destined to sink all boats.

Contradiction 15: Capital and Nature. Capitalism both harms the environment and saves it—and of course by so doing generates a new market for its services to help use up the surplus capital being generated. Shipping is a major contributor to this artificial market which it both destroys and then recreates benefitting from both actions. The opposition of the shipping industry to environmental protection and reform is hardly surprising as curbing its polluting activities not only generates costs (equipment, training, vessel improvement) but also limits the contribution (and profit) it can make from serving the industries that clean it up.

Contradiction 16: Alienation. Perhaps the least definable of all the contradictions, the development of technology (not least in the shipping industry) reduces costs for the shipowner through lower labour costs and greater efficiency in the use of vessels. However, a major if indirect effect is also to make employment less interesting, to alienate employees and to increase inefficiencies as a result. Trying to achieve a balance between allowing technology to take away the human element (and risk) of work whilst still sustaining enthusiasm, attention and interest is a fundamental contradiction within the industry that continues to need to be addressed.

Conclusion

So that's it then. All sorted. All the problems of maritime governance resolved by a thorough redesign which addresses the inadequacies of static unidimensionality and the need to recognise the relationship between governance and globalisation, making steady progress to accommodate dynamism in policy-making. Unfortunately no; there remains much more to decide let alone accomplish because there remains very little understanding of even the significance of globalisation, and consequently, dynamic governance still remains a dream. And yet in addition to this, there is even more to do.

This is the second of three books that address issues of inadequacy in maritime governance. One way of understanding the problems faced is to take them as representing three dimensions. The first, examined in *Maritime Governance and Policy-Making*, looked at the relationship between a static maritime governance and globalisation, unidimensional in character and in need of broadening and development to move on from the traditional hierarchical, nation-based, institutionally defined approach that was so comfortable for many of the major stakeholders.

This book has introduced the idea that a static approach to governance is inadequate and that a more dynamic framework is the only way to reflect the way the industry operates—constantly changing, adapting to market needs, attempting to avoid the rigours of legislation and policy-making that can be so inconvenient. Issues of form, processes, time and flows have been considered and shown to be essential elements of a new governance framework that recognises not only the role of globalisation but also the dynamic way the industry operates.

The third and final dimension will be central to the next volume which will look at the relationship between policies within a dynamic governance framework and how they affect each other. How consideration of one policy approach has to consider its impact upon others and how adaptations of policy and policy timing will be important in improving their effectiveness. This process of policy juxtaposition would allow a governance framework to be designed specifically for any particular situation. Contextual issues can provide the basis for this process—contexts first identified in the shipping sector by Ledger and Roe (1996: 66) and including operational, organisational, political, economic, spatial, legal, environmental, managerial and social and summarised in Fig. 8.3. Governance of any sector, and not least the shipping industry, has to work within this contextual framework and redesigning governance to be more dynamic and with the faculty to juxtapose across and between policies will benefit from the framework this provides.

Context	**Example**
Operational	Technical impact of change in ship and port operations.
Organisational	Organisational changes within the shipping industry including ownership.
Spatial	The specific geographical context for the shipping sector and activity. For example, port location, supply chain routes, sea lanes, weather etc.
Legal	Legislative impacts across jurisdictions.
Environmental	Impacts of climate change.
Political	The political context for shipping at global (eg IMO), supra-national (eg EU), national, regional and local levels.
Managerial	Levels of management skills and types of managerial designs.
Economic	Methods of business, material prosperity and broad economic change in the maritime sector and elsewhere.
Social	Changes in market desires, conditions of employment, training.

Fig. 8.3 Contexts and the maritime sector. *Source* Adapted from Ledger and Roe (1996)

Finally, two further and highly significant issues will then be considered in the light of governance redesign and adoption of a juxtapositional capability: polycentricity and metagovernance. Both are highlighted by Gritsenko and Yliskylä-Peuralahti (2013) who suggest that maritime governance is essentially a polycentric issue with a multitude of ever-changing foci that have to be accommodated within the governance of the industry if it is to be meaningful. These foci can be considered wholly within a traditional hierarchical context—global, supranational, national, regional and local—or within a newly designed governance framework that accommodates the dynamism that is clearly important to effective policy-making. They provide a polycentric structure for the industry, driving policy-making to understand the multifocal nature of the maritime sector.

Meanwhile, whilst the traditional hierarchy is largely one composed of dominantly spatial and legal characteristics, the environmental context (for example) features a range of other foci, constantly changing and highly polycentric and certainly not dominantly spatial or legal. The same might be said of the management, organisation and operation of the industry where alternative operational, organisational, social or economic foci may well be more important. A polycentric approach to governance has much to offer (see for example Aligica and Tarko 2012).

Polycentricism leads on to metagovernance, referring to the need to have a governance framework that oversees the governance of the industry itself—rather in the way that Juvenal in his *Satires* saw the need for guards to be guarded—*quis custodiet ipsos custodes?* Much has been written on metagovernance—see for example Bell and Park (2006), Kooiman and Jentoft (2009) and Meuleman (2010)—and its significance to effective governance and policy-making should not be overlooked. Both polycentricity and metagovernance will be considered alongside juxtaposition as the third governance dimension for the maritime sector.

But that is for next time.

References

Adams, P. C. (1995). A reconsideration of personal boundaries in space-time. *Annals of the Association of American Geographers, 85*(2), 267–285.

Aldskogius, H. (1977). A conceptual framework and a Swedish case study of recreational behavior and environmental cognition. *Economic Geography, 53*(2), 163–183.

Aligica, P. D., & Tarko, V. (2012). Polycentricity: From Polanyi to Ostrom and beyond. *Governance, 25*(2), 237–262.

Anderson, J. (1971). Space-time budgets and activity studies in urban geography and planning. *Environment and Planning, 3*, 353–368.

Andersson, E., Malmberg, B., & Osth, J. (2012). Travel-to-school distances in Sweden 2000–2006: Changing school geography with equality implications. *Journal of Transport Geography, 23*, 35–43.

Arentze, T. A., & Timmermans, H. (2005). An analysis of context and constraints-dependent shopping behaviour using qualitative decision principles. *Urban Studies, 42*, 435–448.

Baker, A. R. H. (1977). Historical geography. *Progress in Human Geography, 1*(3), 465–474.

Baker, A. R. H. (1978). Historical geography: Understanding and experiencing the past. *Progress in Human Geography, 2*(3), 495–504.

Baker, A. R. H. (1979). Historical geography: A new beginning? *Progress in Human Geography, 3*(4), 560–570.

Baker, A. R. H. (1981). An historical-geographical perspective on time and space and on period and place. *Progress in Human Geography, 5*(3), 439–443.

Banks, M., & MacKiant, S. (2000). Jump in! The water's warm: A comment on Peck's grey geography. *Transactions of the Institute of British Geographers NS, 25*, 249–254.

Baudelaire, C. (1970). *Paris spleen*. New York: New Directions.

Bell, S., & Park, A. (2006). The problematic metagovernance of networks: Water reform in New South Wales. *Journal of Public Policy, 26*(1), 63–83.

Belsey, A. (1992). World poverty, justice and inequality. In R. Attfield & B. Watkins (Eds.), *International justice and the third world* (pp. 35–49). London: Routledge.

Berry, B. J. L. (1994). Let's have more policy analysis. *Urban Geography, 15*, 315–317.

Bigelow, J., Ellis, B., & Pargetter, R. (1988). Forces. *Philosophy of Science, 55*, 614–630.

Buttimer, A. (1976). Grasping the dynamism of lifeworld. *Annals of the Association of American Geographers, 66*(2), 277–287.

Buttimer, A. (1982). Musing on Helicon: Root metaphors and geography. *Geografiska Annaler B, 64*(2), 89–96.

Buttimer, A. (1999). Humanism and relevance in geography. *Scottish Geographical Journal, 115*(2), 103–116.

Byatt, A. S. (1993). *Angels and Insects*. London: Vintage.

Carlstein, T. (1978). Innovation, time-allocation and time-space packing. In T. Carlstein, D. Parkes, & N. Thrift (Eds.), *Human activity and time geography* (pp. 147–161). London: Edward Arnold.

Carlstein, T. (1981a). The sociology of structuration in time and space: A time-geographic assessment of Gidden's theory. *Svensk Geografisk Arsbok, 57*, 41–57.

Carlstein, T. (1981b). *Time resources, society and ecology: On the capacity for human interaction in space and time*. London: Edward Arnold.

Carlstein, T., Parkes, D., & Thrift, N. (1978). Introduction. In T. Carlstein, D. Parkes, & N. Thrift (Eds.), *Human activity and time geography* (pp. 117–121). London: Edward Arnold.

Carlstein, T., & Thrift, N. (1978). Afterword: Towards a time-space structured approach to society and environment. In T. Carlstein, D. Parkes, & N. Thrift (Eds.), *Human activity and time geography* (pp. 225–263). London: Edward Arnold.

Cederlund, K. (1977). Administrativ verksamhet som projekt. *Svensk Geografisk Arsbok, 53*, 69–80.

Chapin, F. S. (1965). *Urban land use planning*. Chicago IL: University of Illinois Press.

Clement, C. (1996). *Care, autonomy and justice: Feminism and the ethic of care*. Oxford: Westview Press.

Converse, P. E. (1968). Time budgets. In D. Sills (Ed.), *Encyclopedia of social science* (pp. 42–47). Chicago IL: Macmillan.

Corbridge, S. (1993). Marxisms, modernities and moralities; development praxis and the claims of distant strangers. *Environment and Planning D, 11*, 449–472.

Couclelis, H. (2009). Rethinking time geography in the information age. *Environment and Planning A, 41*, 1556–1575.

Crang, M. (2003). Rhythms of the city. Temporalised space and motion. In J. May & N. Thrift (Eds.), *Timespace: Geographies of temporality* (pp. 187–207). Abingdon: Routledge.

Cresswell, T. (2006). *On the move, mobility in the modern western world*. London: Routledge.

Danner, U. N., Aarts, H., & de Vries, N. K. (2008). Habits vs intention in the prediction of future behaviour; the role of frequency, context stability and mental accessibility of past behaviour. *British Journal of Social Psychology, 47*, 245–265.

Davies, K. (2003). Responsibility and daily life. In J. May & N. Thrift (Eds.), *Timespace: Geographies of temporality* (pp. 133–148). Abingdon: Routledge.

Deleuze, G., & Guatarri, F. (1988). *A thousand plateaus. Capitalism and schizophrenia*. London: Bloomsbury.
Dewey, J. (1925). *Experience and nature*. Chicago IL: Open Court.
Dijst, M. (2009). Time geographical analysis. In R. Kitchin & N. Thrift (Eds.), *International encyclopaedia of human geography* (Vol. 1, pp. 266–278). Oxford: Elsevier.
Douglas, M. (1966). *Purity and danger: An analysis of concepts of pollution and taboo*. London: Routledge and Kegan Paul.
Downs, J. A., & Horner, M. W. (2012). Probabilistic potential path trees for visualizing and analyzing vehicle tracking data. *Journal of Transport Geography, 23*, 72–80.
Dyck, I. (1990). Space, time and renegotiating motherhood: An exploration of the domestic workplace. *Environment and Planning D, 8*, 459–483.
Eagelton, T. (1996). *The illusions of postmodernism*. Oxford: Blackwell.
Ellegard, K., Hagerstrand, T., & Lenntorp, B. (1977). Activity organization and the generation of daily travel: Two future alternatives. *Economic Geography, 53*(2), 126–152.
Ellegard, K., & Svedin, U. (2012). Torsten Hagerstrand's time-geography as the cradle of the activity approach in transport geography. *Journal of Transport Geography, 23*, 17–25.
Engels, F. (1969). *The condition of the working class in England*. London: Panther.
ERU (1980) Offentlig verkshamet och regional valfard: Perspecktiv pa det politiska besluts-fattandet och den offentliga sektorn, Expertgruppen for forsnkning om regional utveckling (Statens offentliga utredningar, Stockholm), 6.
Fang, Z., Shaw, S., Tu, W., Li, Q., & Li, Y. (2012). Spatiotemporal analysis of critical transportation links based on time geographic concepts: A case study of critical bridges in Wuhan, China. *Journal of Transport Geography, 23*, 44–59.
Forer, P. (1998). Geometric approaches to the nexus of time, space and microprocess: implementing a practical model for mundane socio-spatial systems. In M. J. Egenhofer & R. G. Golledge (Eds.), *Spatial and temporal reasoning in geographic information systems* (pp. 171–190). Oxford: Oxford University Press.
Forer, P., & Huisman, O. (2000). Space, time and sequencing: Substitution at the physical/virtual interface. In D. G. Janelle & D. C. Hodge (Eds.), *Information, place and cyberspace* (pp. 73–90). Berlin: Springer.
Foucault, M. (1966). *The order of things: An archaeology of the human sciences*. London: Vintage.
Friedman, M. (1991). The practice of partiality. *Ethics, 101*, 818–835.
Frihida, A., Marceau, D. J., & Theriault, M. (2004). Development of a temporal extension to query travel behavior time paths using an object-oriented GIS. *Geoinformatica, 8*, 211–235.
Gatrell, J. (2006). Time geography. In B. Warf (Ed.), *Encyclopaedia of human geography* (pp. 489–491). Thousand Oaks, CA: Sage.
Giddens, A. (1984). *The constitution of society: Outline of the theory of structuration*. Berkeley, CA: University of California Press.
Giddens, A. (1985). Time space and regionalism. In D. Gregory & J. Urry (Eds.), *Social relations and spatial structures* (pp. 265–295). London: Macmillan.
Gilligan, C. (1987). Moral orientation and moral development. In E. Kittay & D. Meyers (Eds.), *Women and moral theory* (pp. 19–33). Rowman and Littlefield: Totowa NJ.
Ginzburg, C. (1994). Killing a Chinese mandarin: the moral implications of distance. *New Left Review, 208*, 107–120.
Goodin, R. (1985). *Protecting the vulnerable*. Chicago, IL: University of Chicago Press.
Gregory, D. (1978). *Ideology, science and human geography*. London: Hutchinson.
Gregory, D. (1985). Suspended animation: The stasis of diffusion theory. In D. Gregory & J. Urry (Eds.), *Social relations and spatial structures* (pp. 296–336). London: Macmillan.
Gregory, D. (1994). *Geographical imaginations*. Malden, MA: Blackwell.
Gregory, D. (2000). Time geography. In R. J. Johnston, D. Gregory, C. Pratt, & M. Watts (Eds.), *The dictionary of human geography* (pp. 830–833). Oxford: Blackwell.
Gren, M. (2003). Time geography matters. In J. May & N. Thrift (Eds.), *Timespace: Geographies of temporality* (pp. 208–225). Abingdon: Routledge.

Griffiths, J. (1980). *Three tomorrows: American, British and Soviet science fiction*. London: Macmillan.

Gritsenko, D., & Yliskylä-Peuralahti, J. (2013). Governing shipping externalities: Baltic ports in the process of SOx emission reduction. *Maritime Studies, 12*, 10.

Grosz, E. (1994). *Volatile bodies. Towards a corporeal feminism*. Bloomington IN: Indiana University Press.

Hagerstrand, T. (1967). Bilden av Yttervarlden. *Svensk Geografisk Arsbok Arg, 43*, 65–74.

Hagerstrand, T. (1970a). What about people in regional science. *Papers of the Regional Science Association, 24*, 7–21.

Hagerstrand, T. (1970b). Regional forecasting and social engineering. In M. Chisholm, A. E. Frey, & P. Haggett (Eds.), *Regional forecasting* (pp. 1–7). London: Butterworth.

Hagerstrand, T. (1970c) Tidsanvandning och omgivningsstruktur, *Urbanisering i Sverige: en Geograisk Samhallsanalys*, Stockholm, Statens Offengtliga Utredningar, 14, Allmanna Forlaget.

Hagerstrand, T. (1973). The domain of human geography. In R. J. Chorley (Ed.), *Directions in geography* (pp. 67–87). London: Methuen.

Hagerstrand, T. (1974a). Tidsgeografisk Beskrivning. Syfte och Postulat. *Svensk Geografisk Arsbok, Arg, 50*, 86–94.

Hagerstrand, T. (1974b) Ecology under one perspective, In E. Bylund, H. Linderholm, & O. Rune (Eds.) *Ecological problems of the circumpolar area* (pp. 271–276), Lulea: Norrbottens Museum.

Hagerstrand, T. (1974c). On socio-technical ecology and the study of innovations. *Ethnologica Europaea, 7*, 17–34.

Hagerstrand, T. (1975). Space, time and human conditions. In A. Karlqvist, L. Lundquist, & F. Snickars (Eds.), *Dynamic allocation of urban space* (pp. 1–14). Farnborough: Saxon House.

Hagerstrand, T. (1976). Geography and the study of the interaction between nature and society. *Geoforum, 7*, 329–344.

Hagerstrand, T. (1978a). Survival and arena. In T. Carlstein, D. Parkes, & N. Thrift (Eds.), *Human activity and time geography* (pp. 122–145). London: Edward Arnold.

Hagerstrand, T. (1978b). A note on the quality of life-times. In T. Carlstein, D. Parkes, & N. Thrift (Eds.), *Human activity and time geography* (pp. 214–224). London: Edward Arnold.

Hagerstrand, T. (1982). Diorama, path and project. *Tijdschrift voor economische en sociale geografie, 73*(6), 323–339.

Hagerstrand, T. (1983). In search for the sources of concept. In A. Buttimer (Ed.), *The practice of geography* (pp. 238–256). New York: Longman.

Hagerstrand, T. (1985). Time geography: Focus on the corporeality of man, society and the environment. In S. Aida (Ed.), *The science and praxis of complexity* (pp. 193–216). Tokyo: United Nations University.

Hagerstrand, T. (1986). Den geografiska traditionens karnomrade. *Svensk Geografisk Arsbok, Arg, 62*, 38–43.

Hagerstrand, T. (1988). Time and culture. In C. Kirsch, P. Nijkamp, & K. Zimmerman (Eds.), *The formulation of time, preferences in a multidisciplinary perspective* (pp. 33–42). Aldershot: Gower.

Hagerstrand, T. (1989a). Reflections on 'what about people in regional science'. *Papers of the Regional Science Association, 66*, 1–6.

Hagerstrand, T. (1989b). Globalt och lokalt. *Svensk Geografisk Arsbok, Arg, 65*, 9–20.

Hagerstrand, T. (1991). Tillkomsten av nationalparker i Sverige. En ides vag fran 'andskap'till landskap. *Svensk Geografisk Arsbok, Arg, 67*, 83–96.

Hagerstrand, T. (1992). Geografins innehall och historiska utveckling. *Svensk Geografisk Arsbok, Arg, 68*, 9–18.

Hagerstrand, T. (1995a). Action in the physical everyday world. In A. D. Cliff, P. R. Gould, A. G. Hoare, & N. J. Thrift (Eds.), *Diffusing geography: Essays for Peter Haggett* (pp. 35–45). Oxford: Blackwell.

Hagerstrand, T. (1995b). Landscape as overlapping neighbourhoods. In G. B. Benko & U. Strohmayer (Eds.), *Geography, history and social sciences* (pp. 83–96). Dordrecht: Kluwer Academic Publishers.

Hagerstrand, T. (2004). The two vistas. *Geografiska Annaler B, 86*(4), 315–323.

Hallin, P. O. (1991). New paths for time-geography? *Geografiska Annaler B, 73*(3), 199–207.

Hannah, M. (1997). Imperfect panopticism: Envisioning the construction of normal lives. In G. Benko & U. Strohmayer (Eds.), *Space and social theory: Interpreting modernity and postmodernity* (pp. 344–359). Oxford: Blackwell.

Harvey, D. (1974). What kind of geography for what kind of public policy? *Transactions of the Institute of British Geographers, 63*, 18–24.

Harvey, D. (1990). *The condition of postmodernity*. London: Blackwell.

Harvey, D. (2014). *Seventeen contradictions and the end of capitalism*. London: Profile Books.

Hawley, A. (1950). *Human ecology: A theory of community structure*. New York: Ronald Press.

Hayek, F. A. (1945). The use of knowledge in society. *American Economic Review, XXXV*(4), 519–530.

Held, D. (1993). Liberalism, Marxism and democracy. *Theory and Society, 22*, 249–281.

Hesse, M. (1999) Globalocal. On the significance of space for time, speed and mobility. In J. Beckmann (Ed.), *Speed—A Workshop of Time, Space and Mobility* (pp. 31–38). Copenhagen: The Danish Transport Council.

Holly, B.P. (ed) (1976) *Time-space budgets and urban research: A symposium, Department of Geography*. University of Kent, UK. Discussion Paper 1.

Hoppe, G., & Langton, J. (1986). Time geography and economic development: the changing structure of livelihood positions on arable farms in nineteenth century Sweden. *Geografiska Annaler B, 68*(2), 115–137.

Hoppe, G., & Langton, J. (1994). *Peasantry to capitalism: Western Ostergötland in the nineteenth century*. Cambridge: Cambridge University Press.

Hudgins, E. (2002). *Space: The free market frontier*. Washington DC: The Cato Institute.

Huggett, R., & Perkins, C. (2004). Landscape as form, process and meaning. In J. A. Matthews & D. T. Herbert (Eds.), *Unifying geography: Common heritage, shared future* (pp. 224–239). London: Routledge.

Huntingdon, E. (1926). *The pulse of progress*. New York: Scribner.

Huntingdon, E. (1931). The Matamek conference on biological cycles. *Science, 74*, 229–235.

Huntingdon, E. (1945). *Mainsprings of civilisation* (pp. 459–617). New York: Mentor Books.

Ingaray, L. (2002). *Heidegger's forgetting of air*. London: Continuum.

Jackson, P., & Smith, S. J. (1984). *Exploring social geography*. London: Allen and Unwin.

James, W. (1940). *Pragmatism*. London: Longman Greens.

Jensen-Butler, C. (1981) *A critique of behavioural geography. An epistemological analysis of cognitive mapping and of Hagerstrand's time-space model*, Working Paper, Geographical Institute, Aarhus University, Denmark.

Johnston, R. (2005). Torsten Hagerstrand (1916–2004). *Progress in Human Geography, 29*(3), 327–328.

Johnston, R., Gregory, D., Pratt, G., & Watts, M. (2000). *The dictionary of human geography*. Oxford: Blackwell.

Kafka, F. (1946). *The great wall of China. Stories and reflections*. Berlin: Schoken Books.

Khalid, F., & O'Brien, J. (1992). *Islam and ecology*. New York: Cassell.

Kissinger, H. (1972). *Years of upheaval*. London: Simon and Schuster.

Klinkowitz, J. (1998). *Vonnegut in fact*. Columbia, SC: University of South Carolina Press.

Kooiman, J., & Jentoft, J. (2009). Meta-governance: Values, norms and principles, and the making of hard choices. *Public Administration, 87*(4), 818–836.

Kubler, G. (1962). *The shape of time: Remarks on the history of things*. New Haven, CN: Yale University Press.

Kwan, M.-P. (1998). Space-time and integral measures of individual accessibility. A comparative analysis using point-based framework. *Geographical Analysis, 30*, 191–216.
Kwan, M.-P. (2000). Interactive geovisualization of activity travel patterns using three dimensional geographical information systems. A methodological exploration with a large data set. *Transportation Research C, 8*, 185–203.
Kwan, M.-P., & Hong, X.-D. (1998). Network-based constraints-oriented choice set formation using GIS. *Geographical Systems, 5*, 139–162.
Landau, U., Prashker, J. L., & Alpern, B. (1982). Evaluation of activity constrained choice sets to shopping destination choice modelling. *Transportation Research A, 16*, 199–207.
Law, R. (1999). Beyond 'women and transport': Towards new geographies of gender and daily mobility. *Progress in Human Geography, 23*(4), 567–588.
Ledger, G., & Roe, M. (1996). *East European change and shipping policy*. Aldershot: Avebury.
Lee, H.-B. (1968). From ecology to time: a time orientation approach to the study of public administration. *Ekistics, 25*, 432–438.
Lenntorp, B. (1974). Grupperingar och arrangemang av odelbaba enheter. *Svensk Geografisk Arsbok, 50*, 95–113.
Lenntorp, B. (1978). A time-geographic simulation model of individual activity programmes. In T. Carlstein, D. Parkes, & N. Thrift (Eds.), *Human activity and time geography* (pp. 162–180). London: Edward Arnold.
Lenntorp, B. (1981). Ett fang figurer—tillbakablickande kommentarer till nagra exemplar ur den tidsgeografiska bild floran. *Svensk Geografisk Arsbok, 57*, 80–91.
Lenntorp, B. (2004). Publications by Torsten Hagerstrand 1938–2004. *Geografiska Annaler B, 86*(4), 327–334.
Lenntorp, B., Tornqvist, G., Warneryd, O., & Oberg, S. (2004). Torsten Hagerstrand 1916–2004. *Geografiska Annaler B, 86*(4), 325–326.
Li, R. P. Y., & Thompson, W. R. (1975). The 'coup contagion' hypothesis. *Journal of Conflict Resolution, 19*(1), 63–88.
Liepitz, A. (1997). Warp, woof and regulation. In G. Benko & U. Strohmayer (Eds.), *Space and social theory: Interpreting modernity and postmodernity* (pp. 250–284). London: Wiley.
Lowenthal, D. (1961). Geography, experience and imagination: Towards a geographical epistemology. *Annals of the Association of American Geographers, 71*, 95–107.
Lundberg, G., Komarovski, M., & McInerny, M. (1934). *Leisure: A suburban study*. New York: Columbia University Press.
Lynch, K. (1972). *What time is this place?*. Cambridge, MA: MIT Press.
MacIntyre, A. (1981). *After virtue: A study in moral theory*. London: Duckworth.
MacMurray, T. (1971). Aspects of time and the study of activity patterns. *Town Planning Review, 42*, 195–209.
Maréchal, K. (2010). Not irrational but habitual: The importance of 'behavioral lock-in' in energy consumption. *Ecological Economics, 69*, 1104–1114.
Martensson, S. (1974a). Primargrupp, eskundargrupp och forbrukningen av persontid. *Svensk Geografisk Arsbok, 50*, 126–135.
Martensson, K. (1974b) Drag i hushallens levnadsvillkor, *Ortsbundna levnadsvillor*, Stockholm: Statens Offentliga Utredningar, 2, Allmanna Forlaget.
Martensson, S. (1977). Childhood interaction and temporal organization. *Economic Geography, 53*(2), 99–125.
Martensson, S. (1981). Femton ars erfarenhet av att presentera tidsgeografi. *Svensk Geografisk Arsbok, 57*, 130–140.
Martin, E. (1991). The egg and the sperm: How science has constructed a romance based on stereotypical male-female roles. *Journal of Women in Culture and Society, 16*(3), 485–501.
Marx, K. (1970). *Das kapital*. London: Lawrence and Wishart.
McQuoid, J., & Dijst, M. (2012). Bringing emotions to time geography: the case of mobilities of poverty. *Journal of Transport Geography, 23*, 26–34.
Meentemeyer, V. (1989). Geographical perspectives of space, time and scale. *Landscape Ecology, 3*(3/4), 163–173.

Merleau-Ponty, M. (1962). *The phenomenology of perception*. London: Routledge and Kegan Paul.

Meuleman, L. (2010). The cultural dimension of metagovernance: Why governance doctrines may fail. *Public Organization Review, 10*, 49–70.

Miller, R. (1982). Household activity patterns in nineteenth-century suburbs: A time geographic exploration. *Annals of the Association of American Geographers, 72*(3), 355–371.

Miller, H. J. (1991). Modelling accessibility using space-time prism concepts within geographical information systems. *International Journal of Geographical Information Systems, 5*, 287–301.

Miller, H. J. (1999). Measuring space-time accessibility benefits within transportation networks: Basic theory and computational methods. *Geographical Analysis, 31*, 187–212.

Miller, H. J. (2004). Activities in space and time. In D. A. Hensher, K. J. Button, K. E. Haynes, & P. R. Stopher (Eds.), *Handbook of transport geography and spatial systems* (pp. 647–660). London: Pergamon.

Miller, H. J. (2005a). A measurement theory for time geography. *Geographical Analysis, 37*, 17–45.

Miller, H. J. (2005b). Necessary space-time conditions for human interaction. *Environment and Planning B, 32*, 381–401.

Miller, H. J., & Bridwell, S. A. (2009). A field-based theory for time geography. *Annals of the Association of American Geographers, 99*(1), 49–75.

Morrison, E. E. (1966). *Men, machines, and modern times*. Cambridge, MA: MIT Press.

Nadel, S. F. (1957). *The theory of social structure*. London: Cohen and West.

Neutens, T., Schwanen, T., & Witlox, F. (2011). The prism of everyday life: Towards a new research agenda for time geography. *Transport Reviews, 31*(1), 25–47.

Nilsson, G. M. O. P. (1985). Tidsgeografins mojligheter och restriktioner. *Svensk Geografisk Arsbok, 61*, 255–265.

Nystuen, J. D. (1963). Identification of some fundamental spatial concepts. In B. J. L. Berry & D. F. Marble (Eds.), *Spatial analysis* (pp. 35–41). Englewood Cliffs, NJ: Prentice-Hall.

O'Neill, O. (1996). *Toward justice and virtue: A constructive account of practical reasoning*. Cambridge: Cambridge University Press.

O'Sullivan, D., Morrison, A., & Shearer, J. (2000). Using desktop GIS for the investigation of accessibility by public transport: an isochrone approach. *International Journal of Geographical Information Science, 14*, 85–104.

O'Tuathail, G., Herod, A., & Roberts, S. M. (1998). Negotiating unruly problematics. In G. O'Tuathail, A. Herod, & S. M. Roberts (Eds.), *An unruly world* (pp. 1–24). London: Routledge.

Oberg, S. (1969) Tandvard och befolkningsunderlag en studie i tandlakartathetensgeografiuska variationer. Lund, Urbaniseringsprocessen, 28.

Oberg, S. (1981). Perspektiv pa levnadsoden. *Svensk Geografisk Arsbok, 57*, 255–266.

Oberg, S. (2005). Hagerstrand and the remaking of Sweden. *Progress in Human Geography, 29*(3), 340–349.

Olander, L.-O. (1977). Foretagsadministration som aktivitets system i tid och rum. *Svensk Geografisk Arsbok, 53*, 55–64.

Olander, L.-O., & Carlstein, T. (1978). The study of activities in the quaternary sector. In T. Carlstein, D. Parkes, & N. Thrift (Eds.), *Human activity and time geography* (pp. 198–213). London: Edward Arnold.

Openshaw, S., & Veneris, Y. (2003). Numerical experiments with central place theory and spatial interaction modelling. *Environment and Planning A, 35*(8), 1389–1403.

Ottensman, J.R. (1972) *Systems of urban activities and time: An interpretive review of the literature*. University of North Carolina, Centre for Urban and Regional Studies, Urban Studies Research Paper.

Pacione, M. (1999) Guest editorial. *Scottish Geographical Journal*, 115(2):x–xiv.

Parker, M. (2009). Angelic organization: Hierarchy and the tyranny of heaven. *Organization Studies, 30*(11), 1281–1300.

Parkes, D., & Thrift, N. (1979). Spacemakers and entrainment. *Transactions of the Institute of British Geographers NS, 4*(3), 353–372.

Parkes, D., & Thrift, N. (1980). Time geography: The Lund approach. In D. Parkes & N. Thrift (Eds.), *Times, spaces, places* (pp. 243–278). London: Wiley.

Peck, J. (1999). Grey geography? *Transactions of the Institute of British Geographers NS, 24*, 131–135.

Peck, J. (2000). Jumping in, joining up and getting on. *Transactions of the Institute of British Geographers NS, 25*, 255–258.

Pepper, S. (1942). *World hypotheses*. Berkeley, CA: University of California Press.

Persson, O., & Ellegard, K. (2012). Torsten Hagerstrand in the citation time web. *Professional Geographer, 64*(2), 250–261.

Planck, M. (1948) *Wissenschaftliche Selbstbiographie. Mit einem Bildnis und der von Max von Laue gehaltenen Traueransprache* (p. 22). Leipzig: Johann Ambrosius Barth Verlag.

Polanyi, K. (1957). *The great transformation: The political and economic origins of our time*. Boston: Beacon Press.

Pollard, J., Henry, N., Bryson, J., & Daniels, P. (2000). Shades of grey? Geographers and policy. *Transactions of the Institute of British Geographers NS, 25*, 243–248.

Pred, A. (1977a). The choreography of existence: Some comments on Hagerstrand's time geography and its effectiveness. *Economic Geography, 53*, 207–221.

Pred, A. (1977b). Urbanisation, domestic planning, problems and Swedish geographic research. *Progress in Geography, 5*, 1–76.

Pred, A. (1978). The impact of technological and institutional innovations on life content: Some time-geographic observations. *Geographical Analysis, 10*, 345–372.

Pred, A. (1981a). Of paths and projects: Individual behavior and its societal context. In K. R. Cox & R. G. Golledge (Eds.), *Behavioral problems in geography revisited* (pp. 231–255). London: Methuen.

Pred, A. (1981b). Production, family and free time projects: A time-geographic perspective on the individual and societal change in nineteenth century US cities. *Journal of Historical Geography, 7*, 3–36.

Pred, A. (1982). Social reproduction and the time geography of everyday life. In P. Gould & G. Olsson (Eds.), *A search for common ground* (pp. 157–186). London: Pion.

Pred, A. (1984). Place as a historically contingent process: Structuration and the time geography of becoming places. *Annals of the Association of American Geographers, 74*, 279–297.

Pred, A. (1985a). The social becomes the spatial, the spatial becomes the social—enclosures, social change and the becoming of places in the Swedish province of Skane. In D. Gregory & J. Urry (Eds.), *Social relations and spatial structures* (pp. 337–365). London: Hutchinson.

Pred, A. (1985b). Interpenetrating processes: Human agency and the becoming of regional spatial and social structures. *Papers of the Regional Science Association, 57*, 7–17.

Pred, A. (2005). Hagerstrand matters: Life (path) and death matters—some touching remarks. *Progress in Human Geography, 29*(3), 328–332.

Proudhon, J.-P. (1840/2008). *What is Property? Or, an inquiry into the principle of right and of government*. London: Forgotten Books.

Raubal, M., Miller, H. J., & Bridwell, S. (2004). User-centred time geography for location-based services. *Geografiska Annaler B, 86*(4), 245–265.

Reeve, S. & Foden, G. (2001) *A new breed of terror*, September 12th. www.guardian.co.uk/wtccrash/story.

Reichenbach, H. (1958). *Modern philosophy of science: Selected essays*. Westport, CN: Greenwood Press.

Rheingold, H. (1993). *The virtual community: Homesteading on the electronic frontier*. Cambridge, MA: MIT Press.

Rich, A. (1986). *Blood, bread and poetry: Selected prose 1979–1985*. New York: Norton.

Roe, M. S. (2013). *Maritime governance and policy-making*. London: Springer.

Rorty, R. (1989). *Contingency, irony and solidarity*. Cambridge: Cambridge University Press.

Rose, C. (1977). Reflections on the notion of time incorporated in Hagerstrand's time-geographic model of society. *Tijdschrift voor Economische en Sociale Geografie, 68*(1), 43–50.

Rose, G. (1993). *Feminism and geography: The limits of geographical knowledge*. Cambridge: Polity Press.

Roy, A. (1997). *The god of small things*. London: Flamingo.

Russell, B. (1918). The philosophy of local atomism. *The Monist, 28*(4), 495–527.

Sasaki, Y., & Box, P. (2003). Agent-based verification of von Thünen's location theory. *Journal of Artificial Societies and Social Simulation, 6*, 2.

Schutz, A. (1973). *Structures of the life world*. Evanstown, IL: Northwestern University Press.

Schwanen, T. (2007). Matters of interest: Artefacts, spacing and timing. *Geografiska Annaler B, 89*(1), 9–22.

Schwartz, B. (1974). Waiting, exchange and power: The distribution of time in social systems. *American Journal of Sociology, 79*, 841–870.

Schwartz, B. (1975). *Queuing and waiting. Studies in the social organization of access and delay*. Chicago, IL: University of Chicago Press.

Scott, D. M. (2010). Space-time prism. In K. J. Button, H. Vega, & P. Nijkamp (Eds.), *A dictionary of transport analysis* (pp. 369–371). Cheltenham: Edward Elgar.

Scott, D. M., & He, S. Y. (2012). Modelling constrained destination choice for shopping: A GIS-based time-geographic approach. *Journal of Transport Geography, 23*, 60–71.

Shaw, S.-L. (2012). Guest editorial introduction: Time geography—its past, present and future. *Journal of Transport Geography, 23*, 1–4.

Shekhar, S., & Chawla, S. (2002). *Spatial databases: A tour*. Upper Saddle River, NJ: Prentice Hall.

Sidgwick, H. (1874). *The methods of ethics*. London: Macmillan.

Singer, P. (1972). Famine, affluence and morality. *Public Affairs, 1*, 229–243.

Solomon, R. C. (1995). Justice as vengeance, vengeance as justice: A partial defence of Polymarchus. In J. P. Sterba, T. R. Machan, A. M. Jagger, W. A. Galston, C. C. Gould, M. Fisk, & R. C. Soloman (Eds.), *Morality and social justice: Point/counterpoint* (pp. 251–300). London: Rowman and Littlefield.

Sorokin, P., & Berger, C. Q. (1939). *Time budgets of human behavior*. Cambridge, MA: Harvard University Press.

Spretnak, C. (1997). *The resurgence of the real: Body, nature and place in a hypermodern world*. Reading, MA: Addison-Wesley Publishing Co., Inc.

Strauss, A. L. (1961). *Images of the American city*. New York: Free Press.

Sui, D. (2012). Looking through Hagerstrand's dual vistas: Towards a verifying framework for time geography. *Journal of Transport Geography, 23*, 5–16.

Swift, G. (1983). *Waterland*. London: Heinemann.

Szalai, A. (1972). *The use of time: Daily activities of urban and suburban populations in twelve countries*. The Hague: Mouton.

Taylor, P. (2003). Time: From hegemonic change to everyday life. In S. L. Holloway, S. P. Rice, & G. Valentine (Eds.), *Key concepts in geography* (pp. 151–164). London: Sage.

Terborgh, G. W. (1950). Capitalism and innovation. *American Economic Review, 40*(2), 118–123.

Thrift, N. (1977a). *An Introduction to time geography, cat mog 13*. London: Institute of British Geographers.

Thrift, N. (1977b). Time and theory in human geography 1. *Progress in Human Geography, 1*(1), 65–101.

Thrift, N. (1977c). Time and theory in human geography II. *Progress in Human Geography, 2*, 413–457.

Thrift, N. (1996). *Spatial formations*. London: Sage.

Thrift, N. (1998). The rise of soft capitalism. In G. O'Tuathail, A. Herod, & S. M. Roberts (Eds.), *An unruly world* (pp. 25–71). London: Routledge.

Thrift, N. (2005). Torsten Hagerstrand and social theory. *Progress in Human Geography, 29*(3), 337–340.

Thrift, N., & Pred, A. (1981). Time geography: A new beginning. *Progress in Human Geography, 5*(2), 277–286.
Tilly, C. (1988). Future history. *Theory and Society, 17*, 703–712.
Tornqvist, G. (1979) *On Fragmentation and coherence in regional research*. Lund studies in geography, Series B, No. 45, Sweden: The Royal University of Lund.
Tornqvist, G. (2004). Creativity in time and space. *Geografiska Annaler B, 86*(4), 227–243.
Townsend, D. J., & Bever, T. G. (2001). *Sentence comprehension: The Integration of habits and rules*. Cambridge, MA: MIT Press.
Tronto, J. (1987). Beyond gender difference to a theory of care. *Signs: Journal of Women in Culture and Society, 12*, 645–663.
Tuan, Y. (1977). *Space and place: The perspective of experience*. Minneapolis, MN: University of Minnesota Press.
Unwin, T. (1992). *The place of geography*. London: Longman.
Ustinov, P. (1987). *Ustinov in Russia*. London: Michael O'Mara Books Limited.
Van Paassen, C. (1976). Human geography in terms of existential anthropology. *Tijdschrift voor economische en sociale geografie, 67*(6), 324–341.
Veblen, T. (1898). Why is economics not an evolutionary science? *Quarterly Journal of Economics, 12*(4), 373–397.
Virilio, P. (1986) *Speed and politics. An essay on dromology*. New York: Semiotext(e).
Vonnegut, K. (1990). *The gospel from outer space (or Yes We Have No Nirvanas)*. Rockville, MD: Borgo Press.
Weber, J., & Kwan, M.-P. (2003). Evaluating the effects of geographical contexts on individual accessibility: A multi-level approach. *Urban Geography, 24*, 647–741.
Whitehead, M. (2003). 'In the shadow of hierarchy': Metagovernance, policy reform and urban regeneration in the West Midlands. *Area, 35*(1), 6–14.
Whittesley, D. (1945). The horizon of geography. *Annals of the Association of American Geographers, 35*, 1–36.
Wirth, C. (2012). Ocean governance, maritime security and the consequences of modernity in North-East Asia. *The Pacific Review, 25*(2), 223–245.
Wolch, J. R., & DeVerteuil, G. (2003). New landscapes of urban poverty management. In J. May & N. Thrift (Eds.), *Timespace* (pp. 149–168). London: Routledge.
Wong, D., & Shaw, S.-L. (2011). Measuring segregation: An activity-space approach. *Journal of Geographical Systems, 13*(2), 127–145.
Wright, J. K. (1966). *Human nature in geography, fourteen papers 1925–1965*. Cambridge, MA: Harvard University Press.
Wu, Y.-H., & Miller, H. J. (2001). Computational tools for measuring space-time accessibility within dynamic flow transportation networks. *International Journal of Geographical Information Systems, 9*, 153–168.
Yin, L., Shaw, S.-L., & Yu, H. (2011). Potential effects of ICT on face-to-face meeting opportunities: A GIS-based time-geographic approach. *Journal of Transport Geography, 19*, 422–433.
Yu, H. (2006). Spatial-temporal GIS design for exploring interactions of human activities. *Cartography and Geographic Information Science, 33*, 3–19.
Zubrin, R. (1999). *Entering space: Creating a spacefaring civilization*. New York: Tarcher.

Index

M. Roe, *Maritime Governance*, DOI 10.1007/978-3-319-21747-5

MIX
Papier aus verantwortungsvollen Quellen
Paper from responsible sources
FSC® C105338

If you have any concerns about our products,
you can contact us on
ProductSafety@springernature.com

In case Publisher is established outside the EU,
the EU authorized representative is:
Springer Nature Customer Service Center GmbH
Europaplatz 3, 69115 Heidelberg, Germany

Printed by Libri Plureos GmbH
in Hamburg, Germany